The Welfare of Animals Used in Research

The Universities Federation for Animal Welfare

UFAW, founded in 1926, is an internationally recognised, independent, scientific and educational animal welfare charity that promotes high standards of welfare for farm, companion, laboratory and captive wild animals, and for those animals with which we interact in the wild. It works to improve animals' lives by:

- Funding and publishing developments in the science and technology that underpin advances in animal welfare;
- Promoting education in animal care and welfare;
- Providing information, organising meetings and publishing books, videos, articles, technical reports and the journal *Animal Welfare*;
- Providing expert advice to government departments and other bodies and helping to draft and amend laws and guidelines;
- Enlisting the energies of animal keepers, scientists, veterinarians, lawyers and others who care about animals.

> *Improvements in the care of animals are not now likely to come of their own accord, merely by wishing them: there must be research...and it is in sponsoring research of this kind, and making its results widely known, that UFAW performs one of its most valuable services.*

Sir Peter Medawar CBE FRS, 8 May 1957
Nobel Laureate (1960), Chairman of the UFAW Scientific Advisory Committee (1951–1962)

UFAW relies on the generosity of the public through legacies and donations to carry out its work, improving the welfare of animals now and in the future. For further information about UFAW and how you can help promote and support its work, please contact us at the following address:

Universities Federation for Animal Welfare
The Old School, Brewhouse Hill, Wheathampstead, Herts AL4 8AN, UK
Tel: 01582 831818 Fax: 01582 831414 Website: www.ufaw.org.uk
Email: ufaw@ufaw.org.uk

UFAW's aim regarding the UFAW/Wiley-Blackwell Animal Welfare book series is to promote interest and debate in the subject and to disseminate information relevant to improving the welfare of kept animals and of those harmed in the wild through human agency. The books in this series are the works of their authors, and the views they express do not necessarily reflect the views of UFAW.

The Welfare of Animals Used in Research: Practice and Ethics

Robert C. Hubrecht

The Universities Federation for Animal Welfare,
Wheathampstead,
Herts,
UK

WILEY Blackwell

This edition first published 2014
© 2014 by Universities Federation for Animal Welfare
Series Editors: James K. Kirkwood and Robert C. Hubrecht

Registered Office
John Wiley & Sons, Ltd, The Atrium, Southern Gate, Chichester, West Sussex, PO19 8SQ, UK

Editorial Offices
9600 Garsington Road, Oxford, OX4 2DQ, UK
The Atrium, Southern Gate, Chichester, West Sussex, PO19 8SQ, UK
1606 Golden Aspen Drive, Suites 103 and 104, Ames, Iowa 50010, USA

For details of our global editorial offices, for customer services and for information about how to apply for permission to reuse the copyright material in this book please see our website at www.wiley.com/wiley-blackwell

Library of Congress Cataloging-in-Publication Data

Hubrecht, R., author.
  The welfare of animals used in research : practice and ethics / Robert C. Hubrecht.
    p. ; cm.
Includes bibliographical references and index.
ISBN 978-1-119-96707-1 (paper)
  I. Title.
  [DNLM: 1. Animal Welfare.   2. Animals, Laboratory. QY 54]
  R853.A53
  179′.4–dc23
                          2014000050
A catalogue record for this book is available from the British Library.

Wiley also publishes its books in a variety of electronic formats. Some content that appears in print may not be available in electronic books.

Cover image: Source: Shutterstock © Vasiliy Koval (image no.: 87245485).
Cover design by Aubergine Design

Set in 10/12.5pt Sabon by SPi Publisher Services, Pondicherry, India

1   2014

Contents

Foreword

As a longstanding (that means quite old) biomedical researcher who has worked with cells, animals and patients throughout my research career, I am delighted to provide the Foreword to this important and long overdue book.

Animals are used in research to understand biological and disease processes, to develop new markers of disease progression, medical devices and new drugs, and to then test these discoveries before they are tried in humans. The latter is a regulatory requirement by governments across the world.

Experimental and regulatory use of animals has been core to the development of new medicines for many decades, and overall it has been very successful. But it is a highly sensitive issue on which views are diverse and often radically polarised. This is the perfect recipe for entrenchment. Those most strongly opposed to research using animals can participate in quite extreme acts of terror, while the scientists involved in the research retreat behind their bunkers, fail to engage in the research and worse still, do not participate in the critical debate about how to limit the use of animals and ensure the welfare of those that are used.

Thus, there is an urgent need for a text that covers the range of issues related to animals used in research. This book provides such a text and its author Robert Hubrecht commands great respect for his unswerving support for animal welfare, his balanced and open approach and the fact that he has facilitated delivery of major benefits to those animals on which experiments are performed.

The book covers the range of important issues-why and how animals are used, public opinion and concerns, the arguments for and against the use of animals (including the practical, the philosophical and the emotional), how we judge the harm to animals versus the likely benefits of research or testing, how we control and regulate the use of animals and how we move forwards-in partnership.

Robert Hubrecht notes how attitudes have changed. This is true not just in research but more widely in our relationship with animals. As a society we are now much more concerned about the welfare of farm animals, pets and of course animals in research. While some of this can be ascribed to a shift in general societal values, the change in welfare standards for animals used in research owes much to animal welfare scientists such as Robert Hubrecht, who has argued the case while maintaining the respect of scientists who use animals in their research. No mean task.

For me, as a researcher, the most important chapter is that which deals with the '3R's'- the replacement, reduction and refinement of animals in research. First proposed over 50 years ago, Russell and Burch's ideas gained rather modest interest for some time. The 3R's are now absolutely mainstream thinking, at least in the UK and are fundamental to every researcher who uses animals.

Most importantly this book is sensitive to the diversity of opinion, and the arguments are based on evidence-which should sit well with any scientist.

Robert Hubrecht's book will certainly be required reading for any student of staff member who works with me and deploys animals in their research. I am sure it will attract a much wider audience.

I hope that within my life time we will have made advances such that we no longer need to use animals to make major discoveries and to develop important new medical advances. In the meantime this book will help us to think very carefully about why, how and when we use animals. Rightly so.

Professor Dame Nancy Rothwell DSc, FRS, FMedSci, FSB
President and Vice-Chancellor The University of Manchester

Preface

Since prehistory, humans have exploited animals, first hunting them as a source of food and clothing but in more recent years to satisfy other desires. We keep animals as companions, modifying them in the process through domestication or through surgical interventions; we rear them in farms for food and for other products; we use them in sport, as purpose-bred animals such as racehorses, greyhounds or pheasants; and they continue to be used, in small numbers, for entertainment in circuses, stage acts, and in films. Finally, we use them in research as models of disease to test the efficacy of drugs, to test the safety of products, or for the advancement of knowledge. The use of animals in research is probably the most controversial of all these practices. On one side of the argument, pressure groups and organisations, some of which count doctors and veterinarians and philosophers in their number, campaign against animal research, both on ethical grounds and arguing that their use is scientifically flawed. On the other side, there are organisations, often counting in their membership doctors, veterinarians, philosophers and distinguished scientists, that are dedicated to protecting animal research because of the benefits that are seen to flow from it. Most of us in the developed world will depend at some time on medicine, and many of us owe our lives or health to it and thus bear at least some responsibility for the use of animals in research. In addition, in a democracy scientists carry out their work with the permission of the public and if there was a sufficient feeling that animal research could not be justified, then it would be outlawed. So all of us, whether involved in research on animals or not, and whether we agree with it or not, bear part of the responsibility for the continuation of the practice.

The views held by those either for or against animal experimentation are often extremely strongly held and I do not intend to add to the heat of that debate. Unfortunately, it is probably an impossible task to write a book on the welfare of animals used in research that is completely unbiased, so to help readers judge any bias that I might have, it seems right that I should briefly describe my background. I was trained as a research scientist and spent many years studying the behaviour, and sometimes the physiology, of animals in laboratory settings and in the field. I then moved to the Universities Federation for Animal Welfare, an organisation that uses science to advance our understanding of how best to improve the welfare of sentient animals whenever they are harmed as a result of human activities. Undoubtedly my career has affected my views but it has also provided me with relevant experience.

Animals are used in research to benefit humans, and sometimes also other animals, but their use can cause harm to the animals involved, and on occasion is likely to cause considerable suffering. It is therefore right that, whenever animals are used in research, the case for their use should be critically examined to ensure that there are no alternatives, that the numbers used should be kept as small as possible, and that any harm caused to the animals involved should be kept to a minimum. In other words, that the 3Rs principles of replacement, reduction and refinement should be fully implemented, which can only be achieved with a good understanding of the issues.

My aim has been to produce a resource that will be a useful introduction to the issues involved in laboratory animal welfare for those who intend to pursue, or are pursuing, a biological or medical career that involves the use of animals in research. I hope that it may also prove useful to prospective animal care staff and animal welfare scientists, and to those involved in ethical review or otherwise working to improve the welfare of animals used in research. It may also help inform debate amongst those who are not involved in experimentation but who are interested in the issues. Laboratory animal welfare is a very wide-ranging subject with many areas of specialist interest, and to deal with each topic in depth would require more space than I have here. I have therefore tried to provide an informative introduction to the welfare and ethical issues that arise from using animals in research, in the process drawing attention to innovative ideas and research and to sources of information. The animal welfare issues involved in research on animals cross national boundaries but the implementation of measures to address these varies. I have therefore not attempted to produce a guide that explains how research is controlled and should be carried out in one particular jurisdiction, but rather tried to concentrate on principles, where necessary illustrated by national examples. I hope that that this has resulted in a book that will help improve the welfare of animals used in research wherever that occurs.

Robert Hubrecht, UFAW
September 2013

Dedication

To Diana

Acknowledgements

Over the years I have absorbed many ideas during discussions with others working in the field of animal welfare science. Those that I have not acknowledged will, I hope, forgive me. I am very grateful to those who have given me advice on various sections or chapters of the book, or provided me with material, and in particular to Joanne Zurlo, Michael Balls, Ngaire Dennison and Jeff Everitt. I should also like to express my thanks to those who gave permission to use images, including those who did not wish to be acknowledged. I am indebted to Liz Carter who took a great weight off my shoulders by dealing with permissions for images and other material, to Eleanor Hubrecht who read and commented on early drafts and provided encouragement, and to Wendy Goodwin for her help with references. I am particularly grateful to James Kirkwood and Manuel Berdoy for their generosity in providing very detailed comments and suggestions on all of the chapters.

Introduction

In this chapter I outline how animals have been used in history to advance human knowledge, how animals are used in research today, where research is carried out, how many animals are used, and the extent of various types of harm caused to them. This leads on to the need on ethical grounds to reduce harm to a minimum, public attitudes to research on animals, and the public's role in permitting research on animals.

1.1 Reasons for Using Animals in Research

The history of the use of animals to advance human knowledge is long. Even in prehistory, the butchering of animals must have provided some insights into human anatomy and disorders for those who were wise enough to see. However, our earliest records of animal studies date back to the ancient Greeks. Aristotle pioneered the experimental method and carried out dissections some 300 years BC, but he was certainly not an experimental scientist as we would recognise one today, his biological works being described by the Nobel Prize-winning scientist Peter Medawar as 'a farrago of hearsay, imperfect observation, wishful thinking and credulity amounting to downright gullibility'[1]. Alcmaeon of Croton, while in Alexandria (305–240 BC), dissected a living animal to demonstrate the importance of the optic nerve for vision[2] and Erasistratus, a prominent physician in Alexandria

---

[1] Medawar and Medawar (1985).
[2] Maehle and Tröhler (1987).

The Welfare of Animals Used in Research: Practice and Ethics, First Edition. Robert C. Hubrecht.
© 2014 Universities Federation for Animal Welfare. Published 2014 by John Wiley & Sons, Ltd.

(third century BC), used vivisection to distinguish between the sensory and the motor nerves. In the second century AD, Galen of Pergamum, a famous physician who became doctor to the emperors Marcus Aurelius and Commodus, used dissection to study the continuity of the nervous system. The experience that he gained from these studies on animals led him to diagnose loss of feeling in the fingers of a patient as being caused by an injury to the spine. This was probably the first time that it was realised that neural problems could be referred from the actual point of injury.

With the arrival of the Renaissance and its associated flowering of scientific endeavour there was a renewed interest in animal experimentation that has continued to the present day. The following are just a few historical examples of the use of animals in studies on anatomy and physiology. William Harvey used living animals (including shrimp, eels, fish, pigeons, dogs and other mammals) to demonstrate the circulation of blood and, in 1661, Marcello Malpighi saw the capillaries as predicted by Harvey in dissected preparations of the frog lung and urinary bladder. In the 1800s, Claude Bernard studied glycogen and its relationship to diabetes, and Sir Charles Bell and Eduard Hitzig studied the nervous system. Incidentally, Bell was extremely reluctant to carry out his experiments, which, like others of the time, must have resulted in extreme animal suffering as this was before the discovery of anaesthesia[3]. More recently, animals have been used in research into the immune system, and in the development and treatment of diseases such as anthrax, poliomyelitis, influenza, asthma and tuberculosis, blood transfusion, various cancer treatments, muscular dystrophy and neurological disorders such as Alzheimer's and Parkinson's disease amongst many others[4]. In addition, animals have been, and are, used in a wide range of fundamental research including studies to gain knowledge about animal or ecological systems, and ways of improving animal health, welfare, productivity or performance[5]. Animals are also used in the safety testing of pharmaceutical and household products as well as environmental safety testing of chemicals, the legal requirement for which, in Europe, depends on the tonnage of the chemical produced per annum[6].

Today published statistics provide an overview of the types of research in which animals are used. For example, UK statistics on animal use for 2011[7] show that fundamental biological research accounted for 35% of the total procedures carried

[3] Lapage (1960); Sechzer (1983).

[4] Lapage (1960); Smith and Boyd (1991), p. 27. See also the Understanding Animal Research website www.understandinganimalresearch.org.uk, accessed 8 May 2013.

[5] A comprehensive list of various uses to which animals are put as models for various diseases can be found in Hau and Schapiro (2011).

[6] The European Community REACH regulation (2007) deals with the Registration, Evaluation, Authorisation and Restriction of Chemical substances and includes regulations on the testing of chemicals, see http://ec.europa.eu/environment/chemicals/reach/reach_intro.htm. For the USA, see The Toxic Substances Control Act (1976) at http://www.epa.gov/lawsregs/laws/tsca.html, accessed 8 May 2013.

[7] Home Office (2012).

out on animals[8], applied human medicine 13%, applied veterinary medicine 5%, and protection of humans, animals or environment 3%. Only 1% of procedures were used in the direct diagnosis of conditions while 43% involved animals in breeding programmes, a category that includes harmful mutant animals and genetically modified animals[9]. The development of genetic modification and mutant techniques has resulted in greater numbers of animals, particularly mice and fish, being used in fundamental research aimed at elucidating gene function and the control of genetically mediated disease. This has been a contributing factor to the reversal of the downward trend in the use of animals in research in the UK seen in the mid 1990s[10]. However, the UK statistics have recorded all animals bred with a genetic modification unless the researcher can prove over two generations that there is no welfare impact. In practice this means that all have been recorded, even though some are simply used for breeding purposes, are not used directly in research and may not show any ill effects (possess a harmful phenotype). Some have argued that this practice artificially increases the statistics of animal use, but others have pointed to the various harms caused in the production of genetically modified animals. However, implementation of European Directive 2010/63/EU will change the reporting requirements so that those shown not to possess a harmful phenotype will not need to be reported[11].

1.2 Where Animal Research is Carried Out

There are various types of institution in which animal research is carried out. Universities and non-profit organisations use animals in fundamental studies, or work in collaboration with pharmaceutical companies. Academic research includes areas such as neurobiology, gene function, and metabolism, but animals are also used in more applied settings such as studies on Parkinson's or Alzheimer's disease. Academic research also includes studies of behaviour or animal welfare that may sometimes be carried out outside the laboratory. Pharmaceutical companies use animals in the research and development of medicines. In these studies animals are

---

[8] Regulated procedures prior to January 2013 were defined as 'any experimental or other scientific procedure applied to a protected animal which may have the effect of causing that animal pain, suffering, distress or lasting harm'. The implementation of European Directive 2010/63/EU resulted in a revision of this definition (see Glossary and https://www.gov.uk/research-and-testing-using-animals, accessed 8 May 2013).

[9] See Glossary for definitions of genetically altered (GA) and genetically modified (GM) animals.

[10] The UK statistics show a fairly regular increase in the number of genetically modified and harmful mutant animals from 1995 to 2010. However, the 2011 figure was slightly lower than that in 2010 due to a 6% reduction in the breeding of harmful mutants, partially offset by a 3% increase in the numbers of genetically modified animals.

[11] Commission Implementing Decision of 14 November 2012 establishing a common format for the submission of the information pursuant to Directive 2010/63/EU of the European Parliament and of the Council on the protection of animals used for scientific purposes, available from http://ec.europa.eu/environment/chemicals/lab_animals/home_en.htm, accessed 3 April 2013.

used in trials of efficacy of potential drugs and to assess their likely toxicology. Some of this research, typically the efficacy studies, is usually done in-house by the company developing the medicine, while the toxicology studies necessary to obtain a licence from the drug regulators[12] to market the drug may be carried out by independent contract research organisations (CROs). However, there has been an increasing trend for contracting laboratories to offer more and varied research services to the pharmaceutical companies. In addition, CROs carry out safety and environmental toxicity testing of non-pharmaceutical chemicals. Organisations that breed animals for research may also carry out certain types of research, and have begun to offer some of the services traditionally provided by CROs. A further category of research institution is government or other public research facilities. These include establishments whose function may be to monitor and provide advice on serious health risks to the population, monitor and control the standards and quality of biological products, research into agricultural or pest-related issues, or counter defence threats.

1.3 Numbers of Animals Used

The number of animals used in experiments is not trivial. Statistics from the UK[13] show that, in 2011 for example, over 3.79 million procedures were started that were likely to cause pain, suffering, distress or lasting harm to animals (this figure is more than the 3.71 million animals used as some re-use of animals is permitted); 77.5% of these procedures were carried out on mice, rats or other rodents, while other mammals (a category that includes dogs, primates, cats, ferrets, etc.) accounted for only 2% of procedures and fish were used in 15% of procedures. As we shall see in Chapter 6, despite a fall in animal use in the 1980s and 1990s, the development of genetic modification technologies has resulted in increased use of certain animals, particularly mice.

This, however, is only one country. Unfortunately, as the Nuffield Council on Bioethics[14] points out, statistics for other countries can be hard to come by and are not necessarily equivalent. For example, the Animal and Plant Health Information Service (APHIS) of the United States Department of Agriculture publishes statistics on the numbers of animals used in research in the USA in each state by fiscal year (Table 1.1), but the numbers used seem very small (approximately 1.1 million animals per annum) compared with equivalent statistics for the UK. The discrepancy between the UK and US figures is, however, easily explained. In the USA the Animal Welfare Act excludes birds, rats of the genus *Rattus* and mice of the genus *Mus*

---

[12]   There are various regulatory bodies that license medicines: e.g. in the USA the Food and Drug Administration (FDA); in the UK the Medicines and Healthcare products Regulatory Agency (MHRA); and within Europe the European Medicines Agency (EMA).

[13]   Tables and graphs can be found in Home Office (2012), available from https://www.gov.uk/research-and-testing-using-animals#publications, accessed 3 April 2013.

[14]   Nuffield Council on Bioethics (2005).

Table 1.1 United States Department of Agriculture Animal and Plant Health Information Service (APHIS) Annual Report on Animal Usage by Fiscal Year. Fiscal Year 2010, published 27 July 2011.

Cats	21 578
Dogs	64 930
Guinea pigs	213 029
Hamsters	145 895
Marine mammals	126
Non-human primates	71 317
Other farm animals	38 008
Pigs	53 260
Rabbits	210 172
Sheep	13 271
All other covered species	303 107
Total	1 134 693

Source: http://www.aphis.usda.gov/animal_welfare/efoia/7023.shtml, http://www.aphis.usda.gov/animal_welfare/efoia/downloads/2010_Animals_Used_In_Research.pdf, accessed 13 May 2013.

bred for use in research. As these in the UK account for just over 87% of the total procedures, a more reasonable estimate of the animals used annually in the USA might be 8.6 million. Using available statistics and estimates of this sort, it has been estimated that fewer than 60 million animals are used worldwide in research[15]. Whatever the exact figure, it is clear that a significant number of animals are used for research purposes and that this justifies serious ethical consideration. However, it is easy to be seduced by numbers, especially when you have nothing with which to compare them. So to provide some perspective, let us turn to the food industry. Many of the animals produced for food suffer some welfare compromise in the processes of breeding, production, transport and slaughter, and the number that we use is truly astonishing. To take just one animal that we breed and kill for food: in 2011, provisional figures suggest that 931 million broiler chickens were slaughtered in the UK[16], and many broiler birds suffer welfare problems such as lameness and ascites[17]. Does this then mean that we should ignore the issue of animals in research? I would argue not. Numbers can be a useful tool to target and prioritise resources effectively, but it would be wrong to use the fact that more animals are used in the food industry to suggest that the laboratory animal issue is less important. After all, for each animal, it is the personal experience that is important, not the numbers of its fellow sufferers.

[15] Understanding Animal Research http://www.understandinganimalresearch.org.uk, accessed 28 March 2012. Taylor *et al.* (2008). See also Knight *et al.* (2008) for suggestions that the number of animals used worldwide is much larger.

[16] Department for Environment, Food and Rural Affairs www.defra.gov.uk/, accessed 22 October 2012.

[17] For example Julian (1998); Butterworth *et al.* (2002); Knowles *et al.* (2008).

1.4 Harmful and Harmless Research

A common misconception about animal research is that it inevitably results in animal suffering[18], usually as a result of surgery or as a response to substances administered to the animal. In some cases research will involve these sorts of harms, but many other types of harm (e.g. fear, discomfort, boredom, hunger) can also occur. Potential harms of whatever sort, whether deliberately induced or as an unintended consequence of the research, need to be taken into account alongside the proposed benefits of the research when considering whether the research is justified, as we shall see in Chapter 5.

Not all animal research is likely to result in harm to the animal. Ethologists, for example, are interested in what factors stimulate particular behaviours (why animals do what they do), what evolutionary processes led to the behaviour, how behaviour develops and what is its purpose, or to put it another way, the study of the causation, evolution, development and function of animal behaviour[19]. Ethological studies are generally carried out for reasons of curiosity as to how the world works, rather than with the aim of reducing human or animal suffering[20] so it is a good thing that many studies of behaviour do not result in pain, suffering, distress or lasting harm. Animals may, for example, be observed in zoos with no additional ill effects on the animals whatsoever. Observations can also sometimes be made on animals in the wild with minimal impact, but it is certainly not the case that all ethological studies are neutral in their effect on the animal.

A personal example may be instructive. In the early 1980s I carried out research into the ranging behaviour of free-living marmosets in north-eastern Brazil. I was interested in their natural history, in particular how they used their environment and the structure of their social groupings. In order to do this I needed to be able to track them through dense vegetation and to do this reliably required the use of radio tags. To fit these I had to trap and anaesthetise some animals to allow me to fit the tag, make measurements and take samples. In those days few people discussed the ethical issues involved in such field research but it was obvious that there were both potential and real costs to the animals in this research. The first harm was trapping, which frightened and stressed the animals. The marmosets then had to be removed from the trap, and anaesthetised by injection. This was again stressful, and there was a potential risk of harm from the anaesthetic (fortunately no ill effects were seen on this occasion). A radio tag was then fitted which added a weight to the marmoset, and may have caused discomfort or affected energy

---

[18]   Even today not everybody believes that animals are capable of suffering, but many believe that at least some should be given the benefit of the doubt that they may experience feelings, both good and bad. This is the basis for many for concerns about the welfare of animals. These issues will be discussed in Chapters 3 and 4.

[19]   Tinbergen (1963).

[20]   There is, however, a flourishing science of applied animal welfare science aimed at improving animal welfare that grew out the ethology discipline.

expenditure and subsequent survival. The collar might also have rendered the marmoset more visible to predators, could have snagged on branches or chaffed. Finally, the marmoset was left in a quiet place to recover fully from the anaesthesia before release. Again, waiting in the cage in proximity to humans, even though the cages were covered with a cloth, may well have resulted in fear and stress, and perhaps foraging time was lost. This example may seem to be rather an obvious one as there were manipulations to the animals, but in the past field investigators have often underestimated the impact to the animal of their studies. Even watching animals can have effects, which may be either beneficial for the animals watched (scaring off predators and thus reducing the risk of being eaten), or negative (perhaps through disturbance of the animals or habituation to humans which may be risky if others are not so kindly disposed to the animals).

The detrimental effects described above were not a required part of the study, but occurred as a consequence of the techniques used. However, other sorts of ethological research may cause harm to animals as an integral function of the experimental procedure. Examples include the deliberate manipulation of clutch size, which might be done to study how parents allocate resources to their offspring, or studies on aggression, predation or territorial displays and communication. As in all other areas of research using animals, an ethologist is expected to identify potential harmful effects (even those that may not at first be obvious) in order to try to eliminate or reduce them and to assess whether the work justifies any remaining harm. Organisations such as the Association for the Study of Animal Behaviour have published guidelines to help researchers do this[21].

1.5 How Much Suffering is Caused by Research?

It is hard to obtain good evidence on the suffering experienced by animals in research. The UK has for many years published detailed statistics on animal experimentation, but the only data relating to suffering has come from the severity banding of licences (Table 1.2). These provide a prospective assessment of suffering whereby licences to carry out research are assigned as 'mild', 'moderate', 'substantial' or 'unclassified', based on an assessment of the likely experience of suffering of the average animal[22]. The revised UK legislation required by European Directive 2010/63/EU requires that applicants will have to report the actual severity of procedures from 1 January 2014, and the publication of this data should provide much greater transparency.

My own experience, as a member of various ethics committees, has been that the majority of licence applications are classified as either mild or moderate severity[23],

---

[21] ASAB (2006).

[22] Animal Procedures Committee (2009a).

[23] These severity limits refer to the maximum suffering or harm that the animal may experience, without the researcher referring back to the Home Office, the UK government department responsible for issues relating to the use of animals in science.

Table 1.2   Severity banding of licences for use of animals in scientific procedures in the UK.

Severity banding	Licences in force on 31 December 2011
Mild	936
Moderate	1591
Substantial	55
Unclassified (procedures carried out under terminal anaesthesia)	42
Total	2624

Note that these are a prospective assessment of the experience of the 'average' animal involved in a given project, which may contain many different protocols, and so give no indication of the level of suffering imposed on individual animals. From 1 January 2013 UK licences are no longer banded (given an overall severity rating). Individual procedures are classified into non-recovery, mild, moderate and severe categories. From 2014 onwards, statistics on actual suffering will be published.

Figures from UK Home Office Statistics (Home Office, 2012). © Crown Copyright 2012. Contains public sector information licensed under the Open Government Licence v2.0.

the mild category including procedures that can be as minor as the taking of a blood sample or an injection of saline. I have also visited many animal houses, and most of the animals, at any one time, in these buildings appeared to be healthy and free of pain. However, we should not underestimate the extent of suffering that can occur. Sometimes pain or other harm is an inevitable consequence of procedures; surgery, for example, is always likely to result in some pain, and electric shocks have been used as part of experimental paradigms. In the past, the notorious LD_{50} test required that test compounds be given to animals in increasing amounts until a dose was arrived at which killed 50% of the animals tested. Nowadays, special justification and permission is required to use the LD_{50} test in the UK, and there are alternatives that require far fewer animals. Another example of a test that causes considerable suffering is one in which mice are used to detect toxins absorbed by shellfish when they ingest certain bloom-forming dinoflagellates such as *Gymnodinium breve*. These neurotoxins can be extremely dangerous to humans, in some cases resulting in death. The mouse assay uses lethality as an endpoint and causes considerable suffering, but fortunately humane alternatives that may also be scientifically better are being developed to replace this test[24].

Finally, it is worth noting that some experiments may cause so little harm to the animal that they fall outside legislation. Within Europe, the level at which a procedure requires licensing has been set at 'practices not likely to cause pain, suffering, distress or lasting harm equivalent to, or higher than, that caused by the introduction of a needle in accordance with good veterinary practice'[25]. Studies that therefore involve only observation of the animals or perhaps collection of faeces or urine would normally not require regulation, but if they resulted in significant fear or stress then, even if there is no invasive component, licences would be needed. Some would wish to go much further, arguing that just keeping animals in confinement

---

[24]   See Cefas website www.cefas.defra.gov.uk, accessed 8 May 2013.
[25]   European Directive 2010/63/EU, Article 1.

within research institutions results in harm to the animals that should not be permitted. As we shall see in Chapter 5, animals used in research certainly can, and do, experience harms that are not a planned part of the study (so-called non-contingent harms). These harmful effects may not be trivial and can include stress from handling, transport, unnatural social groupings, or inadequate housing.

1.6 Attitudes to Animal Experimentation

Knowledge derived from animal experimentation comes at a cost to the animals used in the research, which is why many countries have laws to regulate animal experimentation. However, attitudes towards animals have not always been as favourable to their welfare as they are today. Over time, attitudes to animals have generally moved towards treating them in a way that would reduce suffering, although there has always been a spread of opinion regarding their ability to suffer and how this should impact on their use including animal research. There are a number of excellent accounts, referenced below, that provide great detail on the development of these attitudes and the factors that influenced these changes but it is worth providing a short summary here.

Even amongst the ancient Greeks there was a range of beliefs regarding our relationship to animals and the correct way to treat them. For example, Aristotle and the Stoic and Epicurean philosophers excluded animals from considerations of moral concern. On the other hand, the Cynics considered animals to be superior to humans while other philosophers developed the concept of a kinship between humans and other animals[26]. Galen, who lived in the second century AD, was not particularly concerned with the suffering of the animals that he dissected, although he did refuse to dissect the sexual organs of living animals or even dead animals in upright man-like postures, largely on aesthetic grounds, and recommended using pigs or goats instead of primates to avoid seeing the unpleasant expression of the ape when vivisected. Indeed, much early animal research was carried out in the belief that the distinction between humans and animals was such that the only ill effect of causing suffering to animals came from the possibility that these actions might lead to inhumanity to humans (a view taken by Thomas Aquinas, René Descartes and Immanuel Kant)[27]. Descartes, writing in the seventeenth century, has been quoted as considering that as animals lacked the necessary soul, they were unable to feel real pain although they might feel some form of inferior sensation. In fact, his position has probably been overstated, his views being rather that animals did have feelings but that they were not self-consciously aware of those feelings[28]. Unfortunately, some of his followers were firmly of the view that animals had no feelings and went so far as to cause them deliberate and pointless pain, laughing at

---

[26] Grayson (2000), p. 3.
[27] Maehle and Tröhler (1987).
[28] Cottingham (1978); Bekoff and Meaney (1998), p. 131.

those who objected. Indeed, the vocalisations of vivisected dogs dissected to show the circulation of blood (demonstrated by William Harvey in 1628) were interpreted as nothing more than the creaking of the animal 'clockwork'. Animals were considered irrational beings, and as such did not fall within the system of 'natural right' and thus humans had no obligations towards animals[29].

By the beginning of the eighteenth century there were growing concerns about ethical aspects of such experimentation from a number of literary men[30]. Samuel Johnson, for instance, was highly critical of animal experimentation, writing 'he surely buys knowledge dear, who learns the use of lacteals [lymphatic ducts] at the expense of his humanity'. He took particular issue against the repeated and popular demonstrations of vivisections to the public. However, his concern may have been more for the corrupting influence of experimentation on the researcher than for the animals themselves[31]. The physiologist Claude Bernard may have also been expressing some concern about the means required to achieve biological knowledge when he wrote that 'If a comparison were required to express my idea of the science of life, I should say that it is a superb and dazzling hall that can only be reached by passing through a long and ghastly kitchen'. Humphrey Davy, who used animals to study the effects of various gases, similarly became increasingly concerned about the pain he caused them. Charles Darwin, while explicitly supporting the use of vivisection to advance physiological knowledge, also wrote that its use for trivial purposes to satisfy 'damnable and detestable curiosity' made him sick with horror[32]. More practically, some were already considering means of avoiding using animals in research. The Scottish astronomer and instrument maker James Ferguson suggested a non-animal, mechanical alternative to the use of animals in demonstrations of Boyles' vacuum pump experiments.

Concern about animal experimentation did not occur in a vacuum, but was part of growing discomfort regarding various uses and abuses of animals. Animals were property and as such, when the owner perpetrated the abuse, were not subject to any legal protection; both deliberate and unnecessary cruelty such as cock fighting, bull baiting and to food animals was rife[33]. Perhaps most influentially, philosophers such as Rousseau, Primatt and Jeremy Bentham argued that it was the ability of animals to experience feelings, such as pleasure and suffering, that made them valid objects of moral concern, which is essentially an argument based on empathy. Although these philosophers had some influence amongst educated persons, the majority in the UK was still largely unconcerned by the suffering of animals[34], but views were gradually changing. James Wright of Derby produced a series of paintings of scientific demonstrations, and in one, painted in 1768, he vividly depicted the varied attitudes to experimentation at the time. The painting shows a scientist

[29]   Maehle (1994).
[30]   Daly (1989).
[31]   Maehle (1990).
[32]   Browne (2002), p. 421; Holmes (2008).
[33]   Radford (2001).
[34]   Radford (2001).

Figure 1.1 *An Experiment on a Bird in the Air Pump* by Joseph Wright, 1768. Reproduced with permission from The National Gallery, London.

demonstrating the effects of a vacuum pump on a bird (Figure 1.1). The painting is Romantic, in that the light of reason dispels the surrounding darkness, but the painting conveys fear as well as wonder[35]. The watchers display responses that seem to reflect the range of views that we might recognise in today's debates, from the didactic, through interested, to the distress of the girl covering her eyes.

Eventually the disquiet caused by these demonstrations, and by vivisection in general, which until about 1850 was carried out without anaesthesia[36] (Figure 1.2), led in 1870 to the British Association for the Advancement of Science publishing voluntary guidelines. However, these were not sufficient to satisfy public concern and, following the submission of two proposals for bills by members of the Houses of Lords and Commons, the Government announced a Report by the Royal Commission for the Advancement of Science. As a result, the UK, in 1876, passed the very first legislation anywhere in the world that controlled animal experimentation. However, this did not by any means bring an end to the vivisection debate[37].

Today, in the UK it seems that although there is general concern about animal welfare and some question the validity of the use of animals in research[38], there is

[35] Holmes (2008).
[36] Smith and Boyd (1991).
[37] See Grayson (2000), chapters 2 and 3.
[38] It is often not easy for those not expert in a particular research area to assess competing claims regarding the benefits, or otherwise, of using animals. Some argue that all animal experiments for biomedical purposes are scientifically invalid, a position that has not been accepted by various reviews into the subject, e.g. Nuffield Council on Bioethics (2005). However, the Bateson Report (Bateson, 2011b)

Figure 1.2 *A Physiological Demonstration with the Vivisection of a Dog* by Émile Édouard Mouchy, 1832. Reproduced with permission from the Wellcome Library, London.

also some evidence that the public accepts that such use should be allowed as long as it is well justified and regulated. Within the UK, a 2012 Ipsos MORI poll carried out on behalf of the Department for Business, Innovation and Skills (BIS)[39] found that 66% could accept animal experimentation so long as there is no unnecessary suffering to the animals. On the other hand, 37% of British adults objected to animal experimentation, with 21% of those polled agreeing, or tending to agree, with the proposition that the government should ban all experiments on animals for any form of research. Interestingly, 66% agreed with the statement that they accept animal experimentation as long as it was for medical research purposes, when in fact the law in the UK also permits it for the much more general purpose of advancing human knowledge. Until recently, polls carried out by MORI on this subject have tended to be reasonably constant in their findings, as well as being consistent with focus group research[40]. However, this latest poll shows a small but significant decline in public support for research using animals. It will be interesting to see how opinions move in future years.

---

points out that those arguing for animal experimentation sometimes make incomplete cases, in which there is often not enough background information on the importance of the condition, or subject being investigated, nor on progress relating to alternatives to animal use.

[39]   MORI poll, 2012, Views on the Use of Animals in Scientific Research, http://preview.tinyurl.com/cog74fh, accessed 8 May 2013.

[40]   For example Macnaghten (2004).

Not surprisingly, there are cultural differences between countries in their attitudes to animals. A survey of polls in 1994 showed that the UK and some European countries seem to have had a higher level of opposition to animal research than Japan or the USA. Even so, a Gallup poll in 2010, questioning the use of medical testing on animals, indicated that 59% of Americans found the practice morally acceptable, while 34% thought it was wrong. Across cultures, women tend to be more concerned by the issue than men.[41]

However, polls need to be treated with caution as they can deliver widely different conclusions depending on how, when and where they are carried out. An analysis of a range of surveys from different countries showed that 0–27% accept the use of animals in research while 0–68% opposed it[42], which is not a clear answer. It also indicated that much depends on how questions are asked and who asks them. For example, if questions include words like 'pain', then respondents are less likely to support animal research, even though many experiments do not cause the animals much, if any, pain. A MORI poll carried out by *New Scientist* in 1999[43] showed that responses to the question 'On balance, do you agree or disagree that scientists should be allowed to conduct any experiments on live animals?' were affected by whether the respondent had been first told that 'Some scientists are developing and testing new drugs to reduce pain or developing new treatments for life-threatening diseases such as leukaemia and AIDS'. Similarly, the purpose and type of research can also affect its acceptability, so that the public are more willing to accept the use of animals in testing the toxicity of chemicals to humans than to establish the effect on the environment[44]. Another factor to keep in mind when studying polls is that special interest groups frequently use public opinion as a tool to advance their cause in the animal research debate[45], and that polls carried out by these organisations have often not been subject to rigorous peer review that would scrutinise the survey techniques and instruments[46]. It may not be surprising, therefore, that poll results often correlate with the views of the organisations that commission them[47].

If it is hard to find out what people think, then it is even harder to find out why they think it. In assessing the results of such polls it is worth considering whether their views are based on adequate knowledge. Managhten's study[48] of a series of

---

[41]   Pifer *et al.* (1994); Gallup Politics website article by Lydia Saad, 2010, Four Moral Issues Sharply Divide Americans, http://preview.tinyurl.com/26mdavu, accessed 26 April 2013. See also a survey posted by Michael Foust, 2006, Baptist Press, http://www.sbcbaptistpress.org/BPnews.asp?ID=23322, which reported that 61% of those polled felt that medical research on animals was acceptable.

[42]   Hagelin *et al.* (2003).

[43]   *New Scientist* MORI Poll, 1999, Attitudes Towards Experimentation on Live Animals, http://preview.tinyurl.com/blplnjn, accessed 26 April 2013.

[44]   MORI, 2009, Views on Animal Experimentation, http://preview.tinyurl.com/csfxv2n, accessed 26 April 2013.

[45]   Hobson-West (2010).

[46]   Hagelin *et al.* (2003).

[47]   Grayson (2000); Nuffield Council on Bioethics (2005).

[48]   Macnaghten (2004).

focus groups indicated that there appears to be general ignorance amongst the public about the broad issues of animal experimentation, for example relating to the numbers of animals used, how they are used and how they are regulated. This finding is supported by the fact that when respondents are asked what controls should be included in the UK regulatory system, they often list provisions that already exist[49]. On the other hand, a Eurobarometer report in 2005[50] indicated that knowledge of biotechnology in European countries was growing. Perhaps the truth of the matter is that most people do not think very hard or often about the ethics of animal experimentation, except when forced to do so by inclusion in surveys or opinion polls. Certainly, Macnaghten's focus group study suggested that people often prefer to avoid the issue, because, when they think about it, they find themselves in the uncomfortable position of realising that their views on the treatment of animals conflict with their desire to provide appropriate care for humans in distress or need. Many people feel that humans are, in some sense, fundamentally different to other animals and worthy of special consideration. However, Darwin's publication of *On the Origin of Species* in 1859 followed by *The Descent of Man* 12 years later, and backed up by 150 years of subsequent research have made it clear that we share not only a common origin but also many anatomical, physiological and behavioural features with other species[51]. Inevitably, this relationship has had important consequences on our understanding and views as to how animals should be treated.

The scientific establishment now generally accepts that we share with at least some animal species the ability to experience feelings, although we may not experience them in the same way or be sure what the feelings are[52] (this will be covered in more detail in Chapters 3 and 4). The consequence of this belief that complex animals, such as mammals, birds, fish and others, are capable of suffering, combined with a view that some animal experimentation is necessary to advance human fundamental or medical knowledge (both propositions are disputed by some), is that one finds oneself in an ethical dilemma. The deliberate causation of suffering to another being is clearly a wrong, and some feel, like Regan and Singer[53], that if it is ethically wrong to carry out certain experiments that would cause suffering or harm to humans, then it must also be wrong to carry out these experiments on animals that are similarly capable of suffering. On the other hand, if the research is not carried out, then harm to humans and other animals may also ensue: sick people and animals may not be cured, people and animals may suffer unnecessary harm as a result of poisoning from an untested chemical, and we may be less well

[49]   Festing and Wilkinson (2007).
[50]   Gaskell *et al.* (2005).
[51]   At the end of *On the Origin of Species* Darwin implies a common origin. More explicitly, in *The Descent of Man* he writes of humans and other vertebrate animals: 'Consequently we ought frankly to admit their community of descent; to take any other view is to admit that our own structure, and that of all the animals around us, is a mere snare laid to trap our judgement.' Darwin (1871), p. 43.
[52]   Kirkwood and Hubrecht (2001).
[53]   See Chapter 3.

equipped to make crucial decisions that affect our environment and the animals living within it. Therefore, others counter that while it may be wrong to carry out such experiments on animals, it is a greater wrong not to do so, either because the experiments have the potential to reduce future suffering of humans, and possibly some animals, or because the benefit to humans outweighs the cost to the animals. People who take this position base it on a special utilitarian approach that combines aiming to achieve the greatest good for the greatest number, with a view that there are some things that are ethically unacceptable to do to humans but which you can do to other animals[54].

At the extremes of the debate, the pro and anti positions are mutually incompatible and, sadly, there is no way of proving which is right. These polarised positions are based on two different moral frameworks, both of which have an internal consistency and logic, which explains why the debate over the rights and wrongs of animal experimentation have been so heated and long-lived. However, it is important to emphasise that debate is not as rigidly polarised as this. There are a range of middle positions in which it is argued that some experimentation should be allowed, as long as there are proper controls and restrictions on the conduct and types of experimentation[55]. These controls generally include (1) that the potential benefits of the research must be justified by weighing them against the likely harms that will accrue to the animals; and (2) that some types of research, or the use of some species, should never be permitted. Most people's views lie somewhere in this middle area, and legislative controls on research (where they exist) reflect this, although some may feel that more needs to be done to find alternatives or to reduce the welfare costs of research. Such a system requires decisions to be made on prohibitions, such as which species should or should not be used, and requires the benefits and harms of a specific piece of research to be weighed. However, the practicalities involved in making these decisions are not simple and will be covered in more detail in later chapters.

The presumed current public acceptance of restricted and regulated animal experimentation raises the question what is different about other animals that makes it morally acceptable for us to carry out experiments on them. Philosophical arguments that have been put forward to support this use include (1) animals are not able to form moral contracts and therefore are not entitled to equal consideration; (2) humans owe more to other humans than they do to animals; (3) the comparative value of human and animal lives, both to themselves and more generally; and (4) that there is a moral tradition that animals' interests are treated as subordinate to ours[56].

It is likely that the majority of people have not studied these arguments in detail, but just generally feel that humans are in some way superior, more valuable, or more

---

[54] Grayson (2000). Richard Ryder argues against a Utilitarian justification of the use of animals, arguing that the harm of many (in this case humans) cannot be summated to justify animal research. However, see Leuven and Višak (2013) for a critique of Ryder's approach.

[55] Nuffield Council on Bioethics (2005).

[56] See also Smith and Boyd (1991), chapter 11; Bekoff and Meaney (1998), p. 163.

powerful than other animals, and that while experimentation may be regrettable, our human interests come first. The problem with the human superiority position is that justifications based on perceptions of humans as superior to other animals are very easily criticised on ethical grounds. For example, if might is right, then why is it wrong to experiment on powerless humans? If the justification is the mental superiority of humans, perhaps exemplified by our capacity to be self-conscious and able to reflect on our feelings, then why is it wrong to experiment on a fetus or a brain-damaged human, either of which may be less sentient than certain animals? Indeed Ryder and Singer argue that to treat animals as morally inferior to humans is speciesism[57], which, Ryder has suggested, is as unacceptable as racism. It is perhaps important to clarify that Singer does not argue that it may never be right to use animals in research, only that making decisions on the basis of species alone is wrong.

So what factors should be taken into account when we try to make moral decisions about the broad rights and wrongs of animal experimentation? Singer suggests that some organisms (either animal or human) will possess features that make them more valuable than others. So, for example, pain is as bad for an organism that is self-aware as for one that feels pain but is not self aware, while the loss of life matters much more to one that is self-aware (hence the life of a human will, usually but not always, be worth more than that of another animal). Incidentally, some would argue that the use of animals in biomedical experiments is more justifiable than using them for food, because we need not rely on animals for food, while without animal research some people would die. However, is the difference as great as it first appears? Medicine is concerned with making us feel better or speeding our recovery from disease. The reason for doing this is essentially hedonistic, that is to make us feel good and to banish unpleasant feelings, as is the desire of some of us to supplement our diet with meat.

The Nuffield Council on Bioethics[58] in an in-depth review of the ethics of animal experimentation identified five features that they considered to be relevant when comparing the moral status of humans with particular species of animals: sentience, higher cognitive capacities, the capacity to flourish, sociability, and the possession of a life. They explain why these features are important as follows. Sentience is considered important as it is usually considered to be the capacity to feel pain or pleasure. Higher cognitive capacities include abilities such as language or tool use and may be important as they could affect how the animals see themselves in the world. The capacity to flourish is a factor as there can be practical difficulties in providing research housing and husbandry conditions in which animals do well. Sociability is seen by some as important as it places the animal in a broader context where what happens to it may also matter to others, either animals or humans. Finally, there is the question as to whether life itself has a value or whether it is only important when the owner of the life has expectations about it. Clearly, all animals possess some of these features, but some animals are thought to be better endowed in some of these features than others,

---

[57] Ryder (1975); Singer (1990).
[58] Nuffield Council on Bioethics (2005).

and it is these distinctions that are important in deciding which animals should be protected, and to what extent (see Chapter 4). However, philosophers and scientists have debated the nature, importance, and in some cases reason for using these features as the basis for moral concern. Moreover, people's ethical concerns about animal research are not just linked to the issue of suffering but can include other issues, such as concerns relating to the intrinsic value of the animal, or how other people might feel about the issue and whether it is right to mutilate or destroy another organism[59]. Many people feel strong concern regarding the use of techniques such as xenotransplantation (transplantation of tissues or organs from one species to another) and genetic modification. While often expressed as a feeling that it is, in some sense, wrong to meddle with species boundaries and thus with nature, these concerns are often based on rational fears about the fundamental safety of such techniques[60].

Given the variety and complexity of these issues, it is not surprising that there appears to be no easy answer to the fundamental issue as to whether it is right in some circumstances to allow animal experimentation. The difficulties were demonstrated by the fact that while members of the Nuffield Council on Bioethics Working Party were able to identify four different ethical viewpoints (Box 1.1), they simply

---

Box 1.1   Four different positions on the ethics of animal research identified by members of the Nuffield Council of Bioethics Working Party

The 'anything goes' view

If humans see value in research involving animals, then it requires no further justification (no member of the Working Party takes this position).

The 'on balance justification' view

In accepting research on animals one acts with full moral justification, while accepting that every reasonable step must be taken to reduce the costs that fall on animals.

The 'moral dilemma' view

Most forms of research involving animals pose moral dilemmas: however one decides to act, one acts wrongly, either by neglecting human health and welfare or by harming animals.

The 'abolitionist' view

There is no moral justification for any harmful research on sentient animals that is not to their benefit.

Source: reproduced from *The Ethics of Research Involving Animals*, para. 14.12 (pp. 244–245) (Nuffield Council on Bioethics, 2005), with permission from the Nuffield Council on Bioethics. Available at http://nuffieldbioethics. org/sites/default/files/The%20ethics%20of%20research%20involving%20animals%20-%20full%20report.pdf, accessed 26 April 2013.

---

[59] Smith and Boyd (1991), p. 307; Nuffield Council on Bioethics (2005).
[60] Macnaghten (2004).

agreed to differ on this issue. While those who take a polarised view of animal experimentation are unlikely to be satisfied with such an outcome, their decision reflects the diversity of views held on the subject and identification of various viewpoints is at least a basis for much-needed rational discussion. With this aim in mind, the Boyd Group[61] has performed a valuable service in the UK by bringing together organisations and individuals with a variety of views on the use of animals in research with the aim of promoting dialogue and identifying areas of commonality. The group is chaired by the ethicist Professor Kenneth Boyd, and was established after a meeting between Colin Blakemore, Professor of Physiology at the University of Oxford, and Les Ward, then Director of Advocates for Animals, during a media debate on animal experimentation. Such affairs are usually run on the basis that the greater the disagreement, the better the programme; however, after the event Professor Blakemore and Les Ward found that they were able to have a constructive exchange of views. Those that have subsequently joined the Boyd Group have helped it to produce a number of reports and papers that have greatly helped to clarify why different views are held, and have found significant areas of agreement.

1.7 The Moral Imperative

1.7.1 The need to reduce suffering to a minimum

Most would agree that if society allows animal experimentation, there is a moral imperative to do all that is necessary to reduce any harm to the animals to a minimum. The various harmful effects that can occur and means of minimising them through the implementation of the 3Rs (Replacement, Reduction and Refinement) developed by W.M.S. Russell and R.L. Burch[62] will be discussed in Chapter 6, but briefly include harm caused by the procedures carried out on the animals as well as disease, injury and, possibly, a range of adverse mental states such as fear, anxiety and frustration.

Unfortunately, the apparently simple imperative to reduce suffering to a minimum is not as simple as it might at first appear. First, how do we measure something that is a private sensation? For example, while I may believe you if you tell me you have a headache, it could be very hard for me to know that you had one if you didn't tell me. Even if you do tell me, I have no easy way of evaluating the extent of your suffering. Human pain scoring systems have been developed by the medical community to try to address this problem and to provide a standardised system of measurement and I might use one of these. Alternatively, I might measure your voluntary intake of an analgesic such as aspirin, but pain is essentially a private experience, and the response to it by individuals is very variable and very dependent on culture and context[63]. In battle, soldiers have continued fighting with

---

61   The Boyd Group, http://www.boyd-group.demon.co.uk/.
62   Russell and Burch (1959).
63   Rollin (1989), pp. 150–153 discusses the variability of pain experience in humans and likely individual variability in animals.

terrible wounds that would probably have incapacitated them under other conditions, or have required less morphine for some hours after the injury[64]. So, if pain is a private experience for humans, then it is even harder to evaluate in other species[65]. Even veterinarians, for whom animal pain assessment is particularly important, consider their ability to assess pain as inadequate, and are uncertain to what extent pain scoring systems can be generalised from the particular procedures for which they were developed (e.g. laparotomy vs. castration)[66]. Further, humans can experience many unpleasant sensations other than pain, and some animals are likely to experience some of these as well. Examples include loss, anxiety, fear and panic, as well as discomfort such as hunger, thirst, heat, cold, itching, respiratory distress, and nausea[67]. If we humans find it difficult to assess pain in other animals, it is perhaps even harder to know whether, and how much, an animal is suffering as a result of some other emotional state such as boredom or fear.

Second, if our aim is to reduce suffering to a minimum, is it better that fewer animals should suffer a lot, or that many animals should suffer less? An example of this sort of dilemma might arise where animals used in an experiment need to be housed apart from one another. We generally consider that social animals suffer if housed singly, but if the alternative is to supply a companion, then the total numbers of animals used would have to be doubled. We shall return to this problem again in Chapter 6, when we consider conflicts between the desire to reduce animal numbers and the desire to refine studies so as to reduce suffering.

Third, some would argue that it is better, on animal welfare grounds, to use some species rather than others in an experiment, based on their capacity to suffer. As we shall see in Chapter 4, although the animal kingdom is very diverse, including very simple as well as complex animals, it is often very difficult to tell whether any particular species has a greater capacity to suffer than another.

1.7.2 How important is death?

Many people would argue that, in addition to the moral imperative to reduce suffering to a minimum, there is another to avoid killing animals. In fact, most animals used in experiments are killed, either because the animal has been judged to have reached a point where euthanasia is the humane option or because the experiment has finished and there is no more need for the animal. Indeed, new EU legislation restricts the re-use of animals, so as to prevent animals being repeatedly used in experiments that might cause suffering. In addition, there has always been a certain proportion of animals that are bred which end up not being used in experiments. This is not as culpable as it sounds. A report produced by the UK Laboratory Animal Science Association on the breeding of surplus rodents[68] showed that while

---

[64] Melzack and Wall (1988).
[65] Bateson (1991); Gregory (2004).
[66] Flecknell (2001). Akhtar (2011) discusses how animals' experiences of pain might in some cases be worse than those of humans if they lack, for example, the ability to rationalise experiences.
[67] Gregory (2004).
[68] LASA (1998).

approximately twice as many animals were bred as were actually required, there were a number of reasons for the so-called overproduction and that these animals are not usually wasted (dead mice for example were often used to feed captive birds of prey in zoos or wildlife parks). The reasons for producing a surplus included that there is not always an equal requirement to use males and females, and that in order to produce sufficient numbers of animals of the correct age and weight, it is necessary to breed rather more. Moreover, some techniques require a surplus of animals to be bred. For example, genetic modification technologies are still inefficient[69], and during the breeding process animals are produced that do not possess the transgene and are therefore surplus to requirements. Finally, animals continue to breed, whether scientists are ready to use them or not, and while for mice it is possible to freeze embryos of a particular strain so that there is no need to keep the strain as a breeding line, there will always be occasions when animals are surplus to requirements.

Most of us think of death as being very important, and usually as something to be avoided for as long as possible. As Woody Allen remarked 'I don't want to achieve immortality through my work, I want to achieve it by not dying', but are animals similarly concerned by death? Animals usually do their best to avoid situations that are likely to cause death, although there are exceptions such as parents risking their own lives in defending offspring, and Pacific salmon returning to their natal rivers to put all their resources into spawning before dying. It is also true that all life involves some risk: if an animal is too risk averse, it would not be able to function well when foraging for food or trying to obtain a mate. Generally, however, behaviour that reduces the risk of death is selected for, as animals that are not risk averse are likely to leave fewer, if any, offspring. It is therefore not surprising that when animals detect that they are in a perilous situation, they may show signs of fear and attempt to deal with it through flight or fight. Nonetheless, for most species, with the possible exception of the great apes, there is no good evidence that animals have a concept of death that they can imagine in the abstract and which would cause fear. I tentatively exclude great apes, as Koko the gorilla is reported to have used sign language to discuss the death of both her pet kitten and of another gorilla Michael, but the interpretation of such language studies is not straightforward[70]. It is interesting to note here that in the USA, the only country where invasive experiments on chimpanzees are still permitted, euthanasia of surplus chimpanzees purchased, bred or used in research conducted or supported by the Federal Government is not permitted other than for welfare reasons[71]. Provision therefore has to be made for their retirement, which can be long term and expensive.

From the point of view of the staff charged with carrying out euthanasia, there are psychological costs to killing animals. Most people find killing animals distasteful, particularly so to begin with, and find some methods more acceptable than others[72]. It is therefore ironic that the people who have to kill animals in

---

[69]   Joint Working Group on Refinement (2003a).
[70]   Patterson and Gordon (1993); Taylor (2009).
[71]   Schapiro and Lambeth (2010).
[72]   Animal Procedures Committee (2006b); Wallace (2008); Wolfensohn (2010).

laboratories are usually the animal technicians or caregivers who entered their profession because of their love for animals. When I have discussed this issue with technical staff, they frequently mention the numbing effect of killing large numbers of rodents. Staff can also become attached or concerned about particular animals, particularly when working with larger animals such as cats, dogs and primates, leading to emotional issues when they are euthanased[73].

If animals are killed humanely, without pain or fear, then there is not a welfare issue, though there is an ethical issue. Unfortunately, truly humane killing is an ideal that is not always achieved in practice. Various regulatory agencies and professional bodies have produced codes of practice or regulations that specify particular techniques for particular species to try to ensure that euthanasia should not be a welfare problem, but things occasionally go wrong[74]. Moreover, as we shall see in later chapters, scientific concerns have been raised about some widely used methods of killing such as the use of carbon dioxide to kill rodents, and chemical methods of euthanasia for fish and other aquatic animals.

1.7.3 Provision of a good life

When considering the ethical requirements of keeping animals in captivity, most people would want to go further than simply avoiding subjecting animals to negative experiences. There is surely also a moral imperative to provide the animals not just with the minimum to avoid suffering but also to allow them to experience some positive feelings and to lead a contented life. As Temple Grandin and others argue, animals share with humans the structures that produce emotions and animals like humans are likely to want to experience good emotions[75]. Such ideas are beginning to become more common in codes of practice or advisory documents on animal welfare, so that a recent Farm Animal Welfare Council report[76] argues that each farm animal should have as a minimum what they term 'a life worth living', from the animal's perspective, and that a growing number should experience 'a good life'. Similarly, in the USA, the National Research Council *Guide for the Care and Use of Laboratory Animals*[77] requires that the environment within the cage should include items that enhance 'animal well being'.

However, there are difficulties with the aim of achieving positive emotional states. These include knowing both how far and to what extent it is appropriate to go in achieving such positive emotional states. Happiness in humans tends to be a transitory emotional state associated with either seeking or achieving a desired goal, so attempting to produce a permanent positive emotional state in animals is

---

[73] Herzog (2002).

[74] See section on refining euthanasia in Chapter 6.

[75] McMillan (2005b); Grandin and Johnson (2009), p. 5. Grandin limits her discussion to vertebrate animals. For a description of the structures involved in experiencing emotions in humans (orbitofrontal, insular, and anterior and posterior cingulate cortices), see Dolan (2002).

[76] Farm Animal Welfare Council (2009).

[77] NRC (National Research Council) (1996), p. 37. The latest edition NRC (National Research Council) (2010), p. 41 draws a distinction between health and well-being.

likely to be perverse and doomed to failure[78]. We know that positive emotional states and the best interests of the individual experiencing those states do not always go together, so a drug addict, despite the potential harm involved, will feel a high after taking the drug. Similarly, animals should experience positive emotional states when being fed treats, but the continued use of treats is unlikely to be in the best interests of the animal. There are also real difficulties in assessing whether an animal is in a positive emotional state. It is true that some recent research has tried to identify animals' affective states[79], but these studies are often more focused on identifying negative states than positive ones. If welfare scientists have struggled to discover good measures that reflect poor welfare or suffering, they have hardly begun to scratch the surface of the problem of quantifying the positive feelings an animal might have, and deciding which of these to try to meet and how to prioritise these needs. Nonetheless, there have been considerable improvements in the captive environments provided to animals in research. In some cases, the provisions may well have gone further than those needed simply to provide the animal with its basic requirements.

1.8 Trust in the System

The scientist and novelist C.P. Snow famously and controversially warned of the dangers of a divide between those educated in the sciences and humanities. With the rapid development of scientific knowledge and techniques there are far more potential divides today, not just between the arts and the sciences, but even between scientific disciplines. None of us can be expert in all areas of human endeavour, and so there is no particular reason to expect a general understanding of the scientific reasons for using animals in research, or the ethical dilemmas and decisions that need to be taken. Instead, this understanding and decision-making is largely devolved to those who regulate the research and to the scientists carrying out the work.

It follows that a central issue in the use of animals is the trust that the public have in those who do the research on their behalf. Outright mad or bad scientists/doctors such as Mengele, Shirō Ishii and Shipman are, mercifully, relatively few in number but their actions cast a long shadow. Moreover, while the advances of science lead to many benefits, the public are often wary of new technologies for which scientists are held responsible (e.g. electricity, nuclear weapons, nuclear power, genetic engineering). Popular perceptions of scientists as portrayed in fiction and film include megalomaniac and mad scientists whose tunnel vision blinds them to the consequences of what they are doing (Drs Frankenstein, Jekyll, Moreau and Strangelove). More positive representations of fictional and real scientists do exist, such as the palaeontologist in *Jurassic Park*, forensic scientists in television shows

---

[78] McMillan (2005a) discriminates between being transiently happy and true happiness, which is described as a pervasive sense over time that all is well, and which may even apply to the life as a whole.
[79] See for example Mendl *et al.* (2009).

and the real-life physicist/engineer Barnes Wallace (depicted in the film *The Dam Busters*) but positive representations seem to be fewer or are perhaps less memorable. Even where the motives of fictional scientists are good, they are usually obsessed, or eccentric, and some of the real scientists who present popular science programmes sometimes seem to have been chosen, at least partly, for their slightly unusual looks.

Given all this, it may be considered surprising that the public trusts scientists at all, and yet surveys consistently indicate that they do retain a high degree of trust. A poll carried out for the British Medical Association by MORI in 2001 indicated that two-thirds of the adults polled trusted scientists to tell the truth[80]. Another carried out in 2005 showed that 91% agreed that doctors were the most trusted group followed by teachers (88%), professors (77%), judges (76%) clergymen/priests (73%) and scientists (70%), all these professions, and the police, being seen as more trustworthy than the average man/woman in the street. Similar results were obtained in 2009[81], but not all scientists are equal; university scientists tend to be trusted more than those working for industry[82]. However, scientists should not sit back and relax; trust is a fragile thing and once lost is not easily regained. It seems highly likely that if the public lost their trust in scientists, whether in terms of the quality of the science or the implementation of high standards of animal welfare, public acceptance of animal experimentation would end.

It is therefore also important that the public trust the regulators and the systems that have been set up to provide accountability to the public. In the UK, both the Animal Procedures Committee (now replaced by the Animals in Science Committee) and the Home Office Animals Scientific Procedures Inspectorate provide public annual reports of their activities but, as in many countries, there is considerable secrecy and confidentiality about the details of who carries out animal experiments and what is done. Trust is also based on transparency: if people want to find out what is happening, they can. There are some understandable reasons for secrecy: it may be necessary in cases of national security or to protect intellectual copyright, and after the violent activities of some activists many scientists are concerned for the safety of themselves, their families or colleagues. Nevertheless, secrecy can be destructive, leading to suspicion that worse things are happening than is in fact the case and to further hostility, as well as a range of other disadvantages[83]. Increasing openness, where reasonable and possible, could help to defuse some of these issues. Some countries have introduced freedom of information legislation, for example the Freedom of Information Act (FOIA) and state 'open record laws' in the USA and the Freedom of Information Act (2000) in the UK. Although not specifically aimed at issues to do with the use of animals, this legislation has had some controversial implications for state and public bodies that carry out animal research[84].

[80] Reported in Corrado (2001).
[81] MORI poll, Trust in Professions, 2009, http://preview.tinyurl.com/d8s8jcn, accessed 23 April 2010.
[82] Gaskell *et al.* (2005).
[83] Sandøe (1994).
[84] Cardon *et al.* (2012b).

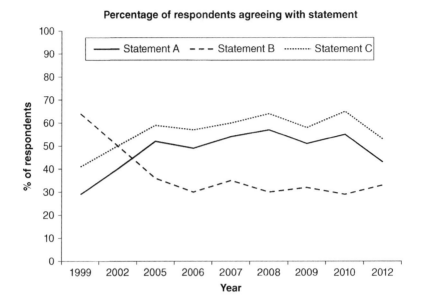

Figure 1.3 Changes in public trust in regulation and enforcement of animal experimentation within the UK 1999–2012.
Statement A: I expect that the rules in Britain on animal experimentation are well enforced.
Statement B: I have a lack of trust in the regulatory system about animal experimentation.
Statement C: Britain probably has tough rules governing animal experimentation.
Redrawn from MORI poll Trust in Animal Experimentation Regulation and Enforcement within the UK 1999–2012, with permission from Ipsos MORI, available from www.understandinganimal-research.org.uk.

Universities, for example, may be required to release information on the research carried out in the establishment, although there are provisions in both jurisdictions to protect information that may compromise the safety of personnel or that would have commercial implications.

Within the UK, the Home Office, in a more targeted step towards greater public transparency, has published online anonymised abstracts of research projects for which licences have been granted[85], and this will become more widespread in Europe as Article 43 of European Directive 2010/63/EU requires the publication of project summaries subject to certain safeguards. The UK Inspectorate has also published reports that include information on visits made to establishments to ensure compliance with legislation. The Animal Procedures Committee (now replaced by the Animals in Science Committee) provided both reactive advice on applications to carry out research, exposés, news reports, etc., and proactive advice on issues identified by the committee. Much of this advice, including a variety of reports, was

[85] Home Office website, https://www.gov.uk/research-and-testing-using-animals, accessed 29 April 2013.

made available to the public on the UK government website. Government research institutions and research funding bodies are also concerned about public account-ability and may use mechanisms such as advisory panels to help them achieve this. Examples from the UK have included the Biotechnology and Biological Sciences Research Council's Bioscience for Society Strategy Panel and the UK Ministry for Defence's Animal Welfare Advisory Committee[86].

The evidence from the 2012 MORI poll (Figure 1.3) indicates that trust in the regulatory and inspection system increased from 1999 to 2005, but also that this trust has fallen away in the last year, and this may suggest that the public requires more information. Certainly this was the view of Sir Mark Walport, Director of the Wellcome Trust, who explicitly drew attention to the link between openness and public trust when he said

> *This poll clearly demonstrates that a majority of the public supports experiments involving animals when these are necessary to advance medical research. But it also reminds us that we must communicate openly with the public if we are to maintain this high level of trust.*

His statement accompanied an agreement by the major UK funders of research to develop principles of openness, practical steps and measurable objectives underpin-ning a more transparent approach to animal research[87]. Undoubtedly this initiative is a step in the right direction, but public trust is also dependent on the existence of appropriate legal controls, and these are covered in the next chapter.

---

[86] Bioscience for Society Strategy Panel http://www.bbsrc.ac.uk/organisation/structures/panels/society/society-index.aspx, accessed 26 April 2013. The Ministry of Defence Animal Welfare Advisory Committee was formally dissolved in 2010.

[87] Declaration on Openness on Animal Research, available from www.understandinganimalresearch.

The Use of Legislative and Other Controls on Animal Research to Meet Public Expectations and Improve Animal Welfare

It is generally accepted that there should be controls on how research on animals is carried out in order to limit harm caused to them. In this chapter I describe various ways in which controls at various levels (e.g. super-national, national, local) can be used to enforce or ensure minimum standards of animal welfare.

2.1 Introduction

Legislative controls are an obvious and attractive option when considering how to best protect the welfare of animals used in research. Legislative controls aimed at minimising harm to animals during their breeding, transport and use in research provide a clear and strong message to those involved of the concern that society has about the practice. Legislation provides a means of monitoring and controlling the activity, and if research is not carried out properly so as to minimise harm, there is the ultimate recourse of penalties and other sanctions. So it is surprising that, in many parts of the world, legislative controls on animal experimentation have been a long time coming. The first law regulating experiments using animals was passed by the UK in 1876; in Australia, Victoria introduced legislation in 1883, but other Australian states did not follow suit until considerably later. Denmark passed animal welfare legislation in 1891, Germany in 1933 and Sweden in 1944[1], but the legislation of many other countries is much more recent.

To provide just a few examples: in the USA, the Laboratory Animal Welfare Act was first passed in 1966, with important additions being made in 1987[2]; Brazil

---

[1] Esling (1981).
[2] Morrison (1981); Rollin (2006); Cardon *et al.* (2012a). See also *ILAR Journal* 52 (Suppl. 2), 2011, pp. 550–551 for a timeline of animal welfare policy, regulation and guidance in the USA.

The Welfare of Animals Used in Research: Practice and Ethics, First Edition. Robert C. Hubrecht.
© 2014 Universities Federation for Animal Welfare. Published 2014 by John Wiley & Sons, Ltd.

passed the Arouca law in 2008; Chile passed the Animal Protection Law in 2009, although at the time of writing in 2012 it was still not yet properly implemented with respect to animal procedures as the National Bioethics Committee had not been constituted[3]; in Japan, despite the large number of animals used in research, there is no specific legislation to protect animals, although the government has produced guidelines; and a proposal for animal welfare legislation in China was produced in 2009 but as yet there is no law. While many countries do now have legal controls, there are significant national differences in both the regulations and their implementation[4]. In some countries, for example, legislation has been specifically enacted to address the issue of the use of animals in research, while in others research is covered as part of more general animal welfare legislation.

Legislation, however, is not a panacea for solving animal welfare and other ethical issues involved in research. By their nature, legislative provisions enforce minimum standards rather than current best practice. Higher standards are often possible, and the decision as to whether to adopt higher standards lies with the researchers, and the institutions in which research is carried out. So, paradoxically, those who carry out research on animals are often in the best position to have a positive impact on the animals' welfare. Researchers in particular are key. They are the ones who make the initial choice to use animals and decide on the types of technique or procedure used on the animals. It follows that they are well placed to take available steps to avoid or reduce animal use, or to minimise or ameliorate any distress, pain or suffering that may result from their research. And researchers often do succeed in making such improvements. There are plenty of examples of animal users who, by developing replacements to animal use or better methods that result in less suffering, have improved the welfare, not just of their own animals, but also of those used by other researchers[5].

However, it is obvious that responsibility for minimising suffering and ensuring good welfare should not lie just with the researcher. Most researchers want to carry out high-quality science while minimising any suffering that may occur, but animal welfare may not be their speciality and they are not disinterested parties. Financial and other constraints can also conspire against the best interests of the animals. External controls, particularly legal controls, can provide animal care staff with persuasive arguments to convince higher management that certain actions are necessary. I once carried out a mini-survey on managers of facilities that bred or carried out research on marmoset monkeys, asking what their priorities were when planning new housing (Table 2.1). The respondents were asked to rank the choices but, as is common with surveys, ignored the instructions and ticked the categories instead. Moreover, all but one ticked two of the categories. While this survey was carried out some considerable time ago now, and undoubtedly could have been done better, it appeared that legislation was a powerful driver of decisions, far

---

[3]   Gimpel in Hawkins *et al.* (2013).
[4]   Bayne *et al.* (2010b).
[5]   The NC3Rs site provides examples, http://www.nc3rs.org.uk, accessed 3 April 2013.

Table 2.1 Priorities of respondents when asked what factors affected their choice of marmoset caging.

Type of institution	Legislative reason	Best use of space	Cost	Best interests of the marmoset
Breeder (n = 16)	14	10	3	4
Research (n = 16)	13	8	9	2
Total	27	18	12	6

Source: Hubrecht (1997).

outranking the respondents' own views as to what might be in the best interests of the animals in their care.

Finally, as discussed in Chapter 1, one very important function of legislative controls is to provide reassurance to the public that research is being carried out with proper checks and balances. These need to be comprehensive, as illustrated in an Australian senate committee report which suggested that if public confidence and support for the use of animals in research was to be maintained, it was essential that the animals' welfare was protected and promoted, that there was transparency of process, public participation in ethical decision-making and that there were mechanisms for accountability[6].

So, legislative controls on the breeding, supply and use of animals in research are useful and arguably essential, but care has to be taken to find a balance between the sometimes conflicting needs of the various parties involved, namely the researchers, the animals used in the research, and any of the general public who have concerns about the research. To achieve this balance, legislation must ensure minimum standards of good welfare and must provide reassurance to the public that things are being done properly, but should not be overly restrictive or bureaucratic.

Other authors have described and contrasted the many different ways that legislation has been used to control animal research in different jurisdictions[7]. Moreover, legal controls are not the only way of ensuring animal welfare and developing higher standards. Codes of practice and non-statutory recommendations have also played a part and deserve to be discussed alongside legal controls. So, in this chapter I will concentrate on the principles of control and oversight of research, looking at what can be controlled and the different means by which it can be achieved, using examples predominantly from the USA and Europe.

---

[6] Rose (2011). This paper also describes how government protection of animals in general has moved from one designed to prevent cruelty to a broader remit to ensure proper animal stewardship.

[7] Bayne *et al.* (2010a,b) provide good reviews of various countries' use of legislative and other means of oversight of the use of animals in research. Matfield (1996) provides a series of articles by various authors on laboratory animal welfare legislation around the globe, and Allen (1998) compares UK legislation under the Animals (Scientific Procedures) Act 1986 and US systems of oversight. A UK House of Lords (2002) Select Committee report also provides a comparison of UK regulation at that time with that of the USA, France and Japan. Note that European Directive 2010/63/EU came into force in January 2013.

2.2 Levels of Control

Controls on animal experimentation can be, and are, applied at a number of different points in the process of planning, carrying out, and reporting of animal experimentation. National or in some cases international controls and regulation can be applied through the legal system. However, other bodies can also play a role in regulating animal research. Funding bodies, regulatory agencies, research institutions, sectors of industry and professional associations can all set their own controls, publishing expectations and requirements on minimum standards, or through systems of accreditation.

2.2.1 Local controls

The ultimate responsibility for the animals used in the research, and thus the first level of local control, lies with the researcher who devises the study. Researchers, like any other group of people, vary in both what they think is ethical and in what they are personally prepared to do. So while the researcher's decisions on the nature of their research are important, many establishments have also developed controls at an institutional, corporate, or group of institutions level, limiting what can be done and specifying how it should be done. Local controls can cover almost any aspect of animal use, from controlling how specific procedures should be carried out to establishing institutional standards on issues such as animal husbandry (i.e. enclosure dimensions and enrichment provision) or staff training.

One might ask why local controls are necessary if the research is carried out in a jurisdiction where there are legislative controls that set minimum standards. In some cases local controls may be required by legislation, but whether they are mandated or voluntary, their importance lies in the fact that they allow establishments to set their own limits as to what is acceptable in their institution. For example, some institutions might decide not to carry out primate research, or perhaps only allow research that causes no more than moderate suffering. It might seem odd that institutions should wish to limit their operations in this way, but there are various reasons why it might be beneficial for them. A desire to operate, and to be seen to operate, ethically is one. An example of a sector where there is a particular need to keep a close eye on public profile vis-à-vis the use of animals is the pet food industry. Pet food manufacturers need to research into and test various aspects of the diets they produce to ensure that they are safe and meet the animals' needs, but for commercial as well as ethical reasons might well wish to limit the types of animal research they do. Similarly, charitable organisations researching cures for human disease depend on public support and thus might wish to keep an eye on the public profile of their research. Even universities may need to balance academic freedom for their research staff with the views of their student clientele. Finally, some organisations operate in areas where there is no legislation, and for this reason may wish to set standards or to harmonise standards across different sites.

One might imagine that local controls are only likely to have a local geographic effect, but when implemented in large institutions the impact is not trivial. The international standards set by multinational organisations, or by client organisations demanding their own minimum standards from suppliers or contract research laboratories can result in local standards being applied across national boundaries where they can influence other users of animals.

2.2.2 National and super-national controls

National and super-national controls tend, by their nature, to be legislative or linked to legislation and, like local controls, may be applied at various levels. In multi-state countries, legislation may be implemented at the federal level, as in the USA, or at state level, as in Canada, while in Australia there is a national code but the legislation is different for each state and territory. These different approaches have probably more to do with history than the merits of the different systems.

In the European Union (EU), regional standards have been implemented through the use of Directives (the first in 1986 and a revised version in 2010). There are various types of European legislation, some more binding than others. Directives are binding, as member states are required to implement the provisions within their national legislation. Some countries already had national legislation prior to the Directives, but as members of the EU they are still required to make sure that their legislation adequately implements the provisions in the Directive. So, regional standards can be valuable in stimulating national legislation in countries that might not otherwise have introduced it and may expand the legislation of those that have it. Further, the harmonisation of regulations achieved by regional standards helps produce a fairer and more standardised legal environment within the region.

2.2.3 The balance between local and national controls

Many areas of oversight, such as ethical assessment of research projects, assurance of competence of staff and compliance with minimum standards, can or must have some degree of local management control. The question then arises, how much should be regulated at a local level rather than through the national regulatory authority? Ethical review is a good example to consider here. Ethical review of projects can be carried out locally, by national or regional regulators, or by some other independent body. National controls may require local review, as in the USA and Australia, or may require both national and local review to address different ethical aspects of the research as in the UK 1986 Animals (Scientific Procedures) Act (ASPA)[8]. Where there are both local and national controls, it is reasonable to ask how much authority should be delegated to the institution. For example, there has been debate within the UK as to whether national legislation should require that minor amendments to research plans should go through the national assessment system or whether they should be approved at local level and subsequently

---

[8] Ethical review is not, in fact, formally a federally mandated function of Institutional Animal Care and Use Committees (IACUCs) but ethical issues are described in NRC (National Research Council) (2010), and IACUCs do review Public Health Service (PHS) funded research using the *Guide* as a basis for evaluation. Bradshaw (2002) describes the Australian system but readers should note that the UK system has evolved since his paper.

reported to the regulatory authority. Questions such as these are not simple. Overly bureaucratic and centralised controls can stifle or delay research and add unnecessary costs. On the other hand, leaving the responsibility entirely with the researcher or institution could cost much more if things go wrong. Good establishments with good management structures may well be able to achieve a very high standard of oversight, albeit perhaps at the cost of some loss of public confidence but poorer establishments pose an obvious risk. Moreover, control and review by a central body, such as the Home Office in the UK, helps to achieve conformity, and the controls are more easily enforced.

So, in achieving an appropriate balance between national and local controls, a balance needs to be struck between enforcement and encouragement. When considering what sort of control is best in any particular situation there are a number of issues that need to be taken into account. These include how quick and easy it is to make changes to the controls and how effective the controls are in achieving the desired high welfare outcome. Some of these are considerations are listed in Table 2.2.

2.2.4 Codes of practice

Throughout the world, codes of practice and guidelines are a common means of either ensuring minimum standards or of working towards minimum standards for many aspects of research. Codes of practice can be produced by government, specialist sectors or by umbrella groups. For example, in Australia[9] documents produced by the National Health and Medical Research Council provide general guidance for those involved in research, while the Animal Research Review Panel (ARRP), a statutory body under the New South Wales Animal Research Act 1985, has produced a series of guidelines on animal care. In the UK, a partnership of UK organisations from the government, charity and commercial sectors that support cancer research have produced guidelines covering all aspects of cancer research with a focus on animal welfare. Other examples include guidelines produced by the Laboratory Animal Science Association (LASA) on the proper transport of animals used in research, the American Veterinary Medical Association guidelines on euthanasia, and the International Association for the Study of Pain Guidelines on the use of conscious animals for pain research.[10]

Codes of practice developed by interest groups can develop momentum and influence legislation. In the USA, the first edition of the *Guide for the Care and Use of Laboratory Animals* (the *Guide*)[11], for example, was developed by a group of veterinarians working in research institutions in the Chicago area, but was subsequently taken on by the NIH and ILAR, giving it national status. Similarly, in the UK, the Home Office's 1989 *Code of Practice for the Housing and Care of*

---

[9] Australian Government National Health and Medical Research Council (NHMRC 2004, 2008), http://www.animalethics.org.au/home, accessed 3 April 2013.

[10] LASA transport guidelines (Swallow *et al.*, 2005); euthanasia (American Veterinary Medical Association, 2007); cancer research (Workman *et al.*, 2010); IASP Guidelines (Zimmermann, 1983).

[11] NRC (National Research Council) (2010), now in its 8th edition.

Table 2.2 Advantages and disadvantages associated with various types of controls to ensure the welfare of animals used in research.

Level of control	Pros	Cons
Local	Changes can be made rapidly Relevant to area/type of research Flexible Promotes sense of personal/ institutional responsibility Advantages relating to setting rules that meet local needs	Good ideas might not be disseminated across the industry Risk of wasted effort in repeatedly developing similar controls Higher controls at some institutions risk loss of competitive edge Additional costs whenever new standards are developed Need for personal responsibility Need for benchmarking Need for buy-in Possibility that controls may not be effectively enforced Possible lack of awareness of current good practice Inconsistencies across institutions
Industry/area of research	Changes can be made moderately rapidly Relevant to area/type of research Flexibility is possible Competitive edge not lost Sense of corporate responsibility Better dissemination Benchmarking easy Enforcement more likely?	Requires collaboration and agreement Issues of intellectual property Possible loss of personal responsibility Possibility that controls may not be effectively enforced
Legislation/ regulation	Can be enforced There is good awareness of how a particular institution or piece of research compares to the norm Best dissemination? Consistency across institutions	Slow to make changes Lack of personal responsibility Inflexible Not relevant to all areas/types of research

Source: courtesy of Ngaire Dennisson, adapted from a presentation given to the RSPCA/UFAW Rodent Welfare Group meeting October 2010.

Animals Used in Scientific Procedures was based on Royal Society and UFAW guidelines published in 1987, and the 1995 *Code of Practice for the Housing and Care of Animals in Designated Breeding and Supplying Establishments* was built on an earlier document produced by the Laboratory Animal Breeders Association.

In New South Wales, Australia the same process occurred. The *Australian Code of Practice for the Care and Use of Animals for Scientific Purposes* (the *Code*) was originally developed by scientists in 1969, and by 1996 all Australian universities had agreed to implement it. Today, it is published by the Australian Government National Health and Medical Research Council, with the Australian Research Council[12], and because it has been adopted by state-based legislation, it is legally binding in all Australian states. Internationally, the Council for International Organizations of Medical Sciences (CIOMS) developed a brief document in 1985 (updated and expanded in December 2012) that provided overarching principles for biomedical research using animals. These formed the basis for the US Government's own principles regarding animals used for research purposes[13]. Another recent international code of practice relating to the use of animals in research is advice produced by the World Organisation for Animal Health (OIE)[14] for its members on the formulation of regulations, or other form of oversight of animal research. The OIE, with the aim of facilitating and promoting international trade, has produced consensus documents (Terrestrial and Aquatic codes), which are intended to be the principal veterinary reference for World Trade Organization members. Initially, the OIE had little to say about the use of animals in research, but in 2010 it introduced a new chapter to the Terrestrial code, specifically on the use of animals in research and education[15]. The chapter is not heavy on detail and is not compulsory, but does signpost important issues and because of the size of the OIE (in 2010 it had 178 member countries[16]) should be important in raising standards worldwide.

2.2.4.1 Codes of practice or mandatory legislation?

When legislation is developed, it is useful to consider what should be provided for in the legislation and what might better be dealt with by codes. In the UK, for example, there are mandatory provisions regarding the use of neuromuscular blocking agents and provisions regarding the re-use of protected animals. In addition, the Secretary of State is required to produce guidance and codes of practice regarding the care and use of protected animals. It would have been quite feasible for these codes, which include housing and husbandry standards and methods of humane killing, to have been incorporated into the legislation, but codes of practice have certain advantages over primary legislation.

---

[12] *Australian Code of Practice for the Care and Use of Animals for Scientific Purposes*, 8th edition, 2004, http://www.nhmrc.gov.au/guidelines/publications/ea28, accessed 17 September 2013.

[13] International Guiding Principles for Biomedical Research Involving Animals (1985), http://www.cioms.ch/publications/guidelines/1985_texts_of_guidelines.htm, accessed 3 April 2013. US Government Principles for the Utilization and Care of Vertebrate Animals Used in Testing, Research, and Training, http://grants.nih.gov/grants/olaw/references/phspol.htm#USGovPrinciples, accessed 3 April 2013.

[14] The OIE is the World Trade Organization's reference organisation for standards relating to animal health and zoonoses.

[15] World Organisation for Animal Health (2010).

[16] OIE membership according to http://www.oie.int/en/about-us/our-members/delegates/, accessed 3 April 2013.

1. Codes can be developed after the legislation, and so allow more time and thought to be put into the content than might be the case if the provisions have to be rushed through with the bill.
2. Because the codes are not part of the primary legislation, they can be updated periodically, as science, expertise and good practice develop, without the expense, time and difficulty of revising primary legislation.
3. Failure to comply with non-mandatory codes is not of itself a breach of the law, which gives regulators a degree of discretion. This can be valuable in cases where an institution with very high standards of animal care does not quite meet the standards in some minor way. It might, for example, possess large enclosures that are a fraction too small in one dimension. If the enclosures are otherwise well furnished, it would be very unlikely that this difference would have any effect on the occupying animal's welfare, but if the dimensions are specified as legal minimum standards then that institution would be breaking the law. On the other hand, another institution with lower overall standards of husbandry and care but which met the letter of the law would be in the clear. Strict enforcement of legal minimum standards might mean wastage or premature euthanasia of animals, not to mention extensive disruption of research. In contrast, codes of practice allow inspectors to exercise their judgement on animal welfare issues, and to take into account the entire husbandry system of an animal unit, rather than forcing them into a box-ticking exercise that may have unintended adverse welfare consequences. You might argue that if codes are not mandatory that is a disadvantage; after all, what is the point of controls if they do not enforce certain standards? However, codes do not have to be toothless. In legal proceedings, failure to comply with relevant provisions within the codes can be taken into account by a civil or criminal court, which shifts the burden of proof towards the licensee with respect to the adequacy of the licensee's provisions for the animals' welfare[17].

Given that codes of practice have these advantages, it is interesting to note that European Directive 2010/63/EU contains mandatory tables of minimum dimensions for the accommodation of animals used in research and these have to be implemented by January 2017 (for those who are interested, Box. 2.1 provides some background as to how this occurred). As EU member states were required to adopt and apply laws, regulations and administrative provisions necessary to comply with this Directive by January 2013, the amended UK legislation must now also incorporate minimum dimensions in its primary legislation. This has been achieved in the UK in a way that allows future changes to housing standards in the Directive to be implemented in the UK without revising primary legislation. The Directive's provisions have the advantage that minimum housing standards are now set across the EU but have the disadvantage that, should new knowledge indicate that different standards are necessary, changes could only be implemented by obtaining agreement within the EU to revise the Directive.

[17] Dolan (2007).

Box 2.1 The development of minimum standards for the care of animals used in research in the UK and Europe

In 1986 the Royal Society and UFAW jointly published guidelines that set out for the first time minimum space recommendations for laboratory animals. In the same year the UK introduced new legislation – The Animals (Scientific Procedures) Act 1986 – and subsequently published codes of practice for the accommodation and care of these animals, the tables of which were based on Royal Society/UFAW guidelines also published in 1986. In the same year, two European organisations published guidelines and legislation on the use of animals in research. The Council of Europe, an organisation concerned with harmonising social and cultural issues across Europe[18], published a Convention for the Protection of Vertebrate Animals Used for Experimental and Other Scientific Purposes (European Convention ETS 123) to which was attached an appendix containing guidelines on animal care[19], while the European Community (which works to facilitate trade across its member states) published Directive 86/609/EEC, to which were attached recommendations on animal care (Annex II to the Directive[20]). The recommended animal housing standards attached to the European Convention and the European Directive were very similar, but the UK 1986 Act's codes of practice for the accommodation and care of animals in some respects exceeded the minimum European recommendations.

The housing and husbandry standards set out in Annex II of Directive 86/609/EEC were advisory and thus it was also permissible for the UK standards of housing to be codes rather than mandatory (member states are required to implement Directive provisions within their national legislation). After 11 years it was felt that animal welfare science had advanced enough to allow housing standards to be looked at again, and so in May 1997 Council of Europe member states that had ratified the Convention (Parties to ETS 123) started the process of revising the Convention's guidelines on animal care set out in Appendix A to the Convention. This difficult process took more than 8 years, but eventually led to considerably more detailed and up-to-date guidelines[21]. As the EU was by this time also a party to the Convention, the EU's standards needed to be revised to take into account the Council of Europe's work. The European Commission therefore published in June 2007 revised recommendations on animal care that were very similar to those produced by the Council of Europe, but these still remained recommendations only[22].

[18] de Leeuw (2004).

[19] Esling (1981) describes the early stages of the development of the Convention.

[20] Directive 86/609/EEC can be found at http://eur-lex.europa.eu/LexUriServ/LexUriServ.do?uri=CELEX:31986L0609:en:NOT, accessed 3 April 2013. Some minor changes were made to the Directive in Amending Act Directive 2003/65/EC.

[21] Revised Appendix A to Convention ETS 123, http://conventions.coe.int/Treaty/EN/Treaties/PDF/123-Arev.pdf. Background animal welfare information and further advice provided by the experts (the Parts B) can be found at http://www.felasa.eu/about-us/library/, both accessed 3 April 2013. Richmond (2004), who chaired the later sessions, describes the process of revision in A Review and Comparison of Processes to Change Regulatory Guidelines: A European Perspective, http://www.ncbi.nlm.nih.gov/books/NBK25387/, accessed 3 April 2013.

[22] Commission Recommendation of 18 June 2007 on guidelines for the accommodation and care of animals used for experimental and other scientific purposes (notified under document number C(2007) 2525), http://eur-lex.europa.eu/LexUriServ/LexUriServ.do?uri=OJ:L:2007:197:0001:0089:EN:PDF, accessed 3 April 2013.

Shortly afterwards, in 2008, the European Commission announced its intention to update its legislation relating to the use of animals in research with the aim of helping to unify standards across Europe (many member states had produced legislation subsequent to the original Directive), and to more fully incorporate the 3Rs principles[23] (see Chapter 6). This revision was completed in 2010[24]. In this revision, minimum enclosure dimensions were incorporated into the body of the text, and became mandatory. However, the Commission recommendations of June 2007 remain as advisory text.

2.2.4.2 How codes can be made effective

Codes of practice can be linked to the legislation, as is the case for animal housing standards in the UK and Australia[25], but there are other ways of ensuring compliance. Financial leverage by funding bodies is often used, both as a carrot and stick, to ensure that due regard is taken of guidelines or codes of practice. For example, in the USA, while institutions using protected animals are required by the Animal Welfare Act to abide by a code of federal regulations, more detailed standards of good practice in animal care and use, which also cover commonly used species not included in the Animal Welfare Act, are provided in the National Research Council's *Guide*[26]. Institutions that do not comply with the *Guide* can have their access to public funding either restricted or cut[27]. Independent funding bodies such as charities can also set conditions. Although the US Animal Welfare Act does not cover research on mice and rats, the National Multiple Sclerosis Society, for example, requires approval by a local committee, the Institutional Animal Care and Use Committee (IACUC)[28], before releasing funds for research using these animals. Within the UK, the National Centre for the Replacement, Refinement and Reduction of Animals in Research (NC3Rs), in association with major UK funding organisations, produces guidelines on primate accommodation, care and use[29] that are

[23]   Commission Proposal 5 November 2008, http://ec.europa.eu/environment/chemicals/lab_animals/proposal_en.htm, accessed 3 April 2013.
[24]   European Directive 2010/63/EU, http://eur-lex.europa.eu/LexUriServ/LexUriServ.do?uri=OJ:L:2010:276:0033:0079:EN:PDF, accessed 3 April 2013.
[25]   Rose (2011).
[26]   NRC (National Research Council) (2010). Regulation of animal use in the USA is controlled by a web of legislation, regulations, guidelines and policies that include the Animal Welfare Act, Animal Welfare Regulations and Public Health Service (PHS) Policy. This latter is a statutory requirement for PHS-funded research. See NRC (National Research Council) (2010).
[27]   Code of Federal Regulations, http://preview.tinyurl.com/cd52nxo, accessed 26 April 2013; Bayne *et al.* (2010b).
[28]   The responsibilities of local committees (IACUC in the USA, Animal Welfare Body under European Directive 2010/63/EU, and Ethical Review Process under ASPA in the UK) vary under different regulatory systems, but may cover issues such as ethical reviews of proposed research, checking or providing advice on animal welfare standards, ensuring appropriate staff training is in place, and even staff safety.
[29]   Guidelines on primate accommodation, care and use, http://www.nc3rs.org.uk/page.asp?id=277, accessed 3 April 2013.

higher than the minima set out in UK legislation. Institutions or researchers, whether in the UK or abroad, who are supported by these funding bodies are expected to comply. The NC3Rs also provides guidance on how organisations may need to upgrade in order to comply, and in exceptional cases the funding bodies may even provide funds to meet the required standards.

Accreditation helps to facilitate commercial interactions and is another way of applying financial leverage to comply with codes or guidelines. The Association for Assessment and Accreditation of Laboratory Animal Care International (AAALAC), a not-for-profit organisation, uses the *Guide* as its primary standard for accrediting institutions around the world, and because of this the *Guide* is influential well beyond the USA's boundaries[30]. If a nation has higher standards than those set out in the *Guide*, then AAALAC requires any institution desiring accreditation in that country to meet those higher standards. On the other hand, in countries where there is no national legislation or the standards are less rigorous or extensive than those set out in the *Guide*, then the institution must, at least, comply with the *Guide*. Some may feel that as accreditation with AAALAC is voluntary, it is a less powerful way of implementing controls than legislation. However, accreditation has proved to be a useful way for many organisations to provide assurance regarding their standards to global partners or clients, and over 800 companies, universities, hospitals, government agencies and other research institutions in 34 countries are AAALAC accredited[31].

2.3 Scope of Controls

The primary aims of controls on the use of animals in research, whether those controls are local or national or are guidance or compulsory, are:

1. to ensure that procedures are only carried out for reasons that can be justified; and
2. if they are carried out, to ensure that the suffering of the animals involved, in terms of numbers of animals affected and the extent of any suffering, is reduced to a minimum.

To achieve these two aims, it is obviously necessary to have controls that ensure that research is competently carried out by trained people in appropriately equipped facilities, and that animal welfare issues are properly addressed. However, there are also areas where it is less clear what should be included or excluded from the control, which raises questions on what is commonly termed 'scope'. The following sections address some of these issues.

---

[30] http://www.aaalac.org/accreditation/resources.cfm, accessed 3 April 2013.
[31] http://www.aaalac.org/about/index.cfm, accessed 3 April 2013.

2.3.1 Species

Given the range of complexity of animals and hence presumed differences in their ability to experience feelings that might involve suffering, any consideration of the scope of legislation must take a view as to which animals need to be protected and which animals, if any, should be prohibited as research subjects. These are complex and challenging issues, and are covered in detail in Chapter 4. Perhaps because the problem is so difficult different jurisdictions have ended up protecting different species. For example, as already mentioned, the US Animal Welfare Act excludes rats, mice and birds (as well as farm animals used in agricultural research). Not surprisingly this exclusion of species that are used in the greatest numbers has been controversial, and in 2000 the US Department of Agriculture agreed to review the issue. There was intense lobbying from both sides of the debate but in 2002 Congress decided to maintain the status quo. In Australia, there is some variation between states as to which species are protected, but generally the use of verte-brates, but not invertebrates, is controlled[32]. In contrast, in Europe, Directive 2010/63/EU applies to live non-human vertebrate animals and one invertebrate group, the cephalopods, while in New Zealand both cephalopods and decapods are protected (see Chapter 4). Given that our knowledge of species capabilities is con-stantly changing, it would seem to be a good general principle, when developing controls on animal experimentation, to build in a simple mechanism to amend the species covered as science provides new information. Indeed, this was the case under the UK's ASPA 1986 legislation prior to 2013 when the Secretary of State had the authority to extend the definition of protected animals[33]. Unfortunately, the 2010 European Directive does not include such a mechanism. Other issues regarding which animals to protect raise equally difficult questions. For example, from what stage in an animal's development should it be protected and should animals bred and then humanely killed for tissue be included in the legislation?

2.3.2 Practices

A fundamental question that developers of legislation or other controls have to address is what practices or type of study should be covered by the control. Again, this issue is more complex than first appears. In the USA, the care and use of any protected animal at a research facility carrying out research, teaching or testing falls under the Animal Welfare Act and requires oversight by an IACUC[34]. However, the term 'use' is a very broad one. We discussed in the previous chapter that there are various types of research that either do not, or are very unlikely to, result in suffering and it would seem reasonable that these should be excluded from controls. In some jurisdictions such exclusions are made. We saw in Chapter 1 that UK and European legislation has drawn the threshold for control at the level of insertion of a hypodermic needle in accordance with good veterinary

---

[32]   Figgis and Griffiths (1998).
[33]   Dolan (2007), p. 22.
[34]   National Institutes of Health (2002).

practice. However, pain is not the only consideration and the legislation makes clear that other practices such as psychological stressors, food restriction or anything likely to cause stress are also included[35]. The impact of an action on an animal can vary enormously depending on the animal on which it is performed. Species, strain, domestication, and the animal's prior experience can all be important factors. Human contact is usually a positive occurrence for domesticated animals, but even something as apparently trivial as a human's approach may need to be regulated if a species or strain is particularly fearful[36]. Hence, legislation should not try to provide an extensive list of controlled procedures but should be written in such a way that it legislates for any procedures that might cause stress to protected animals.

The regulation of techniques or procedures is further complicated by the fact that some of them may also be carried out as part of normal husbandry or for veterinary purposes. A clinician, for example, might take a blood sample for diagnostic purposes. He or she might also carry out exactly the same procedure on the animal and an identical analysis for purely experimental purposes. Although from the point of view of the animal the experience is the same, the purpose of experimental regulation is to control research on animals and thus European Directive 2010/63/EU, for example, specifically excludes from its remit non-experimental agricultural and clinical veterinary practices. Nonetheless, this overlap of procedures common to research and other areas of animal use can lead to anomalies.

One such anomaly is identification. Marking, primarily for the purposes of identification, is specifically excluded in European Directive 2010/63/EU, and ringing (i.e. fitting leg bands to birds) has not been considered a licensable procedure in the UK[37]. Ringing of birds is, however, likely to cause at least as much distress to the animals involved as some procedures that are covered by legislation. Moreover, although bird ringing is primarily carried out for identification, it might well be argued that its purpose is research. Another marking issue relates to transgenic mice where the issue is complicated by the fact that tissue removed for marking purposes can also be used to discover whether that particular animal is carrying the genetic modification of interest (genotyping). Traditionally the tip of the tail of young mice was removed for DNA analysis, but there is increasing evidence that tail tipping may not be the most humane method. Ear clipping is a recognised method of marking rodents[38] and if it is done for this purpose, then it would seem

---

[35]   For example, Guidance on the Operation of the Animals (Scientific Procedures) Act 1986; see also European Directive 2010/63/EU, Article 3.

[36]   Approach by a human has been used as a stressor for hens (Jones *et al.*, 1981), and nervous animal strains have been bred and sometimes used in research, e.g. pointer dogs (Newton and Lucas, 1982).

[37]   However, there are other controls. In the UK, bird ringing is regulated through the British Trust for Ornithology.

[38]   There are a number of methods of marking rodents (microchip, ear-clipping, felt marker, ear-tag) but virtually all of them can have some welfare impact depending on the species, age and strain of the rodent.

to do no further harm to the animal to use the tissue for genotyping. However, the question has been raised as to whether the technique should be licensed as part of a programme of scientific work if the tissue is used for genotyping.

2.3.3 Limits on categories of research

Legislation can limit categories of research to those that are considered acceptable to society by specifically stating what types of research are permissible under the legislation. For example, Box. 2.2 shows the purposes for which animal research can be carried out within the EU (some member states may choose to restrict the categories further).

Legislative controls may also proscribe certain types of research on the grounds that the research is either unacceptable or unnecessary. Examples of such prohibitions in the UK have included cosmetic testing, or the use of animals for training purposes in surgical procedures other than for the training of practising surgeons in microvascular techniques[39].

Box 2.2   Permissible purposes for research using animals under European Directive 2010/63/EU

(a) basic research;
(b) translational or applied research with any of the following aims:
    (i) the avoidance, prevention, diagnosis or treatment of disease, ill-health or other abnormality or their effects in human beings, animals or plants;
    (ii) the assessment, detection, regulation or modification of physiological conditions in human beings, animals or plants; or
    (iii) the welfare of animals and the improvement of the production conditions for animals reared for agricultural purposes;
(c) for any of the aims in point (b) in the development, manufacture or testing of the quality, effectiveness and safety of drugs, foodstuffs and feed-stuffs and other substances or products;
(d) protection of the natural environment in the interests of the health or welfare of human beings or animals;
(e) research aimed at preservation of the species;
(f) higher education, or training for the acquisition, maintenance or improvement of vocational skills;
(g) forensic inquiries.

Source: European Directive 2010/63/EU. © European Union, http://eur-lex.europa.eu/.

---

[39]   Europe has banned the sale of cosmetics products from 2013 containing an ingredient that has been tested for the purposes of the Directive using animals. This ruling is specifically referred to in Directive 2010/63/EU. The UK Cruelty to Animals Act 1876 section 3(6) originally prohibited the use of animals for the purpose of attaining manual skill, and this restriction was maintained under ASPA 1986. The House of Lords Select Committee (2002) criticised this proscription on the grounds that the public would, like them, probably prefer that surgeons gained their skill on anaesthetised animals rather than on them, and the revised ASPA has no such restriction.

Legislation can also limit the maximum suffering that an animal may experience. For example in the UK, under legislation before 2013, personal licensees and project licences contained endpoint conditions describing the conditions under which appropriate humane killing must be carried out. UK guidance also provided an upper limit to suffering beyond which experiments would not be permitted, stating that the 'Secretary of State will not license experiments producing serious injury or severe pain without effective anaesthesia'[40]. The new European Directive 2010/63/EU is not quite as prohibitive, as it requires that member states should not allow procedures that involve severe pain, suffering or distress that is likely to be long-lasting and cannot be ameliorated and there is a clause that allows such procedures in exceptional circumstances. There is no provision limiting the maximum suffering permissible in the US Animal Welfare Act and this issue is left to the regulatory authorities to judge at the time of authorisation. On the other hand, the USA and other jurisdictions do restrict certain procedures that are very likely to result in unnecessary suffering. So, for example, the use of paralysing agents would impair the ability of care staff to judge whether the animal is suffering and their use is banned, by regulation in the USA and by Directive in Europe[41].

Certain types of research, or the use of certain species, may be considered unacceptable, or only acceptable under very special circumstances. For example, European Directive 2010/63/EU permits member states to use great apes in cases of exceptional justification, but there are safeguards. The only acceptable justifications are the preservation of the ape species itself or outbreaks of life-threatening or debilitating clinical conditions in humans. Moreover, their use must be reported to the Commission and other member states giving reasons and evidence for the decision. Similarly, the New Zealand Animal Welfare Act 1999 precludes the use of non-human hominids (great apes), unless the use benefits the individual concerned or its species, and approval is obtained from the Director-general for Ministry of Primary Industries[42]. The UK legislation now absolutely prohibits the use of great apes and also contains special provisions restricting the use of dogs, cats and Equidae. The banning or restricting of certain types of research does not have to be dealt with in primary legislation; it can also be left to local controls. Although some may prefer the certainties of absolute prohibition, an alternative argument is that it is better to deal with these issues during ethical review at the time of authorisation on a case-by-case basis. Another possibility is to adopt policy decisions such as

---

[40]   ASPA 10.2(b), Home Office (2000), Chapter 5 Project Licences, Paragraph 5.17. Some rare procedures are licensed in which death occurs very rapidly and where it may not be possible, or even humane, to kill all animals before death occurs. In cases where anaesthesia may interfere with the outcome, other methods of alleviating distress may be used such as increased monitoring, the provision of appropriate enrichment or heat pads.
[41]   Code of Federal Regulations, Title 9, Chapter 1, Subchapter A, Animal Welfare (known as 9CFR) Sec. 2.31 Institutional Animal Care and Use Committee (IACUC), http://www.aphis.usda.gov/animal_welfare/downloads/awr/awr.pdf, accessed 3 April 2013. European Directive 2010/63/EU, Article 14, 3.
[42]   National Animal Ethics Advisory Committee Annual Report 1 January to 31 December 2011, Ministry for Primary Industries, New Zealand Government, http://preview.tinyurl.com/c5jwhbv, accessed 13 May 2010.

those made in the UK, under ASPA 1986, where guidance has indicated that certain proposals would not be approved, for example the use of great apes, the use of animals for testing or developing alcohol or tobacco products, or for developing or testing offensive weapons.

2.4 Systems for Ethical Review and Authorisation

The previous sections have discussed the provisions that legislation or other codes can provide to ensure that research is carried out in places that have the proper facilities and resources. In addition to these proscriptive controls, there also needs to be a system to critically assess each proposal to use animals before the research is carried out. The process of ethical review will be covered in more detail in Chapter 5, so here I will consider the increasing role of ethical review as a component of control, oversight and authorisation.

One of the purposes of ethical review is to judge whether the research should be carried out at all, and to carry out this task the review must include, as a minimum, an assessment of the likely harms that may occur to the animal and of the likely benefits that may accrue as a result of performing the study (a harm–benefit judgement, see Chapter 5)[43].

Most regulatory systems require some form of review, either by the regulator or by the institution, but a proper harm–benefit decision, which takes into account the procedures to be carried out, the species and numbers of animals to be used, weighing them against the likely benefits of the research, is not always required. Indeed, until the introduction of European Directive 2010/63/EU, the requirement for harm–benefit analysis was found only in UK and German legislation[44].

With the continuing evolution of regulation of animal experimentation, there has been an increase in the extent and scope of review of proposed research. The ethics, as well as the science, of research may now be judged, not only by funding bodies at the time of grant approval, but also before and during the research by the institution's local committees. The growth of review, although sometimes seen by researchers as a drag on their research activities, is undoubtedly a good thing. Scientists may become blasé about procedures that they routinely carry out, and ethical review raises complex questions. Individuals may vary in their views about what is acceptable and because of this it is generally accepted that ethical review and judgements about animal use benefit from the involvement of a number of persons with different backgrounds and experience (see Chapter 5).

---

[43]   There may be factors other than animal welfare that influence this decision. For example, an institution may decide that it does not wish to carry out certain types of research. It is worth noting that the term 'harm–benefit analysis' is misleading. As we shall see in Chapter 5, it is a weighing up of the issues and not in any way a mathematical operation.

[44]   Festing and Wilkinson (2007). Article 38d of European Directive 2010/63/EU requires a harm–benefit analysis.

However, there is considerable variation under different systems of control as to what is required in the review and whether the review is compulsory. IACUCs are required by federal law in the USA but are only a recommendation in Japan under guidelines produced by the Science Council of Japan[45] (in Japan, control of animal experiments is a voluntary responsibility of the research institution). In the USA, IACUCs prospectively review and approve proposed research, ensure that alternatives to animals are used where these exist, and require that discomfort, distress and pain are minimised, but are not required to perform a harm–benefit analysis under either the Animal Welfare Act or Public Health Service Policy. Nonetheless, times are changing and IACUCs are increasingly discussing harm–benefit even if the starting point for the debate is often that if the research is funded then it is justified[46]. IACUCs are, however, required to ensure compliance with regulations and that experiments are carried out properly, as well as advising on 3Rs issues. IACUC members are required to determine that 'procedures involving animals will avoid or minimize discomfort, distress, and pain to the animals' and must possess sufficient ability to 'assess animal care, treatment, and practices in experimental research as determined by the needs of the research facility and shall represent society's concerns regarding the welfare of animal subjects used at such facility'[47]. In Europe, prior to implementation of the 2010 Directive, ethical review has been carried out in a wide variety of ways, at national, regional or institutional level, or even sometimes at more than one level[48]. Harm–benefit analysis is now required across Europe under Article 38d of Directive 2010/63/EU from 2013, but centralised project authorisation by the Home Office has been a requirement in the UK since 1986. The Minister of State makes the decision but except in exceptionally difficult cases the decisions are, in reality, delegated to Home Office officials. The local Home Office inspector plays a key role in providing advice to the Minister

---

[45]   Guidelines for Proper Conduct of Animal Experiments, June 1 2006, Science Council of Japan, http://www.scj.go.jp/ja/info/kohyo/pdf/kohyo-20-k16-2e.pdf, accessed 3 April 2013.

[46]   I have been informed by colleagues who sit on IACUCs that most such committees do not carry out harm–benefit analysis. However, Tannenbaum (2001) suggested that in his experience IACUCs do tend to balance animal pain against the value of the experiment. Further, Rollin (2006) wrote

> Friends of mine at NIH even told me that within five years of the law going into effect in 1987, several committees were discussing cost/benefit issues in terms of animal suffering, even though the law does not mandate such discussions. Though most protocols are not rejected, many are modified to the benefit of animals. The biggest problem remaining with the laws is that primacy is still given to the science being done, not to animal welfare'.

[47]   Animal Welfare Act as of 1 February 2010 §2143. Standards and certification process for humane handling, care, treatment, and transportation of animals. (b) Research facility Committee; establishment, membership, functions, etc., http://preview.tinyurl.com/cyc4va9, accessed 26 April 2013. Code of Federal Regulations Title 9 Animals and Animal Products §2.31 Volume: 1 Date: 2009-01-01. Institutional Animal Care and Use Committee (IACUC). d(1) http://preview.tinyurl.com/bt6v4vv, accessed 26 April 2013. Successive editions of the *Guide* have added to the list of scientific and welfare issues that should be considered by IACUCs during protocol review (Bayne, 2011).

[48]   Smith *et al.* (2005, 2007).

on projects, and as they also inspect the institution for compliance with other standards, such as housing and husbandry, they are well placed to judge the project, the staff and the facilities in total.

Local ethical review is not required by European Directive 2010/63/EU, but in 1999 the UK required licensed institutions to establish an Ethical Review Process (ERP). A specified function of the ERP was to analyse projects with reference to the expected benefits and the likely costs (harms) to the animals and to provide advice to the Certificate Holder (since 2013 this person is now termed the Establishment Licence Holder), in order that he or she could make a decision on behalf of the institution as to whether the harms justified the benefits. The UK has therefore required two levels of harm–benefit analysis, one as part of the process of authorisation, and one as a system of local control within the research institution and has retained local review and advice to the establishment licence holder as a role of the new Animal Welfare and Ethical Review Body (AWERB).

Retrospective review is becoming more common, and is required under Article 39 of European Directive 2010/63/EU for projects classified as severe or which involve non-human primates, as well as any other projects as deemed necessary by the national authorising authority. Retrospective review may occur at the end of the research or at specified stages during it. Institutions may carry out retrospective review of projects, and there may be some degree of retrospective review when the time comes to submit the research to a client or regulatory body. Certain journals have for some time set their own standards for submitted papers. Within the UK some, for example, require that the research be permissible under UK law, and provide ethical guidelines for prospective authors[49]. In the past, many journals did not have policies or publish guidance on the use of animals; however, increasingly journals are now asking for details on the design, procedures and animal husbandry used for studies reported in papers, and the ARRIVE guidelines are helping to formalise these requirements[50]. The US Association for the Study of Animal Behaviour together with the UK Animal Behaviour Society established one of the first journal ethical review systems, in which an ethics committee advises the editor of the journal *Animal Behaviour* on welfare and other ethical issues arising from contentious papers. These reviews are taken into account alongside other scientific reviews and can lead to modification or even rejection of papers. Other journals now operate similar systems, for example the *Journal of Physiology* and *Experimental Physiology* appoint ethics editors to provide advice on issues arising from submitted manuscripts. Guidance to authors states that papers may be rejected on ethical grounds if the experimental procedures 'may reasonably be presumed to

[49]   Smith and Boyd (1991), p. 288; Osborne *et al.* (2009). Examples of journal guidelines include those of the Association for the Study of Animal Behaviour (ASAB) and the British Society of Animal Science (BSAS), http://asab.nottingham.ac.uk/ethics/guidelines.php and http://animal-journal.eu/documents/Ethical%20issues.pdf, accessed 12 September 2013.
[50]   Kilkenny *et al.* (2010).

have inflicted unnecessary pain or discomfort upon them', and there is a growing trend amongst biomedical journals to set an ethical policy[51]. Some might think that as any harms to the animals have already occurred by this stage, post-study review of this sort is pointless. However, the comments of the ethics committee help educate researchers on issues that they may not have considered, ensure that relevant information is included with the publication and, as contentious manuscripts can be rejected, provide a powerful motivator for academics to design and carry out experiments ethically.

2.5 Performance and Engineering Standards

The terms 'performance' and 'engineering standards' are used to describe two different approaches to drawing up controls on animal care and use. Both types aim to achieve a minimum standard of animal welfare by providing a clear description of the expected or permissible minimum standards, but the difference lies in the choice available as to how to get there. Performance standards describe the desired outcome, but not how to achieve it; engineering standards specify precise criteria for minimum standards, but not necessarily how these criteria should be implemented so as to attain the desired standards. Both types of standard can be used to regulate many aspects associated with animal research, but this section will largely focus on animal husbandry and staff training to provide examples of how these standards can be applied and enforced.

Engineering standards for an animal facility might prescribe minimum enclosure dimensions, ventilation flow rates and temperature ranges. However, engineering standards are not confined to specifying units of measurement, as they can also prescribe a particular method or means of achieving the desired outcome. For example, an engineering standard might specify a precise curriculum of a training programme or that particular items or types of enrichment should be provided, such as perches or a minimum duration of exercise.

In contrast, performance standards describe outcomes; for example, housing should provide for the animals' well-being. Because performance standards do not specify precise measurement criteria and may be rather general, they might seem a less rigorous method of ensuring animal welfare than engineering standards. However, performance standards do not free the institution from responsibility for taking measurements. As a publication on the implementation of performance standards in the USA put it,

> ...to set performance standards for any piece of equipment or process, it is important to set the goal, to define the measurements which may be taken, establish the frequency of those measurements and then verify that the desired parameters are reached[52].

[51] 61.6% of biomedical journals surveyed required author assurance of compliance with ethical standards (Rands, 2011).
[52] Gonder *et al.* (2001), p. 18.

Table 2.3 Provisions for training under European and US legislation.

European Directive 2010/63/EU Article 23	US Animal Welfare Act
Competence of personnel	§2143. (d) Training of scientists, animal technicians, and other personnel involved with animal care and treatment at research facilities
1. Member States shall ensure that each breeder, supplier and user has sufficient staff on site	
2. The staff shall be adequately educated and trained before they perform any of the following functions:	Each research facility shall provide for the training of scientists, animal technicians, and other personnel involved with animal care and treatment in such facility as required by the Secretary. Such training shall include instruction on:
(a) carrying out procedures on animals;	
(b) designing procedures and projects;	
(c) taking care of animals; or	1. the humane practice of animal maintenance and experimentation;
(d) killing animals.	2. research or testing methods that minimize or eliminate the use of animals or limit animal pain or distress;
	3. utilization of the information service at the National Agricultural Library, established under subsection (e) of this section; and
	4. methods whereby deficiencies in animal care and treatment should be reported

Source: European Directive 2010/63/EU © European Union, http://eur-lex.europa.eu/.
US Animal Welfare Act found at http://www.aphis.usda.gov/animal_welfare/awa_info.shtml, accessed 7 January 2014.

To do all this, taking responsibility for defining what must be achieved as well as appropriate measurements, might even be more onerous than complying with a more proscriptive engineering regulatory system.

Performance standards vary considerably in the degree of detail provided. Table 2.3 provides an example where there is noticeably less detail in European compared with US standards for staff training. Similarly, both Annex III of European Directive 2010/63/EU and the US regulations are rather brief with respect to meeting behavioural and physiological needs:

> *All facilities shall be constructed so as to provide an environment which takes into account the physiological and ethological needs of the species kept in them.*
>
> (Annex III of European Directive 2010/63/EU)

> *Dealers, exhibitors, and research facilities must develop, document, and follow an appropriate plan for environment enhancement adequate to promote the psychological well-being of nonhuman primates.*
>
> (US Regulation §3.81 regarding the psychological well-being of non-human primates)

On the other hand, the latest edition of the NRC *Guide* provides rather more detail regarding the types of behaviour that the housing should allow. For example, the *Guide* recommends that the cage height for monkeys

> *should be sufficient for the animals to comfortably stand erect with their feet on the floor,* and requires that *at a minimum, animals must have enough space to express their natural postures and postural adjustments without touching the enclosure walls or ceiling, be able to turn around, and have ready access to food and water*[53].

Both performance and engineering standards have their merits and disadvantages. It is relatively easy to check an engineering standard. Cages can be measured, temperature and ventilation records examined and the presence or absence of specified procedures or equipment determined. Nothing is ever totally straightforward, however, and there can be rather esoteric disputes such as whether floor dimensions should be measured from above or below the nesting material in rat cages with walls that curve inwards towards the floor. A major disadvantage of engineering standards is that the link between the standard and the desired outcome is not always obvious. A minimum cage size may have been specified and provided, but the animal needs more than just space. There is also the risk that care staff may concentrate excessively on the standard rather than on the implications for the animals themselves, for example spending too much time concentrating on temperature graphs. So engineering standards can lead to a box-ticking exercise that does not necessarily advance animal welfare.

Performance standards have the advantage that they tend to be clear on the desired outcome, which helps to focus minds on the animals' welfare. Performance standards also allow factors such as the animal species and age as well as design of the institution's facilities and staff experience to be taken into account. Hence, performance standards provide the regulated person or institution with greater flexibility over the means to achieve a particular welfare standard and the standard can be met in a way that suits the local circumstances. It is also easier for performance standards to incorporate new research findings regarding animals' needs or behaviour.

On the other hand, performance standards can be difficult to regulate both by the institution itself and by the regulating organisation. Institutions cannot rely on just complying with a list of regulations, but have to decide how they are going to measure success in achieving a performance standard[54]. This can be difficult. How exactly do you show that you have met the ethological needs of an animal? External inspectors or assessors face similar challenges of measurement. While the ability of a primate to stand within a cage could, in principle, be determined, judging this for a large colony of animals of varying sizes is likely to be quite difficult. It would be even harder for an inspector to judge whether all the enclosures in a facility provided enough space to

---

[53] NRC (National Research Council) (2010), pp. 56 and 61.
[54] Klein and Bayne (2007).

allow all the animals to express all their natural postures, even supposing agreement could be achieved on what were, and what were not, natural postures. Moreover, because inspections by a regulating authority or organisation are necessarily comparatively short, and may in themselves affect the behaviour of the animals, there may be more reliance on institutions' records to determine whether specified standards have been met. Hence, there is the risk that the inspection loses some of its independence.

It is probably true to say that performance standards are preferred in the USA, whilst Europe tends towards being more proscriptive. Nonetheless, as both systems have their advantages and disadvantages, it is not surprising that most regulatory systems and codes of practice use a mixture of performance and engineering standards. Both the *Guide* and European Directive 2010/63/EU provide tables listing minimum enclosure dimensions, stocking densities and temperature ranges. However, both also provide text that describes expected performance standards. As an example, the European Directive requires that:

> *All animals shall be provided with space of sufficient complexity to allow expression of a wide range of normal behaviour. They shall be given a degree of control and choice over their environment to reduce stress-induced behaviour. Establishments shall have appropriate enrichment techniques in place, to extend the range of activities available to the animals and increase their coping activities including physical exercise, foraging, manipulative and cognitive activities, as appropriate to the species. Environmental enrichment in animal enclosures shall be adapted to the species and individual needs of the animals concerned. The enrichment strategies in establishments shall be regularly reviewed and updated.*

The US Government Principles, paragraph VII state:

> *The living conditions of animals should be appropriate for their species and contribute to their health and comfort. Normally, the housing, feeding, and care of all animals used for biomedical purposes must be directed by a veterinarian or other scientist trained and experienced in the proper care, handling, and use of the species being maintained or studied. In any case, veterinary care shall be provided as indicated.*

And an example from the US *Guide*:

> *socially housed animals should have sufficient space to allow them to escape aggressions or hide from other animals in the pair or group.*

2.5.1 When performance and engineering standards collide

Whenever standards are revised, the scientific literature is scoured for evidence to support the cases for or against change. Unfortunately, the evidence is often not clear-cut, the reality being that science cannot provide definitive answers to uncertainties such as the minimum space for an animal enclosure. The reason for this is that so much depends on other factors, such as whether the enclosure will be long and thin or short and fat, the complexity and interest of what is placed within the enclosure, whether the animal will have periods outside the enclosure, and so on. Moreover, any suggestion that a minimum space allowance is absolute is vulnerable to the argument that one could decrease one of the dimensions by a few millimetres with no perceptible effect on animal welfare.

Despite these problems, certain space allowances were increased in the 2010 *Guide* and in European Directive 2010/63/EU, the European regulations being based on the advice of an expert group that took the following points into account:

1. the availability of existing enclosures (e.g. rodent housing is manufactured in various standard sizes, so if a minimum cage size is scrapped, it makes sense to move to the next available size);
2. the cage must have a minimum size that can sensibly accommodate both the animals and any essential environmental provisions such as shelters;
3. relevant published information on the animal's behaviour and welfare, and that age can be more important than size (a young playful rat may require more space than a large old sedentary rat).

However, such matters are open to debate, and it is often easier to gain acceptance for performance standards than engineering ones. Stakeholders with interests in carrying out research are likely to accept statements such as 'socially housed animals should have sufficient space to allow them to escape aggressions or hide from other animals in the pair or group', but upward revisions of space allowances may be resisted on the grounds that they would result in substantial cost increases. This difference in acceptability between the two types of standard can lead to internal tensions in documents containing both standards. So, for example, in the same US guide that required that 'socially housed animals should have sufficient space to allow them to escape aggressions or hide from other animals in the pair or group' also has an engineering standard that two macaques of up to 25 kg can be housed in a cage with a floor area of 1.86–2.8 m² depending on their weight. It is hard to see how two macaques housed in this area or, as another example from the same guidance, how two adult dogs housed in 2.4 m² could control their social interactions by escape or hiding as specified in the performance standard. Similarly, in the table of space allowances for cats, the minimum enclosure height is only 24 inches (60.8 cm) and yet the text argues that vertical height is important. Even the table seems to provide conflicting advice, as a comment within it notes that vertical space with perches is preferred and may require additional height.

When engineering and performance standards collide like this, one's sympathies go out to those who are required to comply with the standards or to regulate compliance with them. From an animal welfare point of view, the sensible option is to ensure that all standards, whether performance or engineering, are met, and in practice this may mean that the minimum engineering standards are not sufficient. However, when engineering standards seem to allow less space than the performance standards it may be difficult for regulators to prove that the performance standards really do require the extra space.

2.6 Roles and Responsibility

Any legislation or other control requires some form of personal responsibility. The researcher is responsible for the planning and execution of the project, but in order to ensure institutional compliance with legislative or other controls there also have to be others with responsibility for various aspects of animal use at the institution.

It is generally accepted that it is necessary to have a person or persons with overall responsibility for an institution's facilities: in the USA, this person is termed the Institutional Official (IO); in the UK under ASPA, this person (previously the Certificate Holder, but since January 2013 the Establishment Licence Holder, or where the holder is a legal entity the Named Compliance Officer) has been responsible not only for the facilities but also for other issues, such as ensuring that unauthorised procedures do not occur, that records are kept and that animals are sourced and disposed of correctly[55]. Some institutions can be large and complex affairs with different research establishments and so, under European Directive 2010/63/EU, one or more persons can fill this role. Whatever the legislation, those in this position have to have sufficient seniority and authority to be able to authorise expenditure or instruct employees to behave in ways that ensure compliance. In the commercial world, it is relatively easy to fit this role within the overall management system of the company, but things can be more complex in academia. Academic staff are usually responsible for finding their own funding, and the concepts of academic freedom and independence are deeply ingrained. Hence when legislative requirements are unclear or interpreted in different ways, or where there is a desire to go beyond the minimum requirements of legislation, academics may feel hindered in their research and the stage is set for conflict. Finding the solution, which is of course easier said than done, depends on having a culture of care based on good and appropriate training and communication that permeates the whole institution.

The institution's responsible individual is accountable for putting the infrastructure and management systems into place, but the detail of compliance is dependent on the actions of researchers and of the staff charged with the care of animals. Clearly, veterinary input is critical for monitoring and ensuring the welfare of animals in breeding and research establishments. European and US legislation, codes and guides require veterinary input (attending veterinarian in the USA, designated veterinarian in Europe). Veterinarians are primarily responsible for ensuring the health of the animals in their care, for advising on welfare issues regarding the animals in their care, and the research that is carried out on them. As such they can also be responsible for programmes of animal care and husbandry, helping to make critical decisions on the supply and transport of animals, their subsequent care and the design and running of the animal facilities. Veterinarians also provide valuable input into local institutional committees. US Animal Welfare Regulations and Public Health Service (PHS) Policy require that the attending veterinarian should

[55] Home Office (2000), pp. 18–26.

be a member of the IACUC and responsible to it, either directly or for a delegated programme. Similarly, UK guidance is that the veterinarian should be involved in the functioning of the local committee. In contrast, the 2010 European Directive only specifies that the veterinarian should provide advice to, but not necessarily be a member of, the local Animal Welfare Body (a body similar to the IACUC and ERP but which does not necessarily address ethical issues).

Technical staff also play a vital role in ensuring animal welfare as they carry out the day-to-day care of the animals. In the UK, under ASPA, their importance is recognised by the requirement for a Named Animal Care and Welfare Officer (NACWO), normally a senior member of the technical staff with broad responsibility for ensuring the welfare of animals under the UK's legislation. Article 24 of the 2010 Directive seems to refer to a broadly similar role for technical staff, stating that each breeder, supplier and user should have one or more persons on site who are 'responsible for overseeing the welfare and care of animals in the establishment'. The Directive also requires that there should be a person with responsibility for training competence and supervision at the establishment.

2.7 Legislation and Licences

One of the ways in which legislation can regulate research activities is by providing permissions for various activities through a system of licences. As Radford[56] points out, the licensing, certification or registering of an activity is symbolically significant as it indicates that the activity is of a type that can only be lawfully carried out with the permission of the State. Moreover, licensing ensures that those charged with upholding the legislation are aware of where research is being carried out, can ensure that standards are being met, and where necessary require that improvements are made or ultimately to refuse an application for renewal or withdraw existing authorisation. It can also be used as a means of raising funds to pay for the regulatory system.

Most systems, as a minimum, involve licensing or registration of institutions that breed or use animals in research. This practice ensures that research on protected species is carried out in appropriate places where the facilities meet the minima specified by regulation or codes of practice, where animals are sourced from institutions that have the appropriate standards; that those involved in the research and animal care are appropriately qualified; and that other protections such as establishment of a local ethics/animal welfare committee are in place. In some cases, establishments that breed animals but which do not carry out research may also require some form of institutional licence. In the UK, licensing is regulated through the Home Office[57] but more commonly licensing is carried out by departments with other responsibilities for animal use such as the Animal Welfare Unit in

---

[56] Radford (2001), p. 291.
[57] The Home Office leads within UK Government on issues related to immigration and passports, drugs policy, crime policy, counter-terrorism and policing.

New South Wales, Australia and the United States Department of Agriculture (USDA) in America.

Under the US regulatory system, the institutional licence is the only one required by the Animal Welfare Act. It serves to notify the authorities and places a responsibility on the institution to self-regulate the conduct of animal experiments through a local committee (IACUC). However, in other countries, regulatory systems may require additional licences to provide a central check on individual programmes of research, the persons responsible for the research, and the skills and experience or supervision of those tasked with carrying out the work on the animals.

Most systems of control include a requirement that there should be appropriate staff, such as veterinary surgeons and animal care staff, to ensure the welfare of the animals, and some legislation requires their notification to the regulatory authorities. It is also usual to require that all persons involved in animal care and use have received appropriate training. The legislative provisions for this are generally couched in rather broad terms (see Table 2.3) and so the detail may be provided in other codes of practice, guidelines, or training materials such as those provided by FELASA, AALAS and other organisations.[58]

2.8 Enforcement and Failure to Comply

2.8.1 Local enforcement and self-reporting

Most would agree that it is much better to prevent animal welfare problems resulting from lack of compliance rather than trying to fix them if they do occur. It follows that the best form of ensuring compliance with codes or legislation is that carried out at a local level. Local enforcement may be through management controls set up by institutions. For example, in the USA it is both PHS policy and a government regulation that the IACUC must carry out facility inspections at least once every 6 months and provide a report to the Institutional Official listing deficiencies and plans for correcting them[59]. Ideally, the culture of care should be so rooted in an establishment that standards do not drop below those expected or, if they do, that they are then detected rapidly by routine checks, where necessary reported to higher management, and quickly corrected. In general, where some error has led to a failure to comply with legislation then it is far better to correct the issue and, if necessary, self-report the incident to the authorities than to have it discovered during an inspection, as the latter suggests inadequate monitoring.

---

[58]   The OLAW Institutional Animal Care and Use Committee Guidebook 2002 describes responsibilities for training in the USA, http://grants.nih.gov/grants/olaw/guidebook.pdf, accessed 3 April 2013. See also NRC (National Research Council) 2010, Appendix A. In Europe FELASA provides recommendations for education and training, http://preview.tinyurl.com/pqmmwg9, accessed 16 September 2013.
[59]   For PHS funded research, there is external oversight by a governmental body. The reports must be made available to the Office of Laboratory Animal Welfare (OLAW). OLAW requires institutions to correct confirmed non-compliance and to take appropriate measures to prevent repeated non-compliance.

2.8.2 Whistle-blowing and exposés

Although not part of the animal welfare regulatory system, whistle-blowing and exposés are inevitable, and are therefore relevant[60]. Because of issues such as confidentiality and potential, and perhaps unnecessary, damage to the public profile of the organisation, whistle-blowing is often unpopular with employers, and there can be consequences for the whistle-blower's future employment. Nonetheless, those developing legislation, or running a research institution, need to consider what an employee of an organisation should do if, rightly or wrongly, he or she believes that there is something wrong with either the welfare of the animals or the way in which research is carried out, but is not able to resolve the issue through the normal internal processes. In such circumstances the employee may feel compelled to solve the issue by going outside the organisation, and the question is where.

It is obviously far better to avoid the issue arising, or to put the matter right in-house, and it may be possible to avoid problems arising by ensuring that employees understand what steps they can take within the organisation. Where internal processes fail, it is usually better for the whistle-blower to report the matter to the regulatory authority rather than to publicise the information through an anti-vivisection organisation, in a blog, or by going to the press. It is true that notifying the regulatory authority will put the matter on a formal footing, but regulators are in the best position to enforce the law or codes, and to do so in a way that does not cause unnecessary damage. It is also worth remembering that whistle-blowers can make mistakes or misunderstand the reasons for certain decisions, and if the matter is simply a misunderstanding, it will be much easier to deal with it out of the glare of publicity.

The activities of whistle-blowers and undercover investigators can be damaging for the institution affected but, as the UK Animal Procedures Committee (APC) stated[61], both types of activity have exposed problems that otherwise might not have been discovered for some time. Moreover, revelations of activities that are publically unacceptable can lead to necessary changes in legislation, which Rollin[62] suggests has occurred in the USA. Nonetheless, it can be very difficult for the authorities or other investigating bodies to get to the bottom of issues, as the same APC report drew attention to disputes of fact, incompatibility of agenda between the aims of the research community and an anti-vivisection organisation, and contrasting perspectives, for example, on the standards of welfare that might reasonably be expected for laboratory animals. There are also difficult moral issues. Whistle-blowing and undercover exposés are carried out by people who feel that their actions are the only way to expose a problem, but there may be some moral distance between the two activities. Whistle-blowing is (ideally) an act of last resort by an employee who originally took on their employment in good faith. In contrast,

[60] Some jurisdictions have legislation that covers whistle-blowing in a broader context than animal welfare regulation, e.g. the UK Public Interest Disclosure Act 1998.
[61] The APC Final Report of the Cambridge/BUAV Working Group, 16 June 2005, http://preview.tiny-url.com/cplmlaw, accessed 26 April 2013. Since January 2013 the Animals in Science Committee has superseded the APC.
[62] Rollin (2006).

undercover exposés are carried out by people who become employees with the intention of ultimately going public. Nonetheless, both activities are undoubtedly controversial. Depending on the circumstances of any particular case, whistle-blowing and undercover investigations may be considered justified activities by some. Others feel that, for one or both of them, there are better ways to deal with regulatory failure.

2.8.3 External review/inspection

Internal reporting and management processes are all very well, but review or inspection by an independent external body provides greater reassurance that research is being carried out properly and that animal suffering is minimised. Moreover, external inspection has the advantage that the inspectors or reviewers will have seen many other institutions, whereas staff within a research facility may become rather isolated. Hence external inspectors/reviewers will be well placed to encourage good practice.

2.8.3.1 Review by clients

Organisations that contract out research or who purchase their animals from breeders usually want to reassure themselves about the quality of those establishments in order to ensure that the work or animals for which they have paid meet their scientific and animal welfare standards. Poor animal care not only risks the quality of the science (see Chapter 7) but, if exposed, is likely to reflect almost as badly on the client as on the institution actually carrying out the work. Often the purchasing organisation will seek reassurance by requesting documentary evidence on the institution, for example standard operating procedures and health records; reassurance regarding compliance with local regulations; and perhaps evidence of external validation such as accreditation. In addition, some organisations will make visits to the sites to see the conditions for themselves. Pressure from clients can be a very valuable way of spreading ideas about good practice, sometimes to countries where there is little history of concern for animal welfare. However, review by clients is not enough on its own for the following reasons: (1) the clients and the contracting organisation are linked by a contract and therefore the client is not a truly disinterested party; (2) the reviews are carried out to satisfy the client's needs, not as part of the social compact with the public; (3) clients will only visit, and have influence on, those areas of the institution that are supplying their contract; and (4) clients are not necessarily expert on the legislation and animal welfare issues relating to species used by the contractor and there may be a tendency to concentrate on rather gross measures of welfare such as health and breeding records. There therefore has to be external inspection by a more independent body.

2.8.3.2 External accreditation and inspection

Accreditation by an external body against set standards is one way of achieving this. In Europe, for example, FELASA provides accreditation for education and training in laboratory animal science, as well as for health programmes for units and for those laboratories carrying out health monitoring.[63] At an international level, AAALAC reviews facilities that it accredits to ensure that they comply with local legislation and, where appropriate, other documents and regulation[64]. Such visits are intensive, carried out by a team from the AAALAC Council on Accreditation as well as ad-hoc consultants. However, accreditation is voluntary, it is peer review, there is usually a fee, and the results usually remain confidential to the inspected institution. As such, accreditation may not provide the public with as much confidence as inspections by a government agency. On the other hand, accreditation is not something that an organisation would want to lose or fail to achieve, and the inspection process is a detailed one that, if done by AAALAC and involving a large organisation, can last several days.

Where mandated by legislation, an executive arm of government may carry out inspections: in the USA, the Animal and Plant Health Inspection Service (APHIS), an arm of USDA, enforces the Animal Welfare Act; in the UK, inspectors within the Animals in Science Regulations Unit (ASRU), which is part of the Home Office, visit licensed establishments. In the past, the mere existence of government inspections was considered sufficient to satisfy public concerns, but with the development of ideas about freedom of information there is now greater release of information collected by government. So, in the USA, USDA publishes reports of inspections on its website. There is no equivalent requirement in Europe, but in the UK the Home Office produces an annual report that describes the failures of compliance that have occurred but does not name the organisations involved[65]. Indeed, UK legislation forbids officials from doing so.

Inspections can be either announced or unannounced and there are pros and cons to each. Unannounced inspections, such as those carried out by APHIS in the USA, and the majority of those carried out by the ASRU inspectors in the UK, provide some reassurance that the conditions during the inspection are truly representative. However, unannounced visits may not be as productive as an announced one, and only inspectors with proper legal backing can truly make an unannounced visit. Announced inspections have the advantage that it is possible to ensure that the correct people are on site to speak to the inspector/reviewer and that the appropriate documentation is easily available to view. All AAALAC reviews are arranged in advance with the site to be reviewed, and AAALAC produces guidance as to how institutions can prepare for reviews.

An adequate frequency of high-quality inspections is essential to enforce legislation. Indeed, prior to the revision of the European Directive, one of the debates was whether it would not be better to first ensure that there was adequate enforcement

---

[63]   FELASA, http://www.felasa.eu/accreditation-boards/, accessed 3 April 2013.

[64]   Such as *The Guide*, *The Agricultural Guide*, European Convention (ETS123) Appendix A and B, PHS requirements, US Animal Welfare Regulations and other documents. AAALAC Accreditation, http://www.aaalac.org/accreditation/index.cfm, accessed 3 April 2013. See also Newcomer (2012).

[65]   USDA Animal and Plant Health and Inspection Service http://www.aphis.usda.gov/animal_welfare/efoia/7023.shtml. UK Home Office Animals Scientific Procedures Division and Inspectorate Annual Report, https://www.gov.uk/research-and-testing-using-animals#publications, accessed 3 April 2013.

throughout the member states. In the USA, APHIS inspects institutions covered by the US Animal Welfare Act at least once annually; however, if serious problems are identified at an institution or if there is a public concern about a situation in a facility, the USDA will conduct additional inspections. In the UK there is no minimum frequency for inspections, but the frequency is determined on a risk basis that takes into account the species used, the severity of the work, the numbers of procedures and the previous compliance history of the establishment[66].

Whether reviews are carried out voluntarily or as part of government regulation, enforcement is expensive and so sufficient resources have to be available to allow a reasonable inspection frequency. However, it is not correct to think that fewer inspections will necessarily cost less. A 2011 Home Office consultation document, seeking views on transposition of the Directive into UK legislation, noted that European Directive 2010/63/EU requires as a minimum that one-third of the animal users should be inspected each year in accordance with a risk analysis. It also requires that breeders, suppliers and users of non-human primates should be inspected at least once a year. When examining how this minimum requirement could be implemented in the UK, the Home Office estimated that it could be achieved by performing an inspection once every 3–5 years. This would represent a very large drop in the frequency of inspections, from just under 2000 in 2010 to about 80 per annum. Despite this apparent reduction in workload, it is estimated that there would be little real financial benefit, because of the increased costs of performing the necessarily much more detailed inspections in a reduced programme of inspection. Even worse, a reduced programme of inspections would also reduce the ability of the inspectorate to offer advice (a function discussed in the next section), and so more resources would be necessary to maintain this role.

2.8.3.3 Relationship of inspectors to sites inspected

Regulatory inspection can operate as a stick to enforce compliance, issuing penalties for non-compliance; more constructively, the inspectors can spend time to get to know the facilities and individuals on their patch, helping them to comply, achieve higher standards and increase the likelihood that problems will be identified before they have major consequences. UK experience has been that inspectors with in-depth site knowledge are well placed to advise on project authorisation, assess the risk of animal welfare problems at a site, and prevent a risky institution becoming non-compliant. According to people who have been subject to both UK and US legislation, UK inspections while equally rigorous are more likely to include helpful advice and encouragement. USDA inspections, on the other hand, have been more focused on compliance and enforcement and less on education, and may become more so in the future[67]. The origins for this difference may, at least partly, lie in the fact that USDA inspects against regulations and law whereas, under ASPA,

---

[66]   Cardon *et al.* (2012a). Animals in Science Regulation Unit: Annual Report 2011, http://preview. tinyurl.com/c7g36hz, accessed 3 April 2013.
[67]   Cardon *et al.* (2012a).

UK inspectors have assessed animal housing against codes of practice; as we have discussed earlier, codes allow a more flexible approach, taking into account both positive and negative issues at the site. With implementation of the 2010 European Directive and hence the introduction of legally binding minimum housing standards, this difference will no longer exist, and it may be a challenge for the UK inspectorate to continue to work collaboratively with institutions as their room for discretion is reduced. A potential downside to the traditional UK approach is that the officials might become, or be seen to become, too close to the people that they regulate, with consequences on their ability or inclination to enforce legislation. Inspectors need to maintain an appropriate independence and find a balance between carrot and stick approaches to achieve the best possible outcomes.

2.8.3.4 Ensuring standards of overseas suppliers

Ethical and welfare issues involved in the use of animals by an institution are not just confined to the site where the research is carried out, as it may buy in animals from breeding establishments or contract out research to other organisations. If these external organisations are also in a different country, then there may be problems in ensuring the welfare of animals sourced from or used at these overseas sites. The purchaser of the service can make visits to the sites, but this does not provide independent scrutiny. External accreditation can also provide reassurance, but the problem is more difficult for a national regulatory system as mandatory inspections cannot be made outside the jurisdiction. A UK report[68] outlines some of the methods that have been used by the Inspectorate, and by those carrying out research to assess the standards of overseas suppliers. Essentially, a risk-based approach has been applied using criteria such as the species of animal and what is known about the external supplier. Information is obtained on colony health and husbandry, while for high-profile species such as non-human primates and dogs, inspectors have made overseas visits to the suppliers, albeit only by prior arrangement and with their agreement. In the future, the options available to the Inspectorate will be fewer as under Article 2.2 of European Directive 2010/63/EU, it will not be possible for individual nations to restrict the import of animals from other nations covered by the Directive on the grounds that the exporting state had lower standards.

Codes of practice and other controls are ways of ensuring that standards are generally good, but what are the bases of our decisions on the welfare of animals used in research? The next chapter examines two very different approaches.

---

[68] Animal Procedures Committee 2007. Consideration of Policy Concerning Standards of Housing and Husbandry for Animals from Overseas non-designated Sources, http://preview.tinyurl.com/bukuuh5, accessed 3 April 2013.

Animal Rights and Animal Welfare: Philosophy and Science

3

This chapter outlines the two main philosophical positions regarding the acceptability of using animals in research. There are already good and very comprehensive accounts of philosophical approaches to human exploitation of animals[1], so this chapter is intended only to be a primer on the subject. I draw attention to how the terms 'animal welfare' and 'animal welfarist' are used somewhat differently in philosophy and research. The chapter then introduces the science of animal welfare, defines what scientists mean when they use the term 'animal welfare', and what the field of animal welfare science has to offer in helping us to make decisions on animal care and use.

3.1 Introduction

Philosophy and science have both played a part in developing our ideas as to how we should best deal with animals. Scientists and philosophers have laboured to develop theories on human and animal consciousness, as well as coherent frameworks of thought to guide our actions, while science has provided us with information to ameliorate harm caused to animals and to find alternatives to their use. The use of scientific methodologies to obtain fundamental or applied knowledge is, of

[1] For brief accounts, some of which cover philosophical viewpoints not mentioned in this chapter, such as contractarianism and natural law, see for example Smith and Boyd (1991), Dolan (1999), Monamy (2009) and Olsson *et al.* (2010). A description of campaigns and other political aspects of the animal rights movement is provided by Kean (1998). For a broad-ranging approach, covering many issues, see Beauchamp and Frey (2011).

The Welfare of Animals Used in Research: Practice and Ethics, First Edition. Robert C. Hubrecht.
© 2014 Universities Federation for Animal Welfare. Published 2014 by John Wiley & Sons, Ltd.

course, the reason why animals are used in research, but the scientific approach is also the only means of providing us with factual information on vital issues such as animal sentience, animals' needs, and methods of assessing individual animal's welfare. I think it is fairly obvious why we need information on animal needs and welfare assessment, but some may wonder why we need to address issues of sentience –whether some animals are capable of experiencing feelings. The answer, as we shall discuss in some detail in Chapter 4, is that it is very important that our decisions on which animals to protect are based on sound information. If we get these decisions wrong then there is a grave risk that many sentient species that should be protected are not, with all the concomitant suffering which would entail. Conversely, if non-sentient species are protected scientific endeavour would be unnecessarily impeded and animal welfare resources and time wasted that would be better employed on protecting sentient animals.

Philosophy and science were originally one and the same, but the rapid development of science, its success in shaping the modern world and the knowledge it has given us about how the universe works and our origins has led to something of a gulf developing, as exemplified by a remark attributed to the geneticist Stephen Jones 'Philosophy is to science as pornography is to sex: it is cheaper, easier and some people prefer it'. Unfortunately, despite the commonality of interest, it seems to me that despite the praiseworthy efforts of several eminent philosophers and scientists to forge links, a gap still remains between the two disciplines in the field of animal welfare. The next section therefore outlines two major philosophical positions that are relevant to the use of animals in research and describes a use of the term 'animal welfare' that could lead to misunderstandings.

3.2 Animal Rights and Animal Welfare: Making a Distinction

Ideas about animal rights have been around for some time. John Lawrence, for instance, wrote a treatise on the bad treatment of horses and other animals, and called for legal recognition of animal rights as early as 1791. However, the development of the concept of modern animal rights really gathered pace in the mid 1900s, with growing concerns about various aspects of animal use, notably the food industry, coinciding with Peter Singer's publication of *Animal Liberation* in 1973 and Tom Regan's *The Case for Animal Rights* published in 1983[2]. The animal rights position holds that while the cognitive abilities of animals might be different and less in some respects than those of humans, animals, like humans, have an inherent value and there is a moral equivalence between humans and those animals with cognitive abilities sufficient to have feelings, to remember and have intentionality

---

[2] Ruth Harrisons's (1964) book *Animal Machines* drew attention to welfare issues associated with the growing intensification of agricultural practices. See also Fraser (2008), p. 61 and Sztybel's article on animal rights, pp. 43–45 in Marc Bekoff's *Encyclopedia of Animal Welfare and Animal Rights* (Bekoff and Meaney, 1998). Fraser (2008), p. 19 quotes John Lawrence's text.

(as Regan puts it 'to be the subject of a life'). For those who take this view, it follows that it is wrong to treat animals in ways that would not be ethically right if humans were the subjects.

The major alternative to the animal rights philosophical position is animal welfare or animal welfarism, a utilitarian philosophy in which the aim is to maximise benefits to the relevant parties and minimise any harms to the animals. Indeed, animal welfarism has been described as 'utilitarianism applied to animals'[3]. The trouble with applying a utilitarian approach to the use of animals is that balancing human and animal interests is harder than when just considering how humans should behave towards each other. If we do harm to animals for the benefit of humans, then how should those benefits and harms be apportioned between the human and non-human animals involved? The classical utilitarian animal welfare approach towards animals, espoused by Jeremy Bentham and John Stuart Mill who are sometimes described as the philosopher architects of animal welfare, is one in which the interests of both humans and the animals should be taken into account, although not always equally. In the words of Robert Garner 'humans are morally superior to animals but…because animals have some moral worth we are not entitled to inflict suffering on them if the human benefit is not necessary'[4]. Many utilitarians therefore feel that it can be morally acceptable to use animals for some purposes, such as for food or medical research for which it would not be acceptable to use humans. However, such use needs to be appropriately justified and places on those involved an obligation, or duty of stewardship, to ensure that any animals that are used are treated humanely.

Unsurprisingly, the animal rights and animal welfare philosophical positions lead to different conclusions. The logical outcome of the animal rights argument is that research on animals should be abolished, while those that take the animal welfare position require the development of processes to weigh up the harms and benefits of proposed research, and by persuasion, regulation and education to improve the welfare of animals used by humans. Positions on the death of animals also tend to differ between the camps. Killing animals for human gain is generally seen as a wrong by those in the animal rights movement, while many who take an animal welfare view are more concerned to ensure that any death deliberately caused by humans is carried out in as humane a way as possible.

However, one has to be careful with labels like 'animal rights' and 'animal welfare' as they can mean different things to different people. Peter Singer and Tom Regan are often seen as leading proponents of animal rights but while many of their concerns and conclusions regarding the treatment of animals are similar, the philosophical bases for their conclusions are very different. Both Singer's and Regan's concerns are based on the principle of the similarity of certain animals to

---

[3]   Regan (1998).
[4]   Robert Garner in Francione and Garner (2010), p. 107. Other positions, in addition to the rights and utilitarian/welfare positions, have been adopted by philosophers and writers in this area. Jack Weir in Bekoff and Meaney (1998), p. 358 notes the virtue ethics position adopted by Stephen Clark, Mary Midgely and Bernard Rollin, who suggest that animals should be able to have a life according to their kind.

us, but they come to different conclusions: Singer argues that if we use animals in ways that we would consider unacceptable for humans then that disregards their interests, while Regan feels that it violates their rights[5]. This is a crucial distinction. Regan's position is that higher animals, like humans, have intrinsic value and thus have rights not to be treated in certain ways. Singer's view, broadly speaking, is that the same considerations should be taken into account when considering the interests of animals and humans. It has been suggested therefore that his position is better described as a utilitarian one (where the morally correct action is that which results in maximising hedonism or the satisfaction of preferences). Indeed, Singer describes himself as a utilitarian and has stated that he regrets having allowed the concept of a right to intrude into his work[6]. So Singer's position with respect to animal experimentation is not so very different from the position of those who carry out experiments, differing mainly in degree, in that he doubts that the benefits of most experiments do justify the harms done to the animals[7]. It is also the case that a utilitarian animal welfare approach does not rule out belief that animals should be given certain rights. Rights have been categorised as legal, moral and natural. Legal rights are those laid down in legislation, moral rights are essentially what people think is right, and natural rights are those that arise from the nature of things and which are universal and inalienable[8]. If one is talking about the rights of animals in terms of legal protection, then it can be perfectly logical to take a utilitarian animal welfare position with respect to the use of animals whilst still believing that they should be given certain legal rights[9].

A more serious issue arising from these labels is that the term 'animal welfare' is used to describe an area of science as well as the utilitarian/animal welfare/animal welfarism philosophical position described above, and this could lead to confusion about the motivation of animal welfare scientists. Further, within the philosophical literature the term 'animal welfare' is not just associated with utilitarianism. The terms 'new welfarism' or 'new welfarist' are used as a description of a pragmatic position adopted by some animal rights advocates where the long-term goal is animal rights, but the short-term goal animal welfare. New welfarists believe that animal use by humans should ultimately be abolished but believe that if that is not achievable now, that there are still worthwhile ethical gains to be made by improving the welfare of animals that are used. Some go even further, arguing that welfare reform raises public consciousness, increases costs and thereby drives down animal

[5] Smith and Boyd (1991), p. 301.

[6] Regan (1982), chapter 2; Singer (1990); Francione in Francione and Garner (2010), p. 10. Robert Garner in the same volume argues that utilitarianism and welfarism should not be conflated, on the grounds that welfarists believe that the ethical treatment of humans and animals should be judged in a different way (p. 107). Some prefer to avoid these two camps. Robert Garner prefers to define himself as an animal protectionist and argues that there may be ethical cases for sacrificing animals provided they do not suffer (pp. 104–119).

[7] Varner (1994). Olsson *et al.* (2010) point out that in most western societies utilitarianism is applied giving humans priority; however, under strict utilitarianism this prioritisation is wrong.

[8] Dolan (1999), chapter 8.

[9] Tannenbaum (1989), pp. 104–115.

use. As Francione puts it 'new welfarists believe that there is a causal connection between cleaner cages today and empty cages tomorrow'[10].

Of course, animal welfare scientists, like any other group, will have a spectrum of views on the morality of various types of research using animals. An animal welfare scientist might be a utilitarian animal welfarist or a new welfarist, but working as an animal welfare scientist does not necessarily mean that one is in favour of animal rights or abolition of research using animals. On the other hand, it would also be quite wrong to think that all animal welfare scientists agree with the philosophical animal welfare position 'that humans do nothing wrong when they use nonhuman animals in research...if the overall benefits of engaging in these activities outweighs the harms these animals endure[11]. On the contrary, many animal welfare scientists and researchers on animals do believe that a wrong is done to the animal even if the general consensus is that on balance the research is justified[12]. The bottom line is that an active research career as an animal welfare scientist tells one little about his or her views on animal research for human benefit. Of course those working in the field of animal welfare science are likely to be motivated to improve the welfare of animals used by humans, but many are also supportive of research for the benefit of humans.

So why do I think that the slightly different uses of the term 'animal welfare' in the fields of philosophy and science matter? The reason is that although animal welfare science has become a respected area of research, it is still young and is not always understood by scientists in other areas of biological research. If the motivations of those who carry out animal welfare research are confused with those who have an agenda to abolish animal use irrespective of the consequences, or worse with the activities of animal rights extremists[13], it could have a serious impact on their credibility and ability to work with those who use animals in research. Certainly, I have occasionally experienced mistrust, and have heard clear misunderstanding of motivation. One example was a view expressed that work to replace animals with non-sentient alternatives (such as *in-vitro* tests or computer models) is anti-science, which was a nonsensical statement as replacements may be cheaper and better scientific models as well as a better ethical option (see Chapters 6 and 7). The confusion might also affect the credibility of scientific animal welfare non-governmental organisations (NGOs) and advisory organisations. For example, Regan[14] writes that stopping the use of animals in research, either in whole or in

---

[10] Francione discusses new welfarism in Bekoff and Meaney (1998), p. 45. See also Francione and Garner (2010), p. 25.

[11] Regan provides this description of the animal welfare position in Bekoff and Meaney (1998), p. 42.

[12] In a *Nature* poll of nearly 1000 biomedical scientists, 33% of respondents had ethical concerns about the use of animals in their work (Cressey, 2011a). I personally have heard a scientist referring to his work as a necessary evil. None of this, of course, means that these scientists think their actions are necessarily wrong.

[13] A very small, albeit dangerous, extremist minority of those who believe in animal rights have used illegal methods in an attempt to frighten or coerce individuals and institutions to cease research (Cressey, 2011a).

[14] Regan (1982), p. 65.

part 'has been one of the long standing objectives of what for the sake of convenience, I shall refer to as "animal welfare organisations"'. This may be true of many of them, but an approach adopted by others is to work to reduce the unnecessary use and suffering of animals. Suggesting that all welfare organisations have an abolitionist aim makes it harder for those working constructively with the animal research community.

The terms 'animal welfare' and 'animal welfarist' are firmly established within philosophy so it is unlikely that much can be changed, but there is value in animal welfare scientists and philosophers having a better understanding of each others' fields[15]. More could also be done to educate scientists, beginning their careers in research using animals, to understand that the aim of animal welfare science is to gain factual information to inform decisions on how best to treat the animals in our care. How that information is used is a matter for society as a whole.

3.3 Animal Welfare Science

3.3.1 The need for a science-based approach to animal welfare

There are at least two good reasons that can be put forward for ensuring that the welfare of animals used in research is as good as is reasonably possible. The first of these is the ethical, and often legal, obligation on those who carry out animal research to avoid unnecessary animal suffering and, as far as possible, to keep the animals in a state of good welfare. Examples of such requirements include the following. In the USA, Government Principle IV states that the Institutional Official should ensure compliance with the following principle 'the proper use of animals, including the avoidance or minimization of discomfort distress, and pain when consistent with sound scientific practices'. Similarly, in Europe, Article 4.3 of European Directive 2010/63/EU requires 'refinement of breeding, accommodation and care, and of methods used in procedures, eliminating or reducing to the minimum any possible pain, suffering, distress or lasting harm to the animals'. Annex III, paragraph 3.1 of the same document states that 'Animals shall be checked at least daily by a competent person. These checks shall ensure that all sick or injured animals are identified and appropriate action is taken'. The second reason is that good science and good welfare often go hand in hand (see Chapter 7). It is, of course, true that some experiments unavoidably result in stress or pain for the animal. Examples include research where the aim may be to study or find ways of alleviating pain or stress, research into sepsis or bone cancer, or certain toxicity tests of chemicals or pharmaceuticals. However, even when poor welfare is inevitable, the ethical and legal obligations to reduce the harms to a minimum remain, and the research will be better if the animal is as unstressed as possible at the outset of the study[16].

---

[15] Fraser (1999) describes some of the areas where advances have been made in achieving a better understanding between scientists and philosophers.
[16] See Chapter 7.

To achieve these ethical, legal and scientific advantages, specific animal welfare knowledge is essential for the day-to-day care and assessment of individual animals, for the refinement of techniques, for the development and implementation of humane endpoints (the point at which suffering is alleviated through the use of analgesia, anaesthesia or euthanasia), to carry out research to inform the development of new husbandry systems, to set minimum standards and to report on harms experienced so that ethical judgements can be made about the acceptability of such research.

However, a general understanding of the breadth of welfare knowledge is comparatively new. Up until the 1980s, much of the effort to improve the welfare of animals used in laboratories was focused on issues of health, and rightly so as poor health can result in animal suffering and, even when latent, can bias results or increase variation[17]. As a result research and input on animal welfare issues was largely veterinary, dealing with and preventing infections in colonies, the care of individual animals and improving experimental techniques. Health was not the only serious issue that needed to be addressed. Postoperative analgesia, for example, was not standard, and much of the housing did not meet the animals' needs. Concerns about hygiene and standardisation led to the design of barren animal housing that had the benefit that they could be easily cleaned but provided little for the animals housed within.

Things changed following the growth of concerns in the 1960s about the use of animals in food and research industries. Questions needed to be answered regarding animals' needs, and how they could best be met. As Sir Peter Medawar pointed out, these questions require a scientific approach, not only to study the welfare of animals but also to develop new techniques to address the problems[18]. Increasing public concern about the welfare of animals farmed using new intensive techniques led to the 1965 report by the Brambell Committee to the UK Government, and this helped stimulate the development of animal welfare as a scientific field in its own right using a multidisciplinary approach. William Thorpe, an ethologist and prominent member of the committee, made a particularly important contribution by identifying how the fields of physiology and ethology could be used synergistically to address animal welfare issues. Thorpe also saw the importance of studies of motivation and preference, areas of research that have since become a major component of animal welfare research, which itself now encompasses many fields including veterinary medicine, physiology, neuroscience and ethology[19]. While the

---

[17]   Russell and Burch (1959) pointed to the effects of disease on experimental outcomes. Weisbroth (1996) gives an account of the successive steps that have been taken to reduce the incidence of disease in rodents, and organisations such as FELASA have addressed the issue by producing guidelines for health screening of various species (e.g. Niklas *et al.*, 2002) and others are being developed. Animals used in research are considerably healthier than they were in the past but there is still work to be done.
[18]   Sir Peter Medawar wrote in 1957 that 'Improvements in the care of animals are not now likely to come of their own accord, merely by wishing them: there must be research on methods of research; and it is in sponsoring research of this kind that UFAW performs one of its most valuable services...'.
[19]   Brambell (1965); Fraser (2008).

Brambell report was concerned with farm animals, it was only a matter of time before animal welfare scientists started to address other aspects of animal use including their use in research[20].

Some people might wonder why science is needed to assess animal welfare, arguing that they can tell perfectly well when an animal is in a poor condition. The reasons are twofold. One is that many of the measures that can be taken to assess welfare require some degree of clinical or scientific expertise (we shall come to what these measures are later). The second reason is that while most people are reasonably good at empathising with other humans, it is harder to accurately judge the motivations and needs of other animals. Common sense and empathy can be useful tools[21], and are the basis for our concerns about animal welfare, but it is all too easy to fall into the twin pitfalls of anthropocentrism and anthropomorphism, in which we tend to assume that animals experience the world as we do (anthropocentrism) and feel and desire the same things (anthropomorphism)[22]. Indeed, this tendency is so pronounced that it is sometimes hard to avoid empathising with objects that we know to be inanimate. In a totally irrational response, I once found myself wincing when the cartoon Homer Simpson drove a nail through his hand, even though I know that Homer is an oddly coloured and not very realistic two-dimensional caricature of a human. If this can happen with a cartoon, how much more likely are we to anthropomorphise animals? The trouble is that animals are not exactly like us. Their senses often differ from ours and they have desires and behaviour and possibly experiences that are shaped by very different evolutionary histories. There is therefore a serious risk that we may wrongly assess what another animal species wants or needs, or even whether or how much it might be suffering[23]. That this is true is demonstrated by the way in which we commonly misunderstand animals that are very familiar to us. Dogs often jump up and attempt to lick the face of their owner when greeting them, while cats rub their faces against their owner. Animal behaviour like this strikes chords with us as humans also kiss or rub portions of their face together in greeting, but such behaviour may be better understood by closer examination of the underlying biology. Wild canid puppies greet an adult when it returns to the den by licking around the face, and this behaviour stimulates the returning adult to disgorge meat from its stomach. Cats possess submandibular scent glands beneath their chin, perioral glands at the corners of the mouth and temporal glands on the forehead; hence their greeting ceremony is not just one of touch and vision, they may also be using scent marks from these glands to maintain social bonds. I accept that the dog greeting its owner may not really be

---

[20]   Broom (2011) provides an account of the development of animal welfare science.

[21]   Françoise Wemelsfelder and her co-workers have shown that, despite the perils of anthropomorphism, qualitative assessments of certain animal species by untrained observers can be consistent, within and between observers, and can also correlate with other measures of welfare (Wemelsfelder *et al.*, 2001; Wemelsfelder and Lawrence, 2001; Wemelsfelder, 2007; Stockman *et al.*, 2011).

[22]   Bradshaw and Casey (2007).

[23]   Barnard and Hurst (1996) discuss how animals' experiences may differ from those of humans, and how some species may be adapted to behave in ways that in humans might lead to suffering.

trying to stimulate the owner to vomit, as display behaviours can become divorced from their original function. I also accept that kissing in humans may not be all that far from the cat rub as both involve the exchange of olfactory information, but these cautionary tales do indicate the potential pitfalls of relying too much on common sense to interpret animals and their motivations – science is needed as well.

3.3.2 Different concepts of animal welfare?

In order to decide what measurements should be taken to assess animal welfare, we have to be clear what we mean by animal welfare. This may seem an odd question. After all, we all know what animal welfare is, don't we? The problem is that we can confidently believe that we understand something with very little real understanding of the depth of the problem. We deal with time as if we know what it is, but as Saint Augustine wrote 'What, then, is time? If no one ask of me, I know; if I wish to explain to him who asks, I know not'[24]. Animal welfare may not be quite such a difficult problem but it is still tricky, and animal welfare scientists have differed in their views on what we mean by the term (whether it is to do with normal functioning of the animal, naturalness, or feelings) and consequently how welfare should be measured. It is therefore not surprising that on occasion there have been differences between experts regarding the welfare of animals or what should be provided for them[25].

This diversity of opinion has led some scientists in other disciplines to criticise the scientific rigour of animal welfare but I do not think there is a real problem. I rather like the analogy that David Fraser makes with the concept of safety. He takes room safety as an example, but I prefer to use bridges as I find them scarier. There is no one measure that one can take regarding a bridge's safety, as its safety depends on many factors, such as the quality of the design, the materials used, and its maintenance. People may well differ regarding the relative importance that they attach to each of these factors; nonetheless, we will probably agree that a decrepit poorly designed bridge is unsafe while a new well-designed bridge is much safer. Judging the extremes of poor bridge safety and animal welfare is relatively easy; the hard part comes somewhere in the middle. Amongst those assessing animal welfare or bridge safety there may well be disagreements as to what matters and how much it matters, but this does not mean that either endeavour is not worthwhile. So, what factors should be taken into account when assessing welfare?

3.3.2.1 Stress, coping, health and welfare

Some scientists have defined welfare in terms of the animal's ability to cope with the environment in which it finds itself. Animal species about whose welfare we are concerned (see Chapter 4) are complex organisms with behaviour and physiology that, as a result of natural selection, tend to be rather well adapted to the environmental

---

[24]   The Confessions of St Augustine, Bishop of Hippo, Book XI, Chapter XIV, http://preview.tinyurl. com/cgo2dgs, accessed 26 April 2013.
[25]   Fraser *et al.* (1997); Fraser (2008).

niche in which they evolved[26]. However, adaptation to an environment does not mean that animals in the wild live in a state of perfect welfare, as all environments include stressors (defined as real or perceived perturbations to an organism's physiological homeostasis or psychological well-being)[27]. Any stressor, if sufficiently severe or prolonged, can result in poor welfare, but mild stressors are not always unpleasant. Exercise can be a stressor, and yet animals (and humans) may seek opportunities to exercise; think of a mouse in a running wheel or a marathon runner, both of which activities can produce physiological stress responses.

Animals have evolved to respond to natural stressors with coping mechanisms that include behavioural, hormonal or immune function changes, the purpose of which are, if possible, to return the animal to a normal (unstressed) state. Behaviour is often the initial response to a stressor, but physiological changes such as increases in blood pressure and heart rate or changes in circulating levels of adrenaline may occur along with, or instead of, behavioural changes. Alternatively, animals may eventually cope by, over time, habituating to the stressor. As behaviour is such an important way of dealing with stressors, it follows that providing animals with a degree of complexity in their environment to give them options and some control is important for limiting stress in captivity (see Chapter 6).

Problems arise for the animal if the situation that the animal is in does not allow an appropriate behavioural or physiological coping response or if the animal's coping response is overloaded. In either case, if the situation is prolonged the animal is no longer able to maintain homeostasis, eventually resulting in a state of distress. Unless this is corrected, sooner or later the animal will fall into a state of poor health or perhaps die. Unfortunately, stress is sometimes linked with poor welfare in a way that does not take into account the animal's ability to cope with the stressor, so that, for example, physiological responses to minor stressors are sometimes taken as indicators of poor welfare. It has therefore been suggested that use of the terms 'stress' and 'stressor' should be restricted to conditions and stimuli where predictability and controllability are at stake[28]. It may be difficult to persuade the world to make this change, but this more rigorous approach would certainly improve communication between scientists in animal welfare research and researchers working on stress in other areas of animal and human research.

3.3.2.2 Naturalness and welfare

Some welfare scientists have argued that biological fitness (the ability to survive and reproduce successfully) can be used as an indicator of welfare, on the grounds that animals have evolved to function well in a particular natural niche. However,

---

[26] I use the word 'tend' as evolution is a dynamic process in which animals are constantly adapting to cope with change. A nice recent example comes from changes in Nebraskan cliff swallow wing length in populations alongside roads, which may be due to selection resulting from collisions with road traffic, http://www.cell.com/current-biology/retrieve/pii/S0960982213001942, accessed 16 April 2013.

[27] NRC (National Research Council) (2008), p. 2.

[28] Koolhaas *et al.* (2011) Broom argues that the term stress should be limited to events that are likely to adversely affect fitness (reviewed in Broom 2011).

this definition runs into problems when one considers battery hens, which can lay eggs regularly despite obvious welfare issues. Others suggest that good welfare is demonstrated by an animal having similar behaviour and physiology to that shown in the wild. It is true that animals may be strongly motivated to perform certain behaviours and may be frustrated if they are unable to do so (the Brambell report pointed to the frustrated behaviour of certain migratory birds in captivity). It is therefore sensible to consider the natural behaviour of animals when design-ing animal accommodation, but this needs to be done with thought and care. For example, what do we mean by natural behaviour and what behaviours should we cater for? After all, domestication can result in changes in behaviour, so for a domesticated animal what is natural behaviour? Furthermore, the behaviour of a species can vary depending on the environment in which the animal finds itself. Indeed, there is a whole discipline of behavioural ecology in which scientists have investigated such variation. Another issue is that just because a behaviour is natu-ral does not mean that an animal's welfare is necessarily improved by giving it the opportunity of expressing it. Giving birth is natural but for some animals can result in pain or death. If there are good reasons for thinking an animal will suffer from dystocia (birth problems), then allowing it to mate might well be wrong[29].

3.3.2.3 Feelings and welfare

Although some people make distinctions between feelings, emotions, motivation and drives, they are all descriptions of states of mind that drive behaviour. These mental states are the product of evolution, which, modified by experience, guide us and other animals as we make our way through the world. Feelings such as fear help us to avoid dangerous situations, while the feeling of hunger causes us to eat. If feelings are consciously experienced by some animal species, as is generally accepted, then it is reasonable to assume that they may also experience mental states that are either positive or negative. Consequently, any assessment of welfare must consider the mental as well as the physical state of the animal[30]. In fact, I believe there is a growing consensus that what matters to those animals capable of experiencing feelings, and therefore what should matter most to us, is how those animals feel[31], whether they are experiencing bad feelings (e.g. pain, itching, nau-sea, loss or frustration) or more positive feelings (e.g. those associated with play or satisfying needs such as hunger, thirst or warmth).

---

[29]   Würbel (2009), however, makes the case that most natural behaviour is rewarding when expressed in suitable circumstances. So, while exposing animals to predators simply to allow them to express natural behaviour would be wrong, giving them the opportunity to express appropriate anti-predator behaviour when the animal perceives that it is under threat would be right.

[30]   The Brambell Committee described welfare as 'a wide term that embraces both the physical and mental well-being of the animal. Any attempt to evaluate welfare therefore must take into account the scientific evidence available concerning the feelings of animals that can be derived from their structure and functions and also from their behaviour.'

[31]   Dawkins (1980); Duncan (1993); Kirkwood (2006); Boissy et al. (2007); Würbel (2009); Mendl et al. (2010b); Panksepp (2011).

However, physical and mental states do not always change in synchrony. Ill health will often impact fairly quickly on how we feel, but it is possible to be in a state of exceedingly poor health and yet, for a while, be unaware of the problem. A life-threatening tumour might have started to develop but as yet causes no pain or discomfort. Similarly, one may have an infection but it can take a few days before the symptoms develop. Clearly, if one is concerned about the future well-being of an animal with a diagnosed disease, decisions have to be made regarding what, if any, medical treatment is appropriate, even if the animal shows no signs of psychological distress. So health matters, and few would disagree with the suggestion that an animal must be in a state of good health to have good welfare.

The average person, thinking about their interactions with dogs or cats, will hardly find the idea that welfare is about feelings and their importance to the animal radical, but it has been, and still is, a controversial idea within the scientific establishment. Animals vary greatly in complexity and if we accept that some animals probably do experience feelings, we then have to ask, as we will in the next chapter, which ones. Some scientists have felt that animal welfare should be about the animal's physiology as they believe that physiological measures are more objective or more likely to indicate a major insult to the animal[32]. There has also been debate amongst those who study animal behaviour. Darwin argued for a continuity of experience between humans and animals, and some pre-war ethologists, such as Julian Huxley, felt that gaining some insight into animals' subjective experiences was essential to understanding their behaviour. However, Niko Tinbergen and Konrad Lorenz, the founding fathers of the modern science of ethology, both argued against the study of subjectivity. Tinbergen, in particular, could not conceive how it was possible to use the scientific method to measure an animal's subjective feelings, as these are only experienced by the animal itself, and he strongly felt that the study of feelings was not an appropriate subject for research[33].

Tinbergen may have been acting rather defensively. The study of behaviour was not considered entirely respectable by some of their contemporaries, and Lorenz and Tinbergen were perhaps influenced by their desire to put ethology on a strong scientific footing. However, possibly they were too cautious. It is true that we can only approach animals' mental states (affect) indirectly, and making the leap from data to inferences about animals' mental experiences is difficult and fraught with pitfalls[34], but as we shall see later in the chapter, scientists have developed ingenious techniques to probe how animals might be feeling. While our concern about animal feelings derives from our own experiences of feeling, we must beware of uncritical anthropomorphism. Animals' feelings may be very different in causation and quality from those that we experience. Not only this, but even in humans different measures of emotions (verbal reports, what people find negatively and positively

---

[32] McGlone (1993).

[33] Burkhardt (1997).

[34] So much so that Marian Dawkins, who believes that some animals are capable of suffering but is also very clear that her belief is not scientific (Dawkins, 2008), now suggests that it is better to use reasons other than sentience when making the case for animal welfare Dawkins (2012).

reinforcing, autonomic changes such as heart rate, hormones, brain activity and behaviour) do not always correlate with each other[35]. This is why clever and rigorous science is so important in order to establish what matters to an animal and how much it matters to it. It is also important to remember that animals (and humans) may have desires and motivations that if satisfied might not always be in their own best long-term interests.

3.3.3 Assessing animal welfare

I have suggested that there seems to be increasing acceptance that how an animal feels is what counts, and this is also my own personal view. However, as animal welfare scientists have placed, and probably will continue to place, emphasis on different aspects to assess an animal's well-being, and as the measures that one chooses will depend on the welfare issues that one is concerned about, it is very important that those assessing welfare should make clear what they are measuring and how they use that data to make inferences about the animal's welfare[36]. Fraser makes this point well. He describes a farm animal welfare case in which European and Australian committees with different conceptions of welfare reviewed the same literature on the use of sow gestation stalls. The Australian experts were primarily concerned with health and biological functioning, whilst the European reviewers were also concerned about the animals' affective state and included issues such as frustration, and as a result the two teams came to very different conclusions[37].

Animal welfare assessment is therefore a two-stage process. The first of these is to collect the data, and the second to interpret the results. Interpretation will always be a personal matter, and well-written papers and reports make the process clear so that the reader can understand how the author came to the conclusions. Good welfare assessment usually benefits from recording a number of different measurements, partly to address the welfare definition issue and partly because a range of measures helps control for issues (e.g. activity, time of day) that might affect some of the measures, but mainly because multiple measures simply provide more information that can be weighed to assess the animals' state[38]. Nonetheless, one should beware of assuming that the use of multiple measures guarantees a better assessment of welfare (if the measures used are bad or poor, then their use is not helpful).

3.3.3.1 Resource-based and animal-based measures

So how does one go about assessing animal welfare? There are two major lines of attack. One is to assess the resources available to the animal by measuring, for example, how much space is provided for the animal, the quality of the food and water, whether there is adequate veterinary care provision, and so on. This approach is sometimes termed resource-based or management-based measures of animal

---

[35]   Reviewed in Dawkins (2008).
[36]   Mason and Mendl (1993); Fraser (1995, 2008).
[37]   Fraser (2008), chapter 4, and p. 240 onwards.
[38]   Mason and Mendl (1993).

welfare. An alternative strategy is to assess how animals are actually faring under the system in question, in other words whether individuals are in pain or injured or showing signs of stress or distress. These types of measures are called animal-based or outcome-based measures.

Resource-based measures

Resource-based measures have some significant benefits, particularly for large establishments housing many animals where it would take a very long time to assess each animal. In practice, an on-site assessment might involve many measures and could take considerable time, but determining whether a resource is present or not is, in principle, relatively simple to carry out. Resource-based measures do not require direct or close contact with the animals, which can also be an advantage if the animal is dangerous or likely to be stressed by close contact. On the other hand, resource-based measures can miss important information. Some animals might do better than others and provision of resources does not guarantee that an animal will be able to use them. Competition between animals can result in an animal being unable to access the food trough or drinker, even though both these resources are present. It is therefore important to look at how the resources are managed as well as whether they are provided. Another point relevant to the use of animals in research is that resource-based measures are not useful for assessing the impact that a scientific procedure has had, or is likely to have, on an animal[39].

Animal-based measures

If one wants to know how an animal is faring, then it is necessary to study it. Animal-based welfare measures include a variety of physiological and behavioural measures, the choice of which will very much depend on the reason for the assessment. If the aim is to detect adverse effects during an experiment, then relevant measures may include signs of pain, injury, fever or response to a toxic compound and, in extreme cases, signs of distress. The aim is often to obtain a quick assessment of the animal so long-term profiling of hormonal measures may not be appropriate. However, if the aim is to establish how best to house or keep the animal, longer-term measures may well be more appropriate. In these studies the measures may include more and rather different behavioural indicators, although there is no hard and fast distinction[40].

3.3.3.2 Observation and clinical assessment

Daily observation and, if necessary, clinical examination by a veterinarian are the first lines of defence in identifying health or husbandry issues, and for assessing expected and unexpected responses of the animal to experimental procedures. Daily observation

---

[39] Main *et al.* (2003); Butterworth *et al.* (2011).

[40] As the Brambell report noted 'The scientific evidence bearing on the sensations and sufferings of animals is derived from anatomy and physiology on the one hand and from ethology, the science of animal behaviour, on the other'. Latham (2010) provides a good summary of the variety of methods available to assess welfare.

of animals that are being used in experiments is, rightly, often a regulatory requirement (e.g. specified in Annex III to European Directive 2010/63/EU), but observations may need to be much more frequent depending on the procedure and the adverse effects expected. Post operation, for example, it would be normal to have a period of much more frequent observations (it has been suggested that for rats 5–10 minutes per animal at 1–2 hour intervals may be required)[41], and depending on the legislation or regulation there may be a requirement to specify this in the plans for the experiment.

Experienced animal care staff can often detect problems very quickly, even if it is not always easy for them to describe exactly what is wrong ('this animal just doesn't look right'). However, although years of experience are very helpful, science-based knowledge is also essential. Traditional, obvious signs used to identify pain in rodents, such as piloerection, unkempt coat/rough hair, hunched posture, apathy, aggression, and self-mutilation[42], may be indicative of relatively severe pain or suffering, so if these are used then intervention will be too late. Earlier signs can be more subtle and harder to detect. Researchers at Newcastle have found that even experienced staff can be poor at detecting signs of pain in rodents without training, but that even a short period of tuition using videotapes can yield substantial improvements[43]. One of the problems is that because humans predominantly use the face to communicate, when assessing animals we tend to look at their faces, whereas signs of pain may be displayed elsewhere depending on the reason for the pain. For example, rats and rabbits, following midline abdominal surgery, show characteristic back arching movements; rats after adrenalectomy writhe and press the belly against the ground; and mice after vasectomy show belly pressing and other behaviours[44]. Signs like these are good markers of pain but those who looked at faces may have had some justification, even if they did not know what to look for. Intriguingly, scientists are now discovering that rodents and rabbits do indeed have facial expressions that change with pain[45], but it takes some training to become aware of these signs. Although there are interspecific similarities in signs shown by animals in pain or distress, there are also differences, so it is essential that staff are trained to identify the signs exhibited by the species they work with or care for.

Observations can be carried out in a relatively unstructured way, with the observer simply recording any significant features, but many sites have developed structured score sheets (see Chapter 6) or computerised data management systems, and these, if based on robust research to identify the appropriate signs to be recorded, can be more effective at detecting signs of pain than just relying on clinical impression[46]. Assessment systems help draw attention to particular signs, and

---

[41]   Flecknell and Roughan (2004).
[42]   NRC (National Research Council) (2003), p. 17.
[43]   Flecknell and Roughan (2004).
[44]   Flecknell and Roughan (2004); Wright-Williams *et al.* (2007); Leach *et al.* (2011).
[45]   Langford *et al.* (2010) for mice, Sotocinal *et al.* (2011) for rats. For ongoing research on the use of facial expressions to assess pain in these and other species, see for example Leach *et al.* (2012), Matsumiya *et al.* (2012) and http://www.nc3rs.org.uk/researchportfolio/.
[46]   Roughan and Flecknell (2006).

No Pain   Maximum pain possible

Figure 3.1 Example of a visual analogue scale that might be used to score pain. A cross is drawn on the line to represent an estimate of the pain experienced by the animal. This is converted into a numerical figure by measuring along the line.

because their presence or absence is recorded in a tabular or visual form, make it easier to spot whether a particular animal is improving, remaining steady or gradually getting worse. Typical signs that may be scored include:

1. changes in the appearance of the animal, such as brightness of eyes, appearance of the coat or posture;
2. changes in behaviour and responses to the staff;
3. food and water consumption, body weight and urine/faeces production;
4. presence of vomit or discharges (nasal, salivation) and, with rats, excretion of porphyrin, so-called red tears or chromodacryorrhoea;
5. specific responses to the procedure, such as wound infection;
6. common physiological measures, including body temperature, heart rate, blood pressure and respiration rate.

The signs can be scored descriptively, but tabulation is easier if they are converted into a numerical score (e.g. on a scale of 1–10) or through the use of visual analogue scales in which the scorer is asked to place a point on a short line (Figure 3.1). As always, signs do not provide a direct measure of welfare but need to be interpreted. Changes occur with age, males and females differ, as do different strains, and even within strains there are individual differences. Assessment needs to take into account these factors as well as the animal's previous history, and this is often best done using a team approach.[47]

3.3.3.3 Behavioural measures
Changes to behaviour (often accompanied by physiological changes, which are discussed later) are often the most effective and flexible way of dealing with a stressor. An animal confronted with a dangerous situation or a competitor can

---

[47] There is both European and US guidance: US guidance is provided in NRC (National Research Council) (2008) and NRC (National Research Council) (2009). The European Commission (2012) has produced guidance for project planning, severity scoring and reporting. See also FELASA Working Group on Pain and Distress *et al.* (1994). Morton and Griffiths (1985) is a seminal reference. Hawkins (2002), Morton and Hau (2010), and Joint Working Group on Refinement (2011) review many of the techniques used to assess stress and distress. Mason *et al.* (2004) describe how chromodacryorrhoea can be scored to assess rats. Honess and Wolfensohn (2010) describe an extended grid system for the assessment of, for example, cumulative suffering. See also section on humane endpoints in Chapter 6.

move or run away, or attack and defend itself if necessary. If it gets too hot or cold, it can move to a more comfortable place or it may begin to shiver[48]. Behaviour therefore provides a good early indicator of potential problems, but it is not limited to this. The development of ethology and the subdiscipline of applied animal behaviour science have vastly increased the tools available to scientists to assess the welfare of animals under different husbandry regimes, during procedures, and to inform regulation.

The use of natural behaviour to inform husbandry

Animal behaviour is species specific and has evolved so that animals largely behave in ways that maximise their survival in their natural habitat. Behaviour that enhances the chances of survival and reproduction therefore tends to be strongly motivated. Captive migratory birds show every indication of wishing to migrate by repetitively flying to one side of their enclosure, even though their needs in terms of food are met. Gerbils are strongly motivated to dig and live in safe refuges in captivity even though there is no risk of predation (see Chapter 6). Likewise, the domestic hen will attempt to dust bathe even when there is no suitable substrate, and mink will work hard to gain access to water in which to swim[49]. If strongly motivated behaviours are frustrated, then it is highly likely that the animal's welfare will suffer. So, the production of ethograms (descriptions of the behaviour categories) combined with the study of an animal's behaviour in the wild can provide useful indicators as to what behaviour is likely to be strongly motivated. Of course, compromises are necessary. There are real constraints regarding the practical size of enclosures, ease of cleaning and capture, as well as ensuring that animals are moderately visible for daily checking of their welfare. Nonetheless, information from field studies can usefully inform decisions as to how research environments should be designed and what resources should be placed within the enclosure.

On the other hand, just because an animal performs particular behaviours in the wild does not mean that the animal must perform those behaviours. Behaviour is versatile to allow animals to cope with a variable environment. Some behaviours may be largely elicited by external stimuli, in which case if the stimuli are lacking there can be no motivation and hence no welfare deficit[50]. Another reason for caution in extrapolating from natural behaviour is that most species used in research have been bred by humans for many generations. During that process we have, either deliberately or accidentally, selected for certain traits and bred out others with the result that many animals used in research differ to some degree from their wild ancestors. The dog, for example, has been domesticated for many thousands

---

[48] This is a bit of a generalisation. Some animals are less mobile than others and some go through a sessile stage where movement is restricted (e.g. barnacles), although they can often mount some behavioural defence if threatened. However, rightly or wrongly, we tend to be less concerned about the welfare of such animals (see Chapter 4), and I am here talking about the sorts of animals protected by legislation.

[49] For example, Wiedenmayer (1997); Duncan *et al.* (1998); Cooper and Mason (2001).

[50] Warburton and Mason (2003).

of years, during which time its social and other behaviour has been modified. It is not therefore possible to use wolf behaviour as a reliable guide in understanding domestic dog behaviour or to deduce dogs' needs[51]. Nonetheless, some behaviour is conserved. There is an excellent video available on the internet of laboratory rats released into an outdoor enclosure which shows that, despite many generations of captive breeding and selection for ease of handling, the rats still retained many ancestral behavioural patterns and were able to adapt well to an outdoor life[52].

Another difficulty with using natural behaviour to inform husbandry choices lies in deciding what is natural, even for the undomesticated animal. Substantial behavioural flexibility is a characteristic of many species that allows them to cope with seasonal and other environmental fluctuations and enables them to flourish in a wide range of habitats. There is indeed an area of science – behavioural ecology – devoted to recording and understanding such adaptive responses. Ranging behaviour is an example where behaviour often changes depending on the resources available. If food is the limiting factor then ranges tend to be smaller in places where food is more abundant. Other behavioural changes can include dietary differences, and even changes in social group size and structure.

Behavioural indicators of stress

Species-specific postures or behaviours that are known to be associated with stress or distress (see also section on daily observation) can be used to assess an individual animal's welfare or to compare different husbandry regimens. In some cases behaviours may be a direct attempt to mitigate the problem, such as when animals with abdominal pain press against a cool surface. Animals may also have evolved means of communicating their distress (e.g. dogs may show low posture, body shaking, yawning, vocalisation, paw lifting and various oral behaviours when exposed to either acute or chronic stress)[53].

However, identifying reliable behavioural indicators of stress is not always easy. The signs of distress in some animals are less obvious to us than those in others. It can be hard to decide whether fish or reptiles are feeling pain, and assessing pain in more closely related animals may also require training (see Chapter 6). Another question to consider is whether the signs that an animal exhibits in response to a stressor reliably indicate the animal's degree of stress. For example, it might be in a young animal's evolutionary interests to exaggerate its need for parental care, in which case variation in calling would not be a reliable indicator of the animal's need. On the other hand, honest signalling of need can be adaptive when the receiver of the signal can do something about it and derives a reproductive advantage from doing so, as is the case when piglets that are hungry or in pain vocalise[54].

---

[51] The period is not known for certain but the earliest domestication event took place 20 000 to 140 000 years ago (Vila *et al.*, 1997; Bradshaw *et al.*, 2009; Bradshaw, 2011).
[52] Berdoy (2002).
[53] Beerda *et al.* (1997, 1998, 1999a,b).
[54] Weary and Fraser (1995); Weary *et al.* (1997).

Abnormal behaviours, such as stereotypical behaviour patterns and self-injurious behaviour, tend to develop when animals are housed in barren housing over long periods, particularly during early development, with little opportunity to express many natural behaviours or to exert much control over the environment. As a result they are often interpreted as indicators of poor welfare, but stereotypies in particular need to be interpreted with caution. These are very repetitive behaviours (e.g. repeated circling, jumping, chewing, head flicking) some of which have been, and in some cases still are, common in some laboratory environments. In the 1990s, it was estimated that 50% of mice, which are the most commonly used species in research (in 2004 estimated at some 7.5 million animals worldwide) showed stereotyped behaviour such as bar biting, jumping and looping around the cage.[55]

Unfortunately, much rodent stereotyped behaviour is shown during the animals' active phase, during the night. Therefore staff at a research establishment may well be unaware of the problem unless they use a reversed light cycle or video monitoring. Stereotypies are easier to spot in diurnal animals. Locomotory stereotypies have been observed in dogs in research housing; in one series of studies, 13% of single-housed laboratory dogs spent more than 10% of their time in such behaviour, with some animals spending substantial proportions of the day in stereotyped behaviour[56]. A survey of US primate facilities carried out in 2003 found that although 73% of primates were housed socially, the percentage was only 46% for the indoor housed population, a housing condition which featured strongly in biomedical research[57]. Macaques, the group of primates most commonly used in research, are very social animals so, perhaps unsurprisingly, single-housed macaques are particularly prone to stereotypies or other behavioural abnormalities; 89% of individually housed rhesus monkeys at a research centre showed abnormal behaviour, with pacing occurring in 78% and self-injurious behaviour in 11%[58].

There are similarities between stereotypies and perseverative behaviour in humans with psychological conditions such as autism and schizophrenia; in humans, such behaviour is linked with defects to the basal ganglia of the brain. This, combined with our knowledge of the sorts of conditions in which stereotypies develop, suggests that animals with stereotypies have been subject to some form of welfare insult. However, the link with welfare and current housing conditions is less clear. There has been debate as to whether stereotypies might in some cases be a way of coping with an inadequate environment[59], in which case an animal with stereotypies might be better off than one without them when

---

[55]   Wiedenmayer (1997); Mason and Latham (2004).
[56]   Hubrecht *et al.* (1992); Hubrecht (1995).
[57]   Baker *et al.* (2007). In the UK, social housing for primates has become much more common in biomedical research, including during data collection. Examples include the use of telemetry, in-home cage testing using touch screens, and it is even possible to house macaques together when one has a surgical head implant as long as the one with the implant is higher ranking than its cage mate.
[58]   Self-injurious behaviour is much more common in individually housed than socially housed rhesus macaques (Novak *et al.*, 2006).
[59]   Mason and Rushen (2006), pp. 14–15.

housed under the same conditions. Another difficulty with using stereotypies to assess welfare is that they are extremely persistent, often continuing after animals have been moved to better housing. Stereotypies may also result from clinical problems or, in some species, inadequate diet, so they do not always indicate poor housing. Stereotypies are therefore a warning signal that something is or has been wrong, but are not necessarily a good indicator of the current welfare state of a particular animal or of the quality of the housing. Equally, their absence is not proof that all is well[60].

Collection of behavioural data by observers is the traditional method employed, and has the advantage that the human observer can react to the unexpected. However, the process can be very time-consuming and requires training. Automated systems are beginning to be developed and may become much more important in the future for routine monitoring of animals.

Finally, it is worth noting that a number of standardised techniques, such as the open-field test and the open arm T-maze, have been used to assess stress, particularly in the fields of psychology and pharmacology. In the open-field test an animal such as a mouse is placed in a bare arena, brightly lit from above, and its behaviour including locomotion and defecation recorded. Freezing and increased defecation are taken as showing increased anxiety or fear. However, this interpretation may well be suspect, as it is more likely a measure of vigilance or anti-predator behaviour[61]. It is therefore important to determine the underlying motivation for behaviours performed in these types of apparatus.

Experimental studies of behaviour

Usually the first stage in attempting to improve animal husbandry systems is to observe the animals, recording behaviour and the way and extent to which they use the resources already provided. These sorts of observational studies provide a baseline before attempting any modifications, should highlight issues such as abnormal or undesirable behaviours, and may stimulate ideas as to new resources or better ways of providing existing resources. It can also be helpful to compare the behaviour of animals housed in different establishments. Such studies can be extended by comparing the behaviour of animals before and after a change in husbandry such as providing some form of enrichment (e.g. a more complex social or physical environment; see Chapter 6) in an attempt to meet the animals' needs. Research of this sort has shown that animals will make extensive use of various types of enrichment provisions, such as foraging devices, manipulatable objects and mirrors for primates, and platforms and chews for dogs[62]. Having gained this information, further experiments, described below, can be carried out to assess what animals want and how much they want it, and to try to assess animal's mental state in different husbandry systems.

---

[60]   Mason and Latham (2004).
[61]   Rushen (2000).
[62]   Hubrecht (1993); Lutz and Novak (2005).

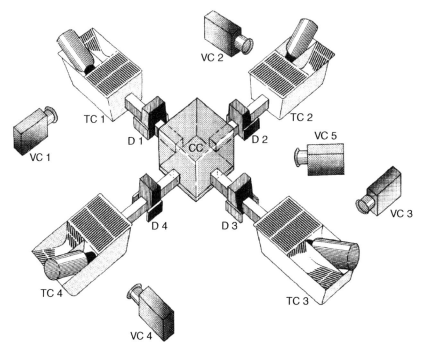

Figure 3.2 Multiple-choice housing system for preference tests with mice. Two or four test cages (TC) can be connected to the central cage (CC) in which the animal is initially placed. Automatic detection units (D) and video cameras (VC) are used to monitor results. Reproduced with permission from H. Blom, Utrecht University.

Often there may be several different enrichment options, and the question is which would the animal most value. For example, is it more important to provide more space or to provide refuges? Preference tests in which the animal can pick from one or more resources and spend some time using them can give some idea as to whether an animal values one resource more than another. In practice this can be tested by recording which chamber of an experimental apparatus an animal chooses or how long it chooses to stay there, where the chambers contain different resources, such as types of nesting material (Figure 3.2), or by recording which resources are used and for how long[63]. However, these tests only give information about the animal's ranking of the resources; they do not tell you how much the animal values the resource. So, if the resources tested were of little importance to the animal, then there might still be a preference between them but the results would not be of much value for informing housing decisions.

One way to address this problem is to assess the strength of an animal's motivation using demand studies. We know from our own experience that if we work very

63   For example, Pullen *et al.* (2010) have shown that dogs prefer toys that are easily destroyed.

hard for something, the chances are that we really want it, so to judge how much an animal wants something, a researcher might arrange things so that the animal has to work for the resource by pressing a lever a certain number of times or by lifting a weighted door, or by requiring the animal to travel varying distances or through a difficult or aversive medium such as water to obtain the resource[64]. By varying the amount of work that has to be done to gain the resource (number of lever presses, weight on the door, distance travelled) and by comparing the work that the animal is prepared to do to gain the resource with some known highly motivated behaviour such as working for food, it is possible to better assess the value of the resource to the animal.

Preference and demand studies have an attraction as ways to measure the motivation of animals for resources. They do, after all, give us an insight into the animal's desire for various resources, and have proved useful in informing husbandry provision for many species. Examples include studies demonstrating rats' motivation for social housing, novelty and solid resting areas; mouse motivation for different types of nesting material as well as running wheels and extra space; rabbits for social contact, and in its absence for cages with mirrors[65]. Preference studies have also been used with hens to examine how other measures might correlate with the animals' preferences. Intriguingly, blood glucose levels seemed to be a much better predictor of the hens' choice of environment than corticosteroid concentrations, although the latter are much more commonly used as a welfare measure[66].

However, both preference and demand studies need to be carefully interpreted, and there has been considerable debate as to how the studies should be designed and what measures should be used[67]. Many factors can influence animals' choices and the strength of their motivation to achieve particular goals, including what resources are given to them and in what context, the animal's previous experience of the resource, how long it is available to them and whether they become satiated. There can also be conflict between the animal's short-term motivation for a resource and its long-term best interests. For example, an animal might work for a tasty high-calorie treat, but if it is given easy access to too many of these, there is increased risk of obesity and disease. Another problem is that for larger animals the apparatus also has to be larger, which may explain why there are rather few publications that have explored the use of these techniques for dogs and primates.

---

[64]   These are termed instrumental or operant methods of assessing motivation.

[65]   Manser *et al.* (1996); Sherwin (1996); Van de Weerd *et al.* (1998); Patterson-Kane *et al.* (2002); Seaman *et al.* (2008); Edgar and Seaman (2010). The mice in Sherwin's studies worked as hard to maintain access to some of the resources as they did for food, giving some indication of the strength of their motivation.

[66]   Nicol *et al.* (2009).

[67]   Behaviour can be assessed or animals can be asked to perform operant tasks in which they have to perform a task to indicate what they want, how much they want it, or even perhaps how they are feeling (Bateson, 2004; Duncan, 2005; Kirkden and Pajor, 2006; Boissy *et al.*, 2007). Würbel (2009) discusses the limitations of demand studies for determining needs.

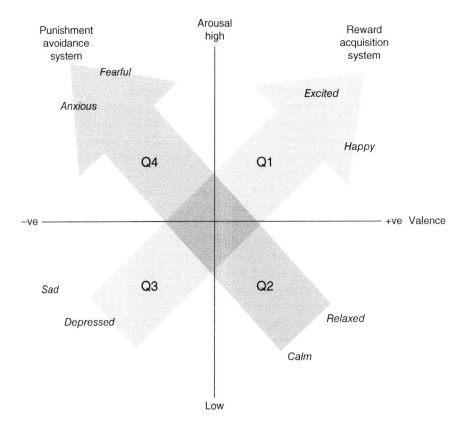

Figure 3.3 Core affect represented in two-dimensional space. Words in italics indicate possible locations of specific reported affective states (including discrete/basic emotions). Positive affective states are in quadrants Q1 and Q2, and negative affective states in quadrants Q3 and Q4. Arrows indicate putative biobehavioural systems associated with reward acquisition (the Q3–Q1 axis of core affect) and punishment avoidance (the Q2–Q4 axis of core affect). Reproduced from Mendl, M., Burman, O.H.P. & Paul, E.S. (2010b) An integrative and functional framework for the study of animal emotion and mood. *Proceedings of the Royal Society of London B: Biological Sciences* 277, 2895–2904, with permission from The Royal Society.

Studies of affect

It would be wonderful if we could measure the valence of animals' feelings or emotions (their affect), that is whether they have generally positive feelings or whether they are more negative, as this is fundamental to welfare (Figure 3.3)[68]. In human terms, we are asking whether the animals are feeling generally 'happy' or 'sad' (or

---

[68] Mendl *et al.* (2010b) suggest that feelings can be explained functionally if described on two axes, one of valence (i.e. whether positive or negative) and the other of arousal (high or low).

'good' or 'bad'), bearing in mind the caveat that different species may experience these emotional states differently to the way we do.

One approach to studying animals' feelings has been to observe the behaviour of humans (who are able to tell us how they are feeling) and then to look for similar behaviour in animals. People who are depressed are more likely to think that future events will turn out badly than those who are not depressed. Based on this knowledge, researchers at Bristol University devised an experiment that asked whether rats, kept in conditions that one might expect would produce a negative state of mind, would be similarly 'pessimistic'. Their experiment proved this was indeed the case, as rats housed in unpredictable conditions (which one might expect would be stressful) tended to predict future events more negatively than rats kept in predictable conditions. The basic experimental method used in this and other studies of cognitive bias is worth describing as it is quite ingenious. The animals were trained to press a lever when they heard a tone that predicted provision of food but to refrain from pressing it when they heard a tone of a different pitch, as otherwise they would receive an unpleasant burst of white noise (a loud hiss). Once it was clear that the animals had learned when and when not to press the lever, the experimenters then used a range of probe tones intermediate in pitch between the two original tones, and recorded whether the animal pressed the lever and, if so, the length of delay before pressing it. The stressed rats took longer to decide and were less likely to respond to probe tones close to one that predicted food, and so appeared to have a pessimistic outlook. The original paper stimulated a range of further studies in other species, using a variety of types of stimuli and rewards. The results have been broadly similar, suggesting that the technique has promise for the study of animal affect[69].

It may be tempting to assume that animals showing cognitive bias must also possess conscious emotional states analogous to those experienced by humans, but this is not necessarily the case. Cognitive bias is likely to have adaptive advantages, as an animal feeding in a dangerous environment would do well to act with caution to any disturbance, whilst one in a safe environment could take more risks. The mechanism that produces cognitive bias does not necessarily have to be one that requires conscious perception of emotions (one can imagine a computer that is programmed to behave in a more cautious manner in a risky environment). Indeed, similar bias has been demonstrated in the honeybee, an animal that we might not guess would possess conscious emotions[70]. Nonetheless, cognitive bias studies have important implications for the use of animals in research. First, even if animals are not conscious, it is better if the complicating effects of stress on experimental results are avoided. Second, if we assume that certain species are conscious, then it is also reasonable to assume that demonstration of cognitive bias tells us something about the state of its conscious mind.

---

[69]   Harding *et al.* (2004) Bateson and Matheson (2007); Mendl *et al.* (2009, 2010a); Brydges *et al.* (2011).

[70]   Bateson *et al.* (2011); Mendl *et al.* (2011).

Researchers have also used other experimental approaches to try to get into the minds of animals. One group of researchers has found that animals exposed to stressors tend to seek out experiences that are more positive. Not only this, but the anticipatory behaviour they show varies with the value of the reward that they have come to expect. In other words, they show more anticipatory behaviour when the reward is a good one. This is exactly the sort of result we might expect, but it also suggests that anticipatory behaviour might be a useful tool to assess both the existing welfare of the animal and the value to the animal of various resources without the bother of having to carry out difficult and tedious preference tests. Other approaches are to study the animal's response to the loss of rewards and the development of anhedonia[71].

3.3.3.4 Physiological indicators of stress

The physiological literature relating to stress is now very large, and here it is only possible to provide a very brief summary. Fortunately, there are several good overviews available for those who require more detail[72].

When mammals are exposed to threatening situations to which the animal has a suitable behavioural response, activation of the sympathetic nervous system (SNS) leads, amongst other things, to increases in heart rate, blood pressure, respiration rate, coagulability of the blood, and secretion of adrenaline (epinephrine) from the adrenal medulla, the so-called sympatho-adrenomedullary (SAM) response (Figure 3.4). This in turn stimulates gluconeogenesis and lipolysis, resulting in increased levels of glucose and fatty acids in the blood thus making energy reserves immediately available to the animal. These changes prepare the animal for activity that may be violent or strenuous and is often termed the fight or flight response. Deactivation of the SNS is associated with a state of relaxation. When active coping behaviour is not possible, the response is somewhat different. For example, while blood pressure still rises following SNS activation, the rise is due to vasoconstriction rather than to cardiac changes. In fact, heart rate and cardiac output are decreased.

A wide range of stressors, such as restraint, infection, or intense heat or cold, can activate the hypothalamic–pituitary–adrenocortical (HPA) axis (Figure 3.4). Activation of the HPA axis is characterised by the secretion of corticotropin-releasing hormone (CRH) and/or vasopressin from the hypothalamus, which stimulates the secretion of adrenocorticotropic hormone (ACTH) from the anterior pituitary gland. This in turn leads to increased secretion of glucocorticoids (cortisol, or corticosterone in rodents) from the adrenal cortex. Other hormones can also be released from the hypothalamus and pituitary glands that affect growth, metabolic rate, lactation and reproduction.

---

[71] Willner *et al.* (1987); van der Harst *et al.* (2003); van der Harst and Spruijt (2007); Burman *et al.* (2008); Dawkins (2012), pp. 166–167.
[72] Toates (1995) Schneiderman and McCabe (1985); Moberg and Mench (2000); NRC (National Research Council) (2008).

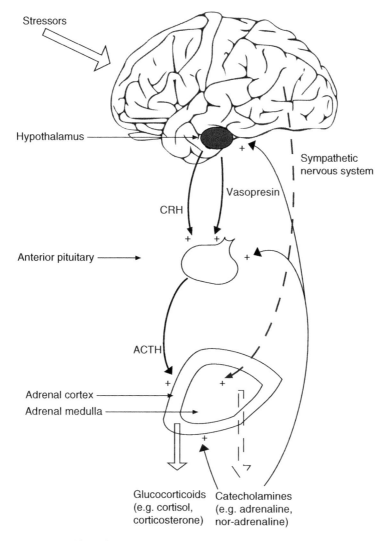

Stressors

Hypothalamus

Sympathetic
nervous system

Vasopresin

CRH

Anterior pituitary

ACTH

Adrenal cortex
Adrenal medulla

Glucocorticoids   Catecholamines
(e.g. cortisol,   (e.g. adrenaline,
corticosterone)   nor-adrenaline)

Figure 3.4 A simplified diagram showing some of the components of the hypothalamic–pituitary–adrenocortical (HPA) axis and the sympatho-adrenomedullary (SAM) response (dashed lines). The pituitary is in fact a small body located under the hypothalamus, which is itself located just above the brainstem. It is here shown much larger and separated for clarity. Activation of the HPA axis occurs when stressors result in the hypothalamus releasing corticotropin-releasing hormone (CRH) and/or vasopressin, which stimulate the anterior pituitary to release glucocorticoids that, in turn, act on various organs and tissues to restore homeostasis. Vasopressin and CRH can to a lesser extent also stimulate glucocorticoid production within the adrenal gland via local cell-to-cell interactions. Stress can also result in stimulation of the sympathetic nervous system, which then stimulates the adrenal medulla to produce catecholamines to ready the body for flight or fight (the SAM response). These can in turn positively feed back to the pituitary and hypothalamus and adrenal cortex. Adapted and redrawn from Matteri, R.L., Carroll, J.A. & Dyer, C.J. (2000) Neuroendocrine responses to stress. In *The Biology of Animal Stress: Basic Principles and Implications for Animal Welfare*, Moberg, G.P. & Mench, J.A. (eds.), pp. 43–76. CAB International, Wallingford. With permission from CAB International, Wallingford, UK.

When mammals are exposed to stressors that are intense and aversive, and where there is no obvious behavioural response that would improve the situation (i.e. those likely to lead to prolonged adverse mental states such as fear or intense frustration), both the SAM and HPA systems are activated. Activation of the SNS may also occur. While animals have evolved to cope with the many normal stressors that they find within the environment, if the stressor is extreme or prolonged or in some other way exceeds their ability to cope with it, then the resultant stress response can be detrimental for the animal. An animal's response to bacterial infection includes an increase in the ratio of neutrophils to leukocytes. As one might expect, there has to be a system to turn off this response (when it has served its purpose), and glucocorticoids do this by downregulating the immune system and, amongst other things, reducing blood neutrophil concentration. Depending on the stressor, acute stress can either increase or decrease various aspects of the immune response, but long-term chronic stress resulting in raised levels of glucocorticoids not only leads to increased susceptibility to infection but also to breakdown of proteins, high levels of blood sugar, and depression including reduced appetite. Obviously, such changes can result in ill health or even death.

All these physiological changes provide a range of measures that can and have been used to either study stress or to try to assess welfare. Hormones such as cortisol or corticosterone, ACTH, luteinising hormone and prolactin have all been used, as have autonomic measures such as temperature, heart rate, heart rate variability, blood pressure, body weight and galvanic skin response. Other measures include changes to immune function, such as ratio of neutrophils to lymphocytes and the production of IgA, and changes in brain receptors such as numbers of opioid receptors.

Physiological measures of hormonal or autonomic changes tend to be rather seductive as potential indicators of welfare. Physiological data are sometimes wrongly seen as 'harder' than behavioural data, perhaps because they tend to come from a machine or assay rather than from human observation. The reality is that, like behavioural data, physiological data need to be interpreted with caution. Like any other measure, physiological data can be invalidated by poor experimental design or technique. Hormone assays, for example, need to be properly validated and carried out, and samples need to taken and stored properly to avoid loss or denaturing of the substance of interest. The hormonal response to stressors can also vary in terms of both quantity and quality depending on the type of stressor and its duration and the response is sometimes counterintuitive[73]. There is also a tendency to view physiological measures of stress as indicators of poor welfare when in fact some stressors can be perceived as being 'good', or within the animal's normal coping ability. In these cases if one were observing the animal one might see behavioural and hormonal perturbations that were simply an adaptive response. Conversely, one does not always see increased HPA activity in animals that one

---

[73] Blecha (2000); Matteri et al. (2000); NRC (National Research Council) (2008). The NRC report points out that the response of the immune system to stress is complex, and depends on which aspect of the immune system is studied, and the duration of challenge with respect to the stressor.

would otherwise consider to be stressed. For example, monkeys with self-injurious abnormal behaviour or with multiple abnormal behaviours are, contrary to expectations, more likely to show lower peripheral cortisol concentrations, perhaps because of a blunting or lowering of the set-point (allostasis) for this response[74]. Finally, levels of circulating cortisol can vary simply due to time of day. So, on their own, elevated stress hormone concentrations could as well indicate improved welfare as reduced welfare; they show a response to stress but not that the animal is in a negative mental state. Equally, an absence of elevated concentrations should not be taken to mean that the animal's welfare is necessarily good.

Another issue to bear in mind is that physiological methods often themselves require some form of intrusion, which itself can be stressful. The animal may need to be captured and restrained in order to collect blood, saliva or other sample. Sensors may need to be placed or implanted and jackets worn to protect the sensors or data pack. Tethers are also sometimes used to transmit the data, although these are increasingly giving way to other methods of remote collection of data such as telemetry, in which implanted sensors wirelessly transmit data to nearby receivers, often allowing the animal to live and behave normally in its home cage with its social partners. However, even telemetry, while it offers many advantages in terms of allowing the animal freedom of movement and access to other animals[75], still requires implantation of the sensors and telemetry device, and repeat surgical operations may be required to replace batteries. Clearly, if stress is the matter of interest, it makes little sense if the measurement itself stresses an animal and so there is considerable effort in this area to refine or improve the techniques[76]. Examples of such refinements include habituating the animal to the techniques – training it to cooperate by, for example, voluntarily moving on to a weighing scale or presenting a limb for a blood sample – or by ensuring that samples are taken before stress hormone levels rise. Telemetry may be an option for some measures such as heart rate, blood pressure and temperature. Whatever the method chosen, it is always worth considering whether there are alternative, less invasive, methods of collecting the same or similar data. For example, it may be possible to analyse hormones in saliva, faeces or urine rather than from a blood sample, or for rats the secretion of red porphyrin from the Harderian glands around the nose[77].

3.4 Funding and Promoting Animal Welfare Research

Winning funds for animal welfare research and its promotion has always been hard. All research funding is extremely competitive, but animal welfare research often suffers as much of it is applied rather than pure research, which disadvantages it in

[74] Novak *et al.* (2013)
[75] Kramer and Kinter (2003).
[76] For example, for telemetry techniques see Joint Working Group on Refinement (2003c).
[77] Mason *et al.* (2004).

the eyes of some funding bodies. Nonetheless, both government organisations and NGOs promote welfare by providing information resources or by funding animal welfare research. Examples of NGOs in the UK include charitable funds such as the Welcome Trust, which supports some welfare research, the Universities Federation for Animal Welfare (UFAW), the Fund for the Replacement of Animals in Medical Experiments (FRAME) and the Lord Dowding Fund. In the USA, the American Association for Laboratory Animal Science (AALAS) and the Association for the Assessment and Accreditation of Laboratory Animal Care (AAALAC) may fund projects aimed at improving the health and well-being of research animals and, additionally, there are numerous private foundations, some of which may fund animal welfare research.

Government sources of funding for animal welfare research in the USA include the National Science Foundation and various Public Health Services (PHS) agencies. In the UK, government funding is routed through research councils such as the Biotechnology and Biological Sciences Research Council (BBSRC), and recently there has been a dramatic increase in the funds that have been made available through the National Centre for the Replacement, Refinement and Reduction of Animals in Research (NC3Rs). Some organisations, such as the Canadian Council for Animal Care in Canada and the NC3Rs in the UK, depend on both government and external funding.

3.5 The Benefits of Animal Welfare Science

There is no question that animal welfare science has led to improvements in the ways in which animals used in research are housed and treated. Postoperative analgesia was rare in the early 1980s, perhaps because analgesia was, and in some cases still is, rarely given to farm animals following procedures such as castration and disbudding. Today, although exemptions can be made on the grounds that the analgesic may influence the science, detection of distress has improved and the use of postoperative analgesia is more common than it was.

Another dramatic change has been in the quality of housing and care. The barren environments that were common in the early 1980s have largely disappeared. My own experience in the UK has been that collaborative research by pharmaceutical companies, shelters and UFAW, together with changing attitudes, has led to enclosures for dogs that are more complex and that contain structures and objects such as platforms and chews that the dogs can use to occupy their time[78]. Primate enclosures for macaques used to be designed for only one or two animals and certainly did not provide sufficient space for them to control their social interactions and a complex and enriched environment. The better facilities now provide enclosures or exercise areas that are reminiscent of good zoo enclosures. Rodent cages are also better. Mice used to be housed in barren 'shoe box' type of cages, not much

---

[78] Hubrecht (1993, 1995, 2002); Joint Working Group on Refinement (2004).

different, in terms of provisions for the animals, from those used in the 1960s, although the material of the base had changed from metal to plastic. As a result of research[79] into the needs of mice and what could be most usefully provided for them, they are now generally housed in enclosures that contain material for nesting in addition to the substrate.

Nonetheless, it is important to understand that, in the words of a sign said to have hung on Einstein's wall, 'Not everything that counts can be counted and not everything that can be counted counts'. Animal welfare science is unlikely to provide precise answers to everything. For example, the minimum space that should be provided for animals is often a controversial issue, usually because little or no scientific evidence can be found to support either existing dimensions or any proposed changes. This is hardly surprising as it would be extremely difficult, and perhaps impossible, to devise a good experiment that resolved the question. So much depends on the quality of the space provided. For a thigmotactic animal (one which prefers to be close or in contact with the edges of enclosures rather than in the open) a smaller, long, thin enclosure might be better than a square enclosure of the same area. The results would also depend on whether the animals are housed in large or small groups, on what enrichment is provided, and on factors such as breed, gender and age. This is why I think it can be unreasonable to require precise specific scientific evidence when considering regulatory changes to minimum dimensions. It can be just as valid to consider how much space is necessary to allow appropriate social housing, structures and objects within the enclosure.

[79] Olsson and Dahlborn (2002); Latham and Mason (2004).

Species Choice and Animal Welfare

In this chapter I review evidence that might be used for making decisions about which species to protect in legislation. For some research there may be opportunities to choose between species on animal welfare grounds. I therefore also look at the evidence for commonly made assumptions that some species might be more capable of suffering than others and describe an approach to making these decisions on a case-by-case basis.

4.1 Introduction

Although the vast majority of vertebrate animals used in research are rodents, many other species have been used. *The UFAW Handbook on the Care and Management of Animals Used in Research* provides some indication of the breadth of current usage, as it contains 33 chapters on individual species such as dogs, cats and rabbits, as well as chapters on groups of species including terrestrial reptiles, amphibians, wild mammals and non-traditional laboratory rodents. The list of animals that have been used in research is not restricted to vertebrates. Invertebrates such as snails, worms, insects, shellfish, cephalopods, crabs and many others are also used in research on subjects as diverse as memory, pain perception, physiology, ecology and behaviour.

We do not, however, treat species equally. Both public opinion polls and legislation reflect the fact that we are more concerned about the use and welfare of some species than others. As we have seen in Chapter 2, when research on animals is covered by national legislation, the protected species tend to be vertebrates. However, some countries protect only some vertebrates, while others also protect

The Welfare of Animals Used in Research: Practice and Ethics, First Edition. Robert C. Hubrecht.
© 2014 Universities Federation for Animal Welfare. Published 2014 by John Wiley & Sons, Ltd.

selected non-vertebrate species, such as cephalopods or decapod crustacea. For those species that are used, there are also differences of opinion as to the degree of protection that we should give them. As we will see later in this chapter, some countries provide extra protection for certain species to further encourage researchers to find alternatives, while other countries do not.

So, what evidence is there for treating different species differently? And are we right in the way in which we tend to apportion our concern for animal welfare across the animal kingdom? This chapter addresses these issues by looking at the evidence base for two species-related decisions: which species to protect, and whether different species should receive different protection. As I wrote in the introduction to the last chapter, if we fail to protect animals that turn out to be sentient, then there are terrible consequences for those animals. Equally, if we wrongly consider that some species have a lesser capacity to suffer, then we run the risk that we may not take sufficient measures to alleviate their suffering.

4.2 Consciousness/Sentience: Evidence for Protecting Particular Taxa

There are a number of reasons why people might have ethical concerns about the use of animals in experimentation, but the most pressing of these relates to the belief that certain of the species used in research may be conscious or sentient (I shall use these terms interchangeably to mean 'capable of experiencing feelings'). I believe that most people naturally assume that animals such as dogs, cats and horses are conscious, but it is also possible that some, or all, animals might be operating as non-conscious automata[1]. These might possess receptors similar to those that respond to stimuli that cause pain in us, and might show similar behavioural responses to these stimuli, but would respond only mechanically and would not *feel* pain. Indeed, it is a good question to ask what is the advantage of evolving the neural mechanisms necessary to experience feelings such as pain[2]. Despite these uncertainties, public views on the matter are such that belief in animal sentience has now been incorporated into a European treaty, the Treaty of Lisbon ratified on 1 December 2009 by European Union member states[3]. Article 13 of the treaty states that:

> the Union and the Member States shall, since animals are sentient beings, pay full regard to the welfare requirements of animals, while respecting the legislative or administrative provisions and customs of the Member States relating in particular to religious rites, cultural traditions and regional heritage.

[1] Solipsists go further, arguing that we cannot be sure that anyone apart from oneself is conscious, but most find this a rather unproductive line of thought.
[2] For a lucid exposition of the issues with respect to fish, see Braithwaite (2010), chapter 2. The issue has also been exhaustively covered by others, for example Dennett (1992); Weiskrantz (1997), chapter 4; Koch (2004), chapter 13; Kirkwood (2006); Dawkins (2012).
[3] Treaty of Lisbon, http://europa.eu/lisbon_treaty/full_text/index_en.htm, accessed 5 April 2013.

As this is legal document it is somewhat surprising that the treaty does not discuss which animal species might be sentient (as worded the treaty includes woodlice and worms for example) or what is understood by the term 'sentience', but perhaps those who drew up the treaty realised the practical problems involved in being more explicit. There are, unfortunately, great difficulties in using the terms 'consciousness' and 'sentience' as they can mean different things to different people[4]. After all, we may say that we are unconscious when we are asleep, or use the word to indicate that we are 'conscious' of some aspect of our environment while we are awake. In fact, one might be conscious of events in a dream while being in another sense asleep and unconscious. Griffin in 1976 defined consciousness as referring to a wide range of states in which there is awareness of thought, memory or sensation, but in order to deal with the variety of conscious states that we can experience, others have split consciousness into subcategories such as phenomenal consciousness, access and monitoring consciousness and self-consciousness (the terms used vary[5]). However, as discussed in the last chapter, those concerned about the welfare of animals generally agree that the most important consideration is whether animals are able to subjectively experience perceptions, thoughts or feelings. This subjective experience is sometimes referred to as phenomenal consciousness or sentience[6].

However, the ability to simply experience something is not enough to raise welfare concerns. Experiences cannot impact on this aspect of an animal's welfare, either positively or negatively, unless the experience also *matters* to the animal concerned. As Marian Dawkins puts it 'For us, one of the most crucial aspects of feeling an emotion is that it matters to us'[7]. As many experiences do matter to us, the

---

[4]   For example, Griffin (1976); Dawkins (2001); Heyes (2008); Broom (2010).
[5]   Braithwaite (2010), pp. 80–81 describes Block's categorisation of consciousness into access, phenomenological and monitoring and self-consciousness. *Access* consciousness is the ability to think about or describe a mental state, either current or associated with a memory. Similar to *perceptual* or *primary* consciousness, the ability to generate a mental image or representation into which you can combine diverse pieces of information and then use integrated knowledge to guide your behaviour and the decisions that you make. *Phenomenal* consciousness is the experience of sensing what is around you and the feelings and emotions generated by what you detect. *Monitoring* and self-consciousness (also known as *reflective* consciousness) is the experience of thinking about your own actions, so that we can reflect on the situation and consider different potential scenarios. See also Griffin (2001), pp. 4–8 and 274–278.
[6]   Terminology is a real issue in discussions of consciousness/sentience. For example, this discussion suggests that phenomenal consciousness is necessary for animals to be able to suffer, although the conscious ability to remember past events (episodic memory) and plan for the future, if present, would also have an important impact on an animal's emotional state (Mendl and Paul, 2008). Carruthers, however, suggests that phenomenal consciousness is unlikely to exist in most animals (Carruthers, 2004) and has argued on this basis that there is no need for concern about their pain (Carruthers, 1989). For a reply see Jamieson and Bekoff (1992). More recently, Carruthers (2005) has suggested that pain can be awful for an animal without phenomenal consciousness, which he defines as the capacity to reflect on personal thoughts/feelings. This is a more restricted use of the term than that used by some others in this field. Broom (2010) also discusses problems of terminology (e.g. with use of the word 'conscious') and provides some suggestions for definitions.
[7]   Dawkins (1993).

question as to whether experiences might matter to an animal might seem an odd one. However, we know that people who have had frontal lobotomies, if pricked with a pin, know that there is pain but are dissociated from the experience – the pain does not matter to them any more[8]. We might feel sympathy for these people because of their condition, but it would be absurd to be concerned about their pain. Similarly, if animals did not possess both the capacity to feel something and for it to matter to them, then while we might value them for reasons such as beauty, monetary value or rarity, there would be no reason to worry about their welfare[9]. Some have tried to separate physical welfare issues from mental welfare, and they are not always linked, as shown by the lobotomy example, but poor physical condition usually impacts on an animal's mental state at some point and if the animal's mental state is good then its other needs have usually been met as well[10].

The problem with placing the emphasis on animals' mental states is that we are then confronted with the problem that the experience of subjective feelings is entirely private. It is not only very difficult (some would say impossible) to infer whether other animals do experience feelings, but also impossible to be sure in exactly what manner they might do so[11]. Nonetheless, despite this lack of certainty, legislative decisions as to which species to protect have to be made.

As one might expect, there is a broad consensus as to which species should be protected, although there are some significant differences between regulatory regimes. In Europe, the 1986 European Directive 86/609 (now superseded) protected non-human vertebrates while, as we saw in Chapter 2, US regulations protect only a subset of vertebrates defined as 'any live or dead dog, cat, monkey (nonhuman primate animal), guinea pig, hamster rabbit, or other such warm-blooded animal as the Secretary (of the Department of Agriculture) may determine'. Curiously, this excludes some of the most commonly used animals in research (rats, mice and birds) from legal protection in the USA, which incidentally is one of the world's biggest users of animals. On the other hand, the major US state research funding agencies, as well as an independent accrediting body (AAALAC), have retained oversight of the use of animals not included in the legislation, and thus help to maintain welfare standards for these species[12].

In most countries, invertebrates are not protected although there are exceptions. Certain invertebrates are protected in New Zealand, Australian Capital Territories and some Scandinavian countries[13], and prior to 2013 the UK national legislation included one invertebrate, a species of octopus, *Octopus vulgaris*. This cephalopod was added to the legislation on the basis of arguments about its neural and behavioural complexity, but the inclusion of just one species did not make much sense on

---

[8] The issue of pain in animals is reviewed in Bateson (1991).

[9] Dawkins (1993, 2001); Kirkwood and Hubrecht (2001); Kirkwood (2006); Heyes (2008).

[10] Duncan and Petherick (1991).

[11] Nagel (1974).

[12] Bayne *et al.* (2010b).

[13] Budellmann (2010); Australian Code of Practice for the Care and Use of Animals for Scientific Purposes, 7th edition, 2004, http://preview.tinyurl.com/dyuzy6a, accessed 26 April 2013.

biological grounds. It was therefore not surprising that suggestions were made in 2005 to either exclude this species or to extend protection to some or all cephalopod species depending on evidence[14]. However, the UK postponed making a decision as Europe had begun the process of revising Directive 86/609, and there were proposals that both cephalopods and decapod crustacea might be included in the Directive's scope. The Commission asked for an expert opinion from its EFSA Animal Health and Welfare Panel, which found in favour of including cephalopods on the basis that:

> *cephalopods have a nervous system and relatively complex brain similar to many vertebrates, and sufficient in structure and functioning for them to experience pain. Notably, they release adrenal hormones in response to situations that would elicit pain and distress in humans, they can experience and learn to avoid pain and distress such as avoiding electric shocks, they have nociceptors in their skin, they have significant learning ability and memory retention, and they display individual temperaments...*

Essentially, the EFSA panel were arguing that cephalopods possess the neurological and physiological apparatus to detect pain, respond to it as a stressor, and to learn from pain. However, the conclusion that all cephalopods should be protected proved controversial, attracting a Royal Society letter that criticised the advice on the basis that 'The evidence presented focuses on the learning, memory skills and sensitive response of invertebrates without clearly demonstrating their ability/inability to suffer'[15]. Perhaps because of the difficulty of directly addressing this point, the issue remains contentious. On the one hand, a review of the behaviour of certain cephalopods suggests that because cephalopods, and especially the octopus, are aware of their location, are capable of learning, and then evaluating and using memory to achieve a particular outcome, they may at least possess primary consciousness[16]. On the other hand, the large size of cephalopod brains may be linked to the fact that they lack an internal skeleton with joints. The greater mobility of cephalopod limbs and body requires more complex musculature than in vertebrates, which in turn requires more neural tissue[17]. Whatever the rights or wrongs of the decision, all cephalopods (including *Nautilus*) are now included in EU Directive 2010/63/EU, but the decision went against the inclusion of decapod crustacea.

---

[14] Animals Procedures Committee Annual Report 2005, http://preview.tinyurl.com/cxql983, accessed 26 April 2013.
[15] 'Opinion of the Scientific Panel on Animal Health and Welfare (AHAW) on a request from the Commission related to the aspects of the biology and welfare of animals used for experimental and other scientific purposes.' *EFSA Journal* (2005), 292, 1–46. Available at http://www.efsa.europa.eu/en/scdocs/scdoc/292.htm, accessed 5 April 2013. Royal Society letter to the European Commission on concerns relating to an EFSA report, http://preview.tinyurl.com/cdlnsex, accessed 5 April 2013.
[16] Mather (2008).
[17] Budellmann (2010).

4.2.1 Criteria used to ascribe sentience/consciousness

So what criteria should be used to ascribe sentience? There is no clear answer to this question but in the absence of direct data relating to the ability to suffer, the criteria that the EFSA panel used to give the benefit of the doubt to cephalopods were not far out of line with previous thinking on the subject. Many years ago, Baker suggested that similarities or homologies in neural structure with that of humans, or other animals believed to be conscious, as well as size, complexity and degree of integration of nervous systems and the intelligence of the animal are all factors that should be taken into account[18]. More recently, a Norwegian report[19] used criteria that included the comparative neuroanatomy of vertebrates and invertebrates as well as their responses to noxious conditions.

For obvious reasons, human consciousness is often taken as a benchmark, but in humans verbal reports of consciousness are the 'gold standard' for ascribing consciousness[20]. Scientists are always coming up with new ways of asking questions of animals, but obtaining something akin to a verbal report of consciousness is, for many species, not likely to be a tenable approach. On the other hand, if there are certain human characteristics that correlate with consciousness, then perhaps we could look for these in animals. Unfortunately, there are perils to this approach as correlation is not causation. Moreover, while there is some agreement between experts regarding the characteristics that we might expect to see in conscious animals, there is certainly not complete consistency. In a critical discussion of the evidence that has been used to assign sentience to vertebrates, Sherwin[21] listed the following criteria that have been used, on the basis that they are thought to be involved in consciousness in humans in some way:

- memory;
- learning;
- spatial awareness;
- deception;
- effects of enrichment;
- learnt preferences/aversiveness;
- operant responsiveness;
- nociception and pain;
- learned helplessness.

Broom suggested the following criteria:

- complexity of life and behaviour;
- learning ability;

---

[18] Baker (1948).
[19] Sømme (2005).
[20] Edelman and Seth (2009).
[21] Sherwin (2001).

- functioning of the brain and nervous system;
- indications of pain or distress;
- studies illustrating the biological basis of suffering and other feelings such as fear and anxiety;
- indications of awareness based on observations and experimental work.

In contrast, Seth and co-authors list three brain properties required for consciousness in humans and a further 14 features of human consciousness associated with these brain properties that, in conjunction with neuroanatomical evidence (see later section on homology), might be used to test the case for consciousness in mammals and perhaps other species. These brain properties are the possession of an appropriate EEG signature, corticothalamic activity and widespread brain activity, while the features of consciousness include accurate reportability, stability of contents, and limited capacity and seriality[22].

Other authors have tried to identify criteria that can be used to determine whether animals are feeling pain rather than simply showing reflex responses to nociception[23].

- Possession of receptors sensitive to noxious stimuli.
- Possession of brain structures analogous to the human cerebral cortex.
- Nervous pathways link the receptors to the higher brain.
- Painkillers modify the response to noxious stimuli.
- The animal consistently avoids noxious stimuli.
- The animal can learn to associate neutral events with noxious stimuli.
- The animal can learn to choose a painkiller when given access to one when pain is otherwise unavoidable.

The differences in these lists suggest that there is no clear consensus about which criteria should be used to ascribe sentience. Moreover, as the authors of these lists would probably agree, none of the criteria, either separately or together, clinch the matter and all have difficulties. Many of the behavioural characteristics, such as learning, aversive responses and even perhaps responses to enrichment, could be duplicated in a robot by a skilled programmer. Indeed, even a plant (*Mimosa pudica*), which folds its leaves together when touched but that clearly does not have any neural machinery, can learn to 'expect' a vigorous shaking when it is associated with darkness[24]. In fact, the aim of Sherwin's paper was to show the inadequacy of commonly used criteria. Sherwin points out that there are invertebrates that

---

[22]   Seth *et al.* (2005); Broom (2007).
[23]   Slightly modified from Gregory (2004). See also Smith and Boyd (1991), p. 63 and Short (1998).
[24]   Armus (1970) cited in Gregory (2004), p. 4. These plants naturally fold their leaves rapidly if shaken but also fold their leaves more slowly as a response to darkness. If they are trained by being shaken when the lights go out, they subsequently fold their leaves more rapidly when exposed to darkness.

possess one or more of the criteria in the list that he provides. It follows, he suggests, that we either need to reconsider whether some or all invertebrates should be included with vertebrates as sentient, or that the argument by analogy from human experiences is a fundamentally flawed process. Nonetheless, our own experience of consciousness is one of the things that we can be most sure about, and so neural complexity, behavioural and cognitive criteria and homologies of brain anatomy, with humans as a benchmark, continue to be used in discussions about animal consciousness[25].

4.2.1.1 Neural complexity

As sentience is produced by neural structures, it seems logical to suggest that sentience might depend to some extent on degree of neural complexity. After all, one or two neurons are very unlikely to be sufficient for consciousness but we humans with billions of neurons are conscious, at least when we are awake. The animal kingdom includes animals with brains that range from the highly complex to organisms such as jellyfish and sea anemones with relatively simple neural nets throughout the body of the animal[26]. Given this variation in neural complexity, if we could show that certain animals simply do not possess the necessary neural architecture to permit sentience, we could exclude them from the list of animals about whose welfare we should be concerned. Unfortunately, we have no idea what degree of complexity is required to permit sentience[27]. Indeed, it has been argued that as very small structures may be involved in various aspects of sentience, the issue is much more likely to depend on organisation of neural structures than complexity per se of the neural system[28].

Nonetheless, it seems obvious that there must be a limit in terms of numbers of neurons below which sentience is impossible, but it is not at all clear what that limit might be. The human brain is undeniably impressive, containing 100 billion neurons (allowing 100 000 billion synapses)[29], but how many is a lot? The octopus brain contains around 168 million neurons[30]. Even the brain of the honeybee contains over 960 000 neurons[31], of which 340 000 are in the mushroom bodies, structures analogous to the vertebrate cerebral hemispheres. Surprisingly, there is at least one vertebrate with a smaller brain than that of the bee. *Thorius narisovalis*

---

[25] Edelman and Seth (2009).
[26] However, there is variation in the complexity of neural structures even of organisms with only neural nets, and simplicity does not necessarily mean a lack of capability. Some species possess a nerve ring or other ganglionic concentrations of nerves. The box jellyfish, for example, possesses eyes similar to those of vertebrates and cephalopods. Its nervous system is capable of processing the information from these eyes and of generating appropriate swimming behaviour in a complex environment (Coates, 2003; Coates *et al.*, 2006).
[27] Kirkwood (2006).
[28] Weiskrantz (1997), pp. 81–82 See also Bermond (2001).
[29] Churchland (1996) cited in Kirkwood (2006).
[30] Wells (1962) cited in Anderson (2006).
[31] Griffin and Speck (2004).

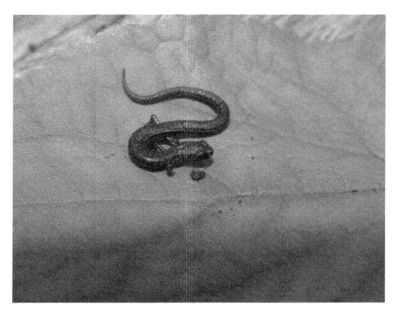

Figure 4.1   *Thorius narisovalis*, a small Mexican salamander. Its brain contains approximately 250 000 neurons compared with about 960 000 neurons in a honeybee's brain. Photograph by Tim Burkhardt, reproduced with permission.

(Figure 4.1) is a very small salamander from Mexico, with a brain the size of a small pinhead, containing only about 250 000 neurons, and there are other verte- brates even smaller than *Thorius*[32]. It is interesting to reflect on the fact that the animal with the smaller brain is protected by European animal welfare legislation, while the insect, with the larger brain, is not.

Early zoologists thought that the size of the brain was likely to be an important predictor of an animal's mental ability, and that consciousness must be a product of this ability, but absolute brain size can be misleading; animals vary in size, and larger bodies generally need more neural tissue just to complete the wiring of sensory organs and muscles, so brain size has to be adjusted to be proportionate to the size of the animal[33]. However, the relationship is not a simple linear one. Measurements of brain and body weight have shown that brain size scales with body weight following a power function of 0.6–0.8. This means that whilst small animals tend to have smaller brains than larger animals, their brains are relatively larger (a human's brain is approximately 2% of body mass compared with a shrew's 10%)[34]. So taking body size into account, the question for any particular animal

[32]   Jerison (2001); Kottelat *et al.* (2006).
[33]   However, Byrne (1996) discusses the relationship between brain size and intelligence in primates, and suggests that, based on Alan Turing's computing principles, absolute size rather than relative size is what matters in terms of cognitive abilities. His view seems to be supported by Deaner *et al.* (2007).
[34]   Roth and Dicke (2005).

might be whether it has a larger brain than would be expected based on its body weight, which might allow it to be sentient.

The encephalisation quotient (EQ) is a widely accepted comparative brain measure that has been used to compare the braininess of species or groups of species, and is the ratio of the observed to expected brain weight calculated from a log/log plot of brain weight versus body weight. There are several ways in which an expected brain weight can be arrived at: a line may be drawn from a specified standard animal on the plot, using a slope of ⅔ (based on the way in which surface area changes with volume for an idealised body); alternatively, a line of best fit can be calculated based on the actual data. While the index results in some outcomes that seem to be reasonable, for example humans attain the highest score, there are also apparent anomalies even amongst relatively closely related species. Capuchins are undoubtedly very clever primates, using tools in the wild and in captivity showing considerable abilities in cognitive tests, but one might not expect them to have a higher EQ score than a chimpanzee, which turns out to be the case[35]. EQs also suffer from the fact that lines with different slopes can be drawn depending on the choice of an appropriate reference animal or group of animals (it makes sense in the analysis to divide animal species by the taxonomic class that they belong to). Moreover, if the assumption is that animals with an EQ of 1 have the correct number of neurons for their body size, based on the numbers needed to provide appropriate connections and processing between sensory organs and muscles, then what does it mean when animals have less than the expected index?[36] Another problem with the concept is that some brain areas may be particularly developed in certain species or groups of species in order to cope with particular sensory or motor systems[37] (I discuss later with respect to non-human primates the notion that certain areas of the brain rather than overall brain size may be the important factor with respect to certain cognitive abilities). Finally, the techniques used to estimate brain size and the correlations that some authors have obtained between either the brain or part of the brain and various types of behavioural complexity or intelligence have been criticised on methodological grounds.[38] The bottom line with respect to EQs and sentience is that it is not clear if or how they might be related, especially when making comparisons between animals as distantly related as insects and vertebrates.

4.2.1.2 Behaviour

If absolute or relative brain size cannot be reliably used to indicate cognitive abilities or sentience, then what about the visible output of the brain, namely behaviour? Behavioural complexity, learning ability and other evidence of intelligence are

---

[35] Roth and Dicke (2005).

[36] Byrne (1996).

[37] For example, the lemur and tarsier have a relatively much larger olfactory bulb than that of humans or chimpanzees (Fleagle, 1988, p. 19), while both dolphins and bats have a well-developed auditory cortex.

[38] Healy and Rowe (2007).

often suggested as factors that would support the attribution of sentience. Some cephalopod species have been described as having behaviour that is no less complex than that shown by some fish or other vertebrates. Octopuses are behaviourally adaptable, can learn how to negotiate mazes, can be trained to distinguish between various stimuli, have personalities (i.e. show individual behavioural differences), and play. Some have learnt how to unscrew jars or childproof containers to obtain the contents, and one even used a jet of water to short out the electrical apparatus of a researcher[39]. However, while it seems likely that sentience must require some degree of intelligence as otherwise there would be no evolutionary function for the capacity, intelligent behaviour does not necessarily demonstrate sentience. To put it alliteratively, cognition is not the same as consciousness[40]. It is true that cognitive capabilities could affect the duration, causes and types of suffering experienced (see later section on intelligence and suffering), but there is no reason to think that intelligence or cognition necessarily correlates with either sentience or the degree of suffering that a sentient animal might experience. So to find criteria for sentience, we need to look at other aspects of animals' behaviour.

Many animals behave in a way that is consistent with the hypothesis that they find certain experiences pleasurable or unpleasant. Unfortunately, as Sherwin pointed out, arguing by analogy with human experiences and behaviour in this way is unreliable. Although our everyday experience shows that many animals respond in a way that is very similar to the way we do to pleasant and unpleasant events or stimuli, it is possible that the animals might be operating as intelligent machines or 'zombies' but without any experiences or feelings. Of course, most people and many scientists do not believe this is true, at least of vertebrates, but the suggestion is not as bizarre as it might sound. After all, even we humans often behave in automatic ways. Simple reflexes are one example, as when we involuntarily withdraw our hand from a hot object. More complex behaviours can also be carried out without full awareness, and some apparently unconscious behaviour can be remarkably complex. Individuals troubled by parasomnias may sleepwalk, or more rarely prepare and eat food, or even perform behaviours as extreme as assault or sexual behaviour while still being completely unconscious of their actions. Similarly, patients with epileptic seizures may perform a series of behaviours some of which may be coordinated but can be unconscious while doing so[41]. It is even possible to be awake and to be unaware of your own behaviour. For example, it is quite a common experience to walk or drive while thinking of some other pressing matter, and then realise that one has travelled a considerable distance apparently on autopilot. It is true that this experience tends to be confined to routine driving, and if something unexpected or dangerous occurs our consciousness snaps back to the urgent decisions that might be necessary, but even driving a car along a known road requires more than simple reflexes. So complex behaviour does not necessarily

[39]   Mather (1992), p. 245; Hanlon and Messenger (1996); Anderson (2006).
[40]   Dawkins (2001).
[41]   Koch (2004), chapter 13; Hughes (2007).

require conscious awareness, and animals could conceivably be operating in this fashion. In fact, it is almost certain that some animals do function in a machine-like way. Sentience must have evolved at some time in our evolutionary history, but nobody knows when this happened or whether it has evolved more than once, nor of the animals alive today which are surely sentient and which are not[42]. Behavioural scientists are devising ever more clever ways of trying to test whether animals are sentient. It has been demonstrated, for example, that fish lose their fear of a novel object when in pain, but this fear is restored when the fish are given an analgesic. The implication is that the pain dominated the attention of the fish and that therefore the fish were aware of the pain[43]. Nonetheless, consciousness is a private experience for which there is no conclusive external behavioural indicator[44] and, despite some recent advances, no one has yet discovered how or where our brain produces this phenomenon of sentience. We do not even know what processes are uniquely involved in the generation of consciousness. Sadly, it is therefore not yet possible to check for sentience in other animals by looking for known neurological correlates of consciousness[45].

4.2.1.3 Homology of brain structure

One pragmatic reason for maintaining the common practice of drawing the line between the sentient and the insentient at the vertebrate/invertebrate boundary is that we humans share with other vertebrates a much closer common evolutionary heritage. The nearest common ancestor between a human and a mouse existed about 75 million years ago while for a cephalopod it is well over 500 million years ago[46]. This means that while there is 150 million years of evolutionary difference between a human and a mouse, there is over 1000 million years between an octopus and a human. The consequence is that although the mouse and the human brain differ in the degree of development of different areas, the basic layout in vertebrates is the same (Figure 4.2). In humans, aspects of consciousness have been thought to be associated with various portions of the cortex and its links with the thalamus. The evidence for this is that damage to the brainstem or thalamus can abolish consciousness, while damage to the sensory cortex results in more specific impairments such as conscious experiences of faces or colour vision[47]. Areas that have been specifically linked with emotions, which, if experienced as feelings that matter, would be important for animal welfare, include the orbitofrontal, insular,

---

[42] Dawkins (1993); Kirkwood (2006).
[43] Lynne *et al.* (2003); Braithwaite (2010), pp. 67–70.
[44] See also section on studies of affect, and discussion on cognitive bias in Chapter 3.
[45] Crick (1994); Griffin (1998); Griffin and Speck (2004); Koch (2004). Block (2005) argues that there may be at least two neurological correlates of consciousness, one for access consciousness and one for phenomenological consciousness.
[46] Dawkins (2004).
[47] Seth and Baars (2005). Additional supporting evidence is provided in Ward (2011). However, functional magnetic resonance imaging of human brains has failed to pin down consciousness to a particular activity or area of the brain (Dawkins, 2012, pp. 76–79).

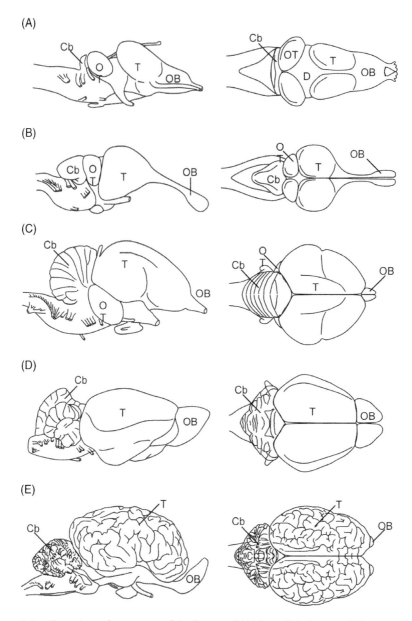

Figure 4.2 The side and top views of the brains of (A) frog, (B) alligator, (C) goose, (D) tree shrew, and (E) horse. Cb, cerebellum; D, diencephalon; OB, olfactory bulb; OT, optic tectum; T, telencephalon. Reproduced from Shimizu, T. (2001) Evolution of the forebrain in tetrapods. *In Brain Evolution and Cognition*, Roth, G. & Wullimann, M. F. (eds.), pp. 135–184. John Wiley & Sons, New York/Spektrum, Heidelberg. With permission from John Wiley & Sons and Springer Science+Business Media. Originally adapted from Romer, A. S. (1970) *The Vertebrate Body*. Saunders, Philadelphia. With permission from Elsevier.

and anterior and posterior cingulate cortices, and these structures are homologous in mammals[48]. In humans, it has been suggested that the dorsolateral prefrontal cortex (DLPFC) may be important for self-consciousness on the grounds that self-consciousness and analytical thought are severely impaired during dreaming and this is associated with deactivation of the DLPFC[49].

The similarities that exist between human brains and other vertebrates, together with behavioural and other evidence, make the hypothesis that other vertebrates are also capable of conscious experience plausible. However, given our lack of certainty it is not surprising that there is ongoing debate. Rose has argued that fish are unable to feel pain as they lack a neocortex, which in humans is essential for sentience. Others, however, have suggested that even if the neocortex is essential for sentience in mammals, it could have taken on this function from other older structures that might serve the same function in fish[50]. Indeed, in 2012, an eminent group of scientists published a proclamation stating that:

> The absence of a neocortex does not appear to preclude an organism from experiencing affective states. Convergent evidence indicates that non-human animals have the neuro-anatomical, neurochemical, and neurophysiological substrates of conscious states along with the capacity to exhibit intentional behaviors. Consequently, the weight of evidence indicates that humans are not unique in possessing the neurological substrates that generate consciousness. Nonhuman animals, including all mammals and birds, and many other creatures, including octopuses, also possess these neurological substrates.[51]

Birds, for example, also do not possess a neocortex but do have structures that can be thought of as having a similar function to the mammalian neocortex and which are derived, like the neocortex, from the pallial structures of the embryonic fore-brain[52]. Developing the idea that structures derived from the pallium are important for sentience, Butler[53] has suggested that consciousness requires an elaborated pallium and dorsal thalamus (or some other comparable structure) resulting in a relatively high brain to body-weight ratio and that it may be associated with a 40-Hz rhythm of neuronal firing in the thalamic neurons. It is interesting to note that such structures and activity have been found not just in mammals but also in other vertebrates: birds, turtle, fish and frogs. If neural structures similar to those required for consciousness in humans are found in other animals, then it may be possible to test for properties or manifestations of consciousness, although identifying these is not likely to be an easy task, and recent research suggests that the neural pathways

[48] Dolan (2002); Panksepp and Northoff (2009).
[49] Muzur *et al.* (2002).
[50] Rose (2002); Braithwaite and Huntingford (2004); Chandroo *et al.* (2004a,b); Rose (2007); Braithwaite (2010).
[51] Francis Crick Memorial Conference, Cambridge, UK, 7 July 2012, http://fcmconference.org/, accessed 5 April 2013. See also Baars (1997), p. 33.
[52] Emery and Clayton (2005).
[53] Butler (2008).

necessary for self-awareness could be more widely distributed in the brain than has recently been thought[54].

Invertebrates have very different neural organisation to vertebrates, which increases the challenge of determining whether they are sentient or not. Decapod crustacea, for example, do not possess a single brain but have a series of clusters of nerve cells called ganglia that are connected by a ventral nerve[55]. The octopus brain is also completely different in layout from those of vertebrates, having no directly equivalent spinothalamic tracts, thalamus or cerebral cortex, all structures thought to be required for consciousness in humans[56]. Instead of the vertebrate hindbrain, midbrain and forebrain, the brain of the common octopus *Octopus vulgaris* (Figure 4.3) consists of a central structure round the oesophagus of the animal, with two lateral optic lobes. The optic lobes are not just involved in vision, but are thought to be the highest of the octopus's brain centres, being involved in memory and learning. As such they may be comparable in their function to the vertebrate forebrain. Other lobes of the octopus brain, such as the vertical and superior frontal lobes, also appear to be involved in the laying down of memories and decision-making[57]. These differences in brain structure between vertebrates and cephalopods mean that if some cephalopods are sentient, then this capacity must arise from different structures than those thought to be involved in consciousness in the humans and other vertebrates. It follows that one would have to postulate that sentience arose at least twice in the course of evolution, once for the vertebrates and once for the cephalopods[58] (it is possible that, like eyes, sentience evolved many times).

Despite the large differences between vertebrate and invertebrate brains, there are also some remarkable similarities between vertebrate craniate brains and the brains of insects. These could be due to convergent evolution or because of genetic similarities. For example, genes necessary for the development of the insect protocerebrum are homologous with genes necessary for the development of the mammalian forebrain. Structural similarities have also been found between the adult nervous systems of vertebrates and arthropods. For example, there are similarities between the craniate spinal cord and arthropod segmental ganglia, and there is a common organisation of neurons in the vertebrate retina and insect optical lobes. There are many other examples, and the intriguing suggestion has been made that perhaps brain areas performing similar functions in mammals and insects require both similar wiring and genetic programmes[59]. Whether these similarities and homologies increase the likelihood of sentience in arthropods is hard to say, but there is behavioural evidence that decapod crustacea respond to anaesthetics in a way that suggests that they experience pain, that they are capable of learning about

[54]   Seth *et al.* (2005); Philippi *et al.* (2012).
[55]   Roth and Øines (2010).
[56]   Baker (1948); Koch (2004); Budellmann (2010).
[57]   Hanlon and Messenger (1996).
[58]   Baker (1948).
[59]   Strausfeld (2001), pp. 392–394; Elwood *et al.* (2009).

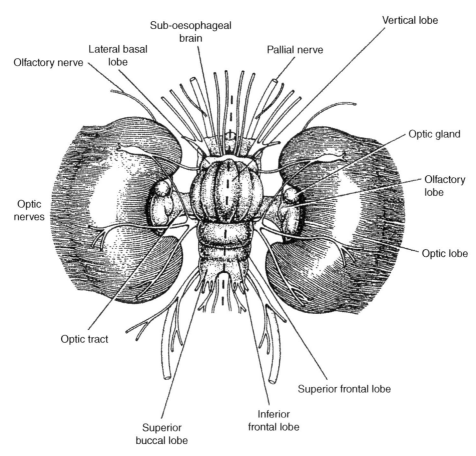

Figure 4.3 Dorsal view of the brain of *Octopus vulgaris*. Note the large optic lobes (visual processing areas and visual memory store), the inferior frontal lobe (together with the subfrontal and superior buccal lobes, forming the tactile memory system), and the vertical and superior frontal lobes (accessory to the visual memory store). From Hanlon, R.T. & Messenger, J.B. (1996) *Cephalopod Behaviour*. Cambridge University Press, Cambridge. Redrawn from J.Z. Young (1964) *A Model of the Brain*. Reproduced with permission from Oxford University Press and Cambridge University Press.

painful experiences and that analgesic opioids affect this learning as they do in vertebrates. It has also been shown that previous painful experiences associated with a particular shell can change the way in which a hermit crab values the shell when it later compares it with others. In other words, the crabs seem to be weighing up various positive and negative aspects of a shell over a period, which is consistent with the idea that they are capable of remembering a painful event rather than simply responding to nociception[60]. However, decapod crustacea are only one

[60]   Barr *et al.* (2008); Elwood and Appel (2009). See also Gherardi (2009) for a review of behavioural indicators of pain.

group among many different invertebrates and for others, such as insects and spiders, there is as yet no clear evidence that they respond to pain as if they are aware of it[61]. The jury is still out, but these are fascinating and tantalising pieces of information, and we need to keep an open mind towards these animals that are so different to us whilst being careful not to over-interpret evidence that might suggest sentience.

The Royal Society's concerns about broadening the scope of legislation protecting animals used in research were, rightly, that the scientific basis for any changes to legislation should be absolutely clear. However, as we have seen, there is at present no agreement on criteria that can be used to prove whether an animal is sentient or not. The problem is that there are major, and possibly insuperable, obstacles to finding out whether any particular species is conscious[62]. We may accept on the balance of probabilities that some closely related species may well be sentient and decide to protect them on a precautionary basis but we have to draw the line somewhere, and as we look at animals that are more distantly related to us the decision becomes harder as homologies become fewer. Invertebrates such as the cephalopods and the decapod crustacea are so distantly related to us that knowing what evidence to use is very difficult indeed. The behavioural studies described above on hermit crabs are an attempt to tackle the problem, but unfortunately such findings can always be explained in ways that do not require the attribution of consciousness to the animals. As I said at the start of the chapter this matters in terms of welfare if we mistakenly conclude that a species is not sentient when in fact it is. On the other hand, the inclusion of new species as protected animals should not be taken lightly, as this might limit research that is beneficial for other animals and humans (and result in substantial financial, bureaucratic and other costs).

4.3 Are Some Species Capable of Suffering More Than Others? Neurological Complexity and Capacity to Suffer

The concept of preferentially using species that are thought to have a reduced capacity to suffer has a long history in animal research. In 1947, a *Lancet* article argued that experiments should be performed 'using the lowest order of animals'[63]. Similarly, Russell and Burch in 1959 described an option in which an octopus could be used instead of a rat in a study of vision[64], but one should not overestimate the opportunities for exercising species choice in practice. In most instances there is

---

[61]   See for example Sherwin (2001); Broom (2007). *ILAR Journal*, 52(2), 2011 is devoted to issues of invertebrate sentience and welfare.

[62]   Dawkins (1993), p. 3. Dawkins (2012) now takes the position that the problems involved in determining whether animals are conscious are so intractable that other evidence should be used for making the case for animal welfare.

[63]   Hall (1847) cited in Richmond (2010).

[64]   Russell and Burch (1959), p. 70.

rather little opportunity for investigators to choose between species, as choice of species is usually driven by the science; that is, the best model available for the study is chosen. In toxicology tests, for example, final species selection depends on the results of earlier pilot studies using a range of species as well as on other information that may be available to the investigator[65]. Moreover, the range of species that might be used by the experimenter is limited as there are constraints of availability and custom. Laboratory-bred mice, for example, are easy to obtain and keep and there is a lot of information available on them. So an investigator carrying out research in an area where mice are usually used might well be reluctant to even consider using another species because of the hurdles to overcome in obtaining and maintaining the species, and in discovering whether it would indeed be a suitable model. Nonetheless, there are instances where there is choice and in some countries there is guidance on how such choices should be made. The difficulty is that both the guidance and the background for the guidance have often been unclear.

The 1986 European Directive contained the requirement that 'In a choice between experiments, those which [...] involve animals with the lowest degree of neurophysiological sensitivity [...] shall be selected'. Because European Directives, once ratified, are enacted into national legislation, a similar provision was made within the UK 1986 Act regulating animal use in research (this provision has now been changed)[66]. Unfortunately, 'neurophysiological sensitivity' is an expression that does not have any clear mutually understood biological meaning. It is not clear how neurophysiological sensitivity could be measured or how it might be distributed through the animal kingdom. The suggestion that species should be judged by placing them on a scale of sensitivity is reminiscent of the *Scala Naturae*, as proposed by Aristotle and subsequently Charles Bonnet. In this classification system, organisms were seen as progressing in a series of steps from inanimate matter through plants, invertebrates, fish, reptiles, birds, mammals, and finally to humans at the top of the stairs. Since the taxonomic work of Cuvier, and Darwin's theory of evolution, we understand that there is no natural progression towards an end, simply selection of variants that are capable of exploiting new or changing niches. While some organisms are certainly simpler than others, there is no reason to think of them as lower; they are just as well adapted to their way of living as more complex animals. The hierarchical way of looking at nature exemplified by the *Scala Naturae* is therefore now thoroughly discredited amongst scientists; nonetheless, many people, if asked to rank the acceptability of using different species groups in biomedical research, might produce a result that is not that different to the *Scala Naturae*.

Public polls certainly show species biases: 65% of people in the UK polled in 1999 were prepared that mice should suffer pain, illness or surgery in experiments

---

[65] Morton (1998).

[66] EEC Directive 86/609 Article 7, http://preview.tinyurl.com/cxcxul4, and Section 5(5b) Animals (Scientific Procedures) Act 1986, http://preview.tinyurl.com/cyqavn8. See also the NC3Rs guidance on species choice, http://www.nc3rs.org.uk/category.asp?catID=78, all accessed 5 April 2013.

to develop a drug to cure childhood leukaemia, but only 52% would let monkeys be used in the same experiments. When asked about a perceived lesser benefit – the development of a new painkilling drug – 47% were prepared to allow mice to suffer pain, but for monkeys the figure fell to 35%. The figures obtained vary by poll and date, but continue to show that people have greater ethical concerns about the use of some species, for example primates[67]. Another survey, not specifically aimed at the use of animals in research, found that people tended to rate chimpanzees, monkeys, dogs, horses and dolphins highly in terms of usefulness, importance, smartness, loveableness and responsiveness, with cats, pandas, sea otters, deer, rabbits, pigs, elephants and sheep scoring slightly lower. Rats, on the other hand, tend to get a very low score in terms of loveableness, although still somewhat higher than those assigned to mosquitoes or earthworms[68]. The reasons for these choices are probably complex, but it is well known that humans tend to favour larger animals, those thought to be more intelligent, and those that are anthropomorphous (i.e. those that to some extent look like humans)[69]. Unfortunately, these attitudes, while not necessarily based on scientific evidence, can bias the choices made as to which species to use in research and how they should be treated.

The 2010 EU Directive, which now affects all member states' legislation, rightly avoids using the term 'neurophysiological sensitivity', instead requiring, along with some other considerations, that where there is a choice between procedures, those that involve 'animals with the lowest capacity to experience pain, suffering, distress or lasting harm…' should be selected. Similarly, the Canadian Council on Animal Care (CCAC) used to recommend 'using species that current scientific evidence indicates have a significantly lower potential for pain perception…'[70]. These approaches move beyond prejudice but require the collection of hard evidence to rank one species over another. This is obviously a difficult task and one that is doomed if sentience proves to be an all-or-none phenomenon such that an organism can either experience feelings or not. On the other hand, it is at least possible, and some feel more probable, that sentience evolved progressively, and may have developed to different extents in different species[71]. If this is the case, then an

[67] *New Scientist* Mori poll, 1999, Attitudes Towards Experimentation on Live Animals, http://tinyurl. com/blplnjn, accessed 26 April 2013. A 2002 poll carried out for the Coalition for Medical Progress showed that while 39% of respondents considered that the use of monkeys in research was acceptable provided that all regulations were enforced, 66% felt that such research was acceptable for rats and 63% for mice.

[68] Driscoll (1995).

[69] Serpell and Paul (1994).

[70] CCAC website as accessed in 2010. Their guidance has now been updated and emphasises the need to carry out a literature search to determine or confirm the best species and model for the scientific objective, and to find the most recent 3Rs information relating to the species and planned procedures, http://preview.tinyurl.com/bphv32m, accessed 5 April 2013.

[71] Kirkwood (2006), for example, writes 'It seems highly likely that there is variation among sentient animals in the range of sensory, emotional or cognitive states that fall within the spotlight of their sentience'. See also Kendrick (1998) who proposes that awareness in sheep may differ from that in primates by being more influenced by current priorities, which suggests more limited powers to reflect on the past or to plan future behaviour.

animal's degree of sentience may, to some extent, depend on the complexity of its neural system, which, if true, suggests that we might be less concerned about harm done to animals with simpler nervous systems.

The difficulty with this idea is that, even for vertebrates, our uncertainties about animals' experiences are so great that there is little scientific data to guide us in making choices. We have already discussed the impossibility of knowing for sure what other animals experience but noted that homologies of structure tend to bias us, without a very firm footing, towards assigning sentience to animals that share a similar neural architecture with ourselves. If a similar homological approach is used to assess the *degree* of suffering likely to be experienced, then that might lead us, where there is a choice, to favour a primate over a mouse, a mouse over a fish, and a fish over an invertebrate. However, is this rational? If we take pain as an example, there is no a-priori reason for assuming that the pain of an injection is likely to be any less in a mouse than in a primate. Nociception and pain are the organism's way of activating processes that either remove the animal from immediate danger or which result in memories that will help the organism avoid danger in the future. Emotions necessary to drive such behaviour are likely to be an earlier rather than a later adaptation and thus should be common (if not identical) in sentient species.

Some might counter that evolution could result in some species becoming more or less responsive to pain than others. For example, it is sometimes suggested that certain prey species may be selected to hide pain, as injured individuals tend to be targeted by predators because they are an easier meal, but if pain is maladaptive then, rather than experience pain and hide it, it would perhaps be simpler, and more likely, that the animal would feel pain less intensely (as can happen with humans at times of stress). However, this is only an hypothesis and it would be dangerous, without much better evidence, to assume that prey animals feel less pain than non-prey animals. The important point to keep in mind is that while we should always be careful about extrapolating our own experiences, even of pain, to those of other animals[72], it seems very unlikely that any organism, capable of feelings, would normally experience intense painful stimuli as anything other than unpleasant.

Although there is little evidence for differences in vertebrate species' capacity to suffer, this does not mean arguing that choice of species is unimportant in terms of animal welfare. The impact of experimental procedures certainly can vary between different species. Species' needs vary according to the niche to which the species has evolved so that there may well be differences in degree or even order of some motivational priorities. Prey species may be particularly motivated to have access to a safe refuge, so could suffer more than some other species if denied such a resource. The social structure of species varies widely so that the effect of different social housing regimens is also likely to vary between species. Even pain might vary for

---

[72] Bateson (1991). Note also that experience of pain, even in humans, is very dependent on the individual and context.

reasons unrelated to comparative sentience. If we return to our example of an injection, the mouse might feel more pain than the primate, as the size of an injection needle and consequent tissue damage is, relatively, considerably larger for the mouse.

4.4 Beastly Bias in Practice

Article 8 of the 2010 EU Directive provides special protection to limit the use of non-human primates in general and to prevent, except in exceptional cases, the use of great apes. UK legislation also includes provisions to prevent the use of cats, dogs and Equidae (horses, donkeys, etc.) unless no other species are suitable for the purpose or because other species that could be used cannot practicably be obtained, and section 5C(5) now bans the use of great apes. In the USA, special provision has been made to retire rather than euthanase chimps no longer needed for biomedical research and the great cost of this undoubtedly discourages research on these animals[73].

Reasons for these special protections lie partly in the perceived public concern for these species and, in the case of non-human primates, because of other ethical and welfare issues discussed later in this chapter. Companion cats and dogs are often loved by their owners, and thought of as members of the owner's family – almost surrogate humans. Horses and donkeys, while traditionally farm and transport animals, are now also often kept as companion animals. The use of these animals in research is clearly a sensitive one, and it is right that this should be recognised in legislation. However, politics and science may well come to different conclusions. We have already discussed the lack of data that would suggest that any particular species would be more likely to feel pain or to feel it in a worse way than another species, but are cats, dogs, Equidae and non-human primates likely to suffer in a different or greater way than other animals when used in research?

4.4.1 Are domesticated animals likely to suffer more?

Cats and dogs and equids are very different animals and therefore likely to react in somewhat different ways to their housing and use in research, but they are all domesticated species and, albeit to different extents, have been selected for an ability to live closely with humans. Dogs were amongst the first domesticated animals; domestication occurred at least 14 000 years ago, and possibly much before[74]. Physically, the effects of deliberate and accidental breeding of dogs for different functions, sizes and forms is so marked that variation within the domestic dog is now greater than that produced by natural selection between canid species. In contrast, the influences of domestication on cats are comparatively minor[75]. Nonetheless,

---

[73]   Chimpanzee Health Improvement, Maintenance, and Protection Act, PL 106–551 (HR 3514), 20 December 2000, http://www.animallaw.info/statutes/stusfdpl106_551.htm, accessed 5 April 2013.
[74]   Clutton-Brock (1995); Vila *et al.* (1997); Miklósi (2007).
[75]   Liberg and Sandell (1988).

cat breeders are catching up with dog breeders in their propensity to produce new and varied physical forms with the result that some breeds look very different from their ancestral form. The shape and form of different breeds is driven by a combination of fashion and, in the case of working animals, the purpose for which the breed is required. However, just as important as shape are changes in behaviour and motivation.

An animal's behaviour or temperament can be critical for its welfare in captivity or when interacting with humans. It is therefore relevant that domestication results in marked behavioural differences as a result of deliberate or accidental selection for various traits. It used to be thought that domestication results in animals that are less cognitively capable than their wild ancestors, but it is now becoming clear that this is not the case and that their abilities are merely different[76]. Domesticated animals are sometimes neotenised (i.e. they have been selected to retain certain infantile physical and behavioural characteristics into adulthood; this is seen in dogs for example). They also tend to be less reactive and more adapted to humans and the human environment. Indeed, some have argued that as a result of domestication dogs may have improved in their ability to communicate with humans[77]. So how might the changes wrought by domestication impact on the welfare of animals?

One possible hypothesis is that animals such as dogs and cats that have experienced selection for home life with humans might suffer more when kept in an institutionalised surrounding. After all, such environments are not likely to provide the close animal–human bonds, or to be as diverse as one they might enjoy as companion animals. However, no evidence has been published that would support a case that domesticated animals do worse in research establishments than non-domesticated animals. In fact, it is more likely that they would generally do better than non-domesticated animals because of their reduced fear of humans. Indeed, because of the difficulties involved in using wild animals in research, which apart from the risks of disease can include the animal's fear, increased stress responses and problems in handling, most of the commonly used animals have been domesticated to one degree or another. Farm animals, such as sheep, goats and cattle, fall into this category, but even the mice, rat, gerbil, hamster and rabbit strains used in the majority of research are the product of many years of selective breeding and are, to a degree, domesticated animals. In some of these cases the founding stock were already domesticated, originating from animals bred by fanciers, but in all cases there has been continuous selection as generations of animals have been bred for research use. Some of the selection is accidental, as animals that survive better or are easier to handle are more likely to be used as breeders, but much is deliberate.

---

[76] Rehkämper *et al.* (2001), pp. 283–285 describes how domestication reduces the size of certain brain areas (striate area in mammals, and the telencephalic sensory areas). Nicol (1996) and Grandin (1998) describe the effects of domestication on behaviour cognitive ability and stress responses with particular reference to farm animals. See also Cooper *et al.* (2003) for the domestic dog.

[77] For example, Hare *et al.* (2002), but see also Udell *et al.* (2010) who point to the importance of life experience when assessing cognitive abilities.

Selection is usually aimed at producing particular strains to meet research needs, but animals can also be bred with the aim of producing behavioural traits that seem to make them more fitted for life in the laboratory. Selection is not always entirely successful and some strains have their problems in captivity. For example, certain strains of mice are noted for their aggressiveness, which makes social housing problematic. Nonetheless, despite these occasional difficulties, there is no doubt that usually it is considerably easier to use purpose-bred mouse strains than wild mice[78].

4.4.1.1 Ensuring the welfare of domesticated species

While animals may be bred for captivity, it is not the case, as is sometimes asserted, that domestication so changes the animal's natural behavioural repertoire that their needs are entirely different from those of non-domesticated relatives. Price points out that behavioural differences between wild and domesticated animal are usually quantitative rather than qualitative and best explained by differences in response thresholds[79]. Laboratory rats, after many generations, still show a wide range of the behaviours characteristic of wild rats, even though while in laboratory housing they would not be able to display all of them[80]. Anti-predator behaviours are particularly likely to be hard-wired, as shown by gerbil digging behaviour and the innate aversion of mice and rats to cat odours[81]. Knowledge of wild behaviour can therefore be very relevant when determining captive husbandry.

When considering the welfare of domesticated animals in research institutions, it is important to remember that the genetic changes imposed by artificial selection are only one component in determining how animals react to stressors; experience counts as well. It is well known that dogs' early experiences are vitally important in determining how they will react to humans, strange environments and stressful events. If dogs are not exposed to complex social and physical environments between the ages of about 3 and 13 weeks, they are much more likely to become stressed when later meeting unfamiliar people or when placed in unfamiliar settings[82]. Similarly, for cats, environment counts with respect to responsiveness, although there are also genetic influences[83].

Laboratory dog breeders are paying increasing attention to the importance of the early environment, providing puppies with varied experiences from an early age. In addition, at least one breeder has produced beagles that appear more

---

[78]   For a discussion of the issues involved in housing wild animals see Inglis *et al.* (2010).

[79]   Price (1998).

[80]   Berdoy, M. (2002) *The Laboratory Rat: A Natural History.* Film (27 minutes) available at www.ratlife.org.

[81]   For rodent aversion to predator odours see Dielenberg and McGregor (2001) and Apfelbach *et al.* (2005). For gerbil digging behaviour see Wiedenmayer (1997). In Chapter 6, I describe how gerbil anti-predator needs can be met.

[82]   Scott and Fuller (1965) drew attention to the importance, for future socialisation, of a critical period from 3 to 13 weeks during puppy development. The implications of this and other work for the use of dogs in research are described in Hubrecht and Buckwell (2004).

[83]   Mendl and Harcourt (1988) review environmental influences on cat personality, and the importance of paternal genetics on kitten 'friendliness'.

relaxed than some other strains when undergoing various research procedures. These dogs seem to be particularly unresponsive when handled, making it easier to take a blood sample or fit a face mask for inhalation toxicology. The behaviour of these dogs is such as to suggest that they 'mind' less when things are done to them. If that were the case and if the behaviour is genetic rather than resulting from socialisation and handling then, on welfare grounds, it would make sense to use such dogs rather than more reactive strains. However, before leaping to the conclusion that it is kinder to use these purpose-bred animals, we should consider an alternative hypothesis. Certain dog breeds often fail to complete the normal behavioural sequences of the ancestral canid. Pointer breeds, for example, when they identify a prey target, tend to freeze in position, which produces the 'point' behaviour, whereas the ancestral canid would have continued to the stalk, chase and kill sequences. Similarly, herding dogs such as collies need to be focused and aggressive enough to chase sheep, albeit in a controlled way, but must not actually attack them[84]. It is possible that the unresponsive laboratory beagles might actually be afraid and want to struggle, but have a behavioural block. If this hypothesis were true, then their welfare could be worse than for other strains. However, I am inclined to the view that the relatively relaxed behaviour of these dogs is genuine (i.e. they really are calm and unstressed) on the grounds that they have been used in research for some time, and that studies would have picked up indicators of stress before now. Nonetheless, it would be useful to have an independent evaluation of the welfare of these dogs.

4.4.2 Non-human primates
4.4.2.1 New-World versus Old-World primates
Prior to January 2013, when a scientist in the UK filled in an application under the 1986 Act to use non-human primates, a request to use Old-World monkeys rather than New-World ones had to be specially justified, suggesting that the regulators considered that Old-World primates were more likely to suffer when used in research. However, if one considers Old- and New-World species other than the great apes, for which licences are not given, this distinction does not make much sense. The New-World monkeys include the marmosets and tamarins, but also the highly intelligent capuchin monkeys that, like chimpanzees, are capable of tool use. Conversely, Old-World monkeys include bush babies lorises and lemurs, species that do not appear to be anything like as cognitively able as the capuchins.

However, in defence of the UK regulators, the legislation reflected the reality of primate use in the UK rather than any perceived differences in capacity to suffer. The most commonly used primates are rhesus macaques, long-tailed macaques and common marmosets while bush babies and capuchins are not commonly used in research. Hence, the UK requirement on the need for special justification for the use of Old-World rather than New-World primates effectively means that common

---

[84]   Scott and Fuller (1965), pp. 78–80. See also Coppinger and Schneider (1995) for a discussion of the interplay between genetics and environment in the selection and development of various breed traits.

marmosets should be used where possible rather than macaques. When phrased like this, the reason for the requirement becomes clearer, as factors related to marmoset supply and housing make it easier to ensure the welfare of these animals.

Most of the macaques used in the UK for pharmaceutical research tend to come from overseas breeding colonies, often located in the Far East. Marmosets, on the other hand, are usually bred in the UK or Europe. There is therefore a considerable difference in transport times, and transport is a potential stressor[85]. Marmosets, moreover, have been bred in captivity for many years without replenishment from wild stocks. This is certainly not currently the case for macaques, where in some overseas breeding and supplying establishments many of the breeding stock are still wild-caught animals[86]. (The 2010 European Directive requires that those using primates in research move towards using F2 or later captive-bred generations.[87]) The welfare benefits of using only animals that have been bred from animals themselves born in captivity are not entirely clear-cut, but wild-caught animal breeding stock will have undergone the stressors of trapping, and it is assumed that wild-caught animals take time to adapt to captivity and the presence of people.

Marmosets have been captive bred for a number of generations and, because they are smaller animals, are easier to keep in conditions that allow more naturalistic social housing and adequate cage size and complexity. However, the argument that marmosets should be preferred because these characteristics mean that their welfare is likely to be better is not as strong as it used to be. Transport can still be an issue for macaques, but in the UK, macaques are increasingly being kept in more extensive systems than in the past and social housing is the norm.

4.4.2.2 Are non-human primates special cases?

It is, perhaps, not surprising that non-human primates, and great apes in particular, have been given special status and protection. After all, humans are great apes, and some have argued that humans and chimpanzees are so similar that they should be reclassified into the same genus[88].

The Boyd Group, in 2002, laid out the arguments for treating non-human primates as a very special moral case deserving exceptional justification[89]. Their case was based partly on the close evolutionary relationship with humans but rested largely on the special cognitive abilities of non-human primates which, it was argued, might lead them to experience more suffering than non-primate species. Unfortunately, the question of primate cognitive abilities and any consequent predisposition to suffer in captivity is complicated by the fact that the primate order is a large and varied one. Primates include the great apes as well as prosimians such as bushbabies and pottos, and primate abilities and brain size vary with species.

---

[85]   Honess *et al.* (2004); Fernström *et al.* (2008).
[86]   Animal Procedures Committee (2006a); Joint Working Group on Refinement (2009).
[87]   F0, wild-caught animals; F1, offspring of wild-caught animals, and so on.
[88]   Wildman *et al.* (2003).
[89]   Boyd Group (2002). See also Prescott (2010).

Moreover, measuring intelligence across species is not straightforward (it is not even straightforward in humans). While there is general agreement among researchers that there are differences in cognitive adaptability among vertebrates, it is not at all clear what intelligence is or how it should be measured across species. It has therefore proven very difficult to compare intelligence across vertebrate species and produce some sort of hierarchy[90]. The difficulties derive from the fact that how animals respond in tests is partially determined by their sensory/motor abilities and the way in which they have evolved to respond to their environment. This may partly explain why supposedly less able animals often succeed at tasks that were thought to be beyond their capabilities. Nonetheless, primate brains are comparatively large with respect to those of other orders[91], and the neocortex, the area of the brain associated with executive function and complex thought, is particularly well developed. The reason for this development is thought to be that larger brains were necessary for the complex social relationships that are found within some primate groups, this hypothesis being supported by the fact that neocortex size correlates with primate group-size, and also with the use of tactical deception of other group members, suggesting that neocortex size is related to social skills[92].

4.4.2.3 Intelligence and suffering

So primates tend to be clever animals, but there is certainly no evidence to support the notion that primates feel pain more than other non-primate vertebrate species. As previously discussed in this chapter, neither sentience nor capacity for suffering necessarily correlate with intelligence. On the other hand, the cognitive abilities of animals used in research are relevant to considerations of suffering, because increased abilities could permit the animal to experience certain *types* of suffering that would not otherwise be possible. In hominids, the neocortical development associated with large group size may have permitted the coevolution of language to allow more efficient interactions, as well as the ability to attribute mental states (such as beliefs or desires) to others (theory of mind or ToM)[93]. This ability allows us to imagine how another is feeling or to understand another's point of view, and we can use this either to empathise with them or perhaps to deceive them. In ourselves we know that our capacity for reflexive thought, a sense of self and the ability to understand others' mental states, while having a positive side, can also result in unpleasant experiences. We are able, for example, to worry about or anticipate the future, think about good and bad relationships with others and dwell on the duration of good and bad events. Some of these capacities are probably

---

[90] Macphail (2001) views learning, memory and decision-making as central processes that are responsible for intelligent behaviour and reviews evidence that there are quantitative differences but no evidence of qualitative differences between non-human species in these processes.

[91] Hofman (2001) points out that brain size also correlates with development of the senses and that this might partly explain the high encephalisation of primates, many of which live in diverse and unpredictable environments.

[92] Dunbar (1995); Byrne (1996).

[93] Dunbar (2003).

dependent on our possession of language, but it is conceivable that the cognitive abilities of apes or monkeys might also provide them with opportunities for suffering or distress that would not be experienced by less able animals[94].

4.4.2.4 Apes and monkeys: primate cognition and distress

Many feel deep discomfort at the thought of using animals in research that are as closely related to ourselves as the great apes and, reflecting this discomfort, progressively fewer countries permit their use. New Zealand banned research on great apes in 1999 and although their use in Europe is not banned, they have not been used recently and their use requires special justification, and notification of the EU Commission. Some European countries have gone further: the UK and Spain effectively ruled out their use in 1997 and 2008 respectively, while other European states also have restrictions on their use. Austria and Japan also have restrictions, and in the USA the Great Ape Protection Act H.R. 1326 (GAPA) was reintroduced to the House of Representatives in March 2009[95]. As for other non-human primates, ethical concerns about great ape use are based on a number of issues, not least that they are all endangered species, but also because of the concern that some of the mental abilities of great apes, particularly with respect to awareness of self and of others, might be similar to those of humans.

However, the evidence that great apes possess these abilities is not unambiguous. Evidence in favour includes observations of behaviour that when seen in humans are thought to require reasoning about the minds of other individuals, including studies of apparent deliberate deception, and instances of apparent empathy[96]. Against this, there is debate amongst researchers about the meaning of these observations. Many, but perhaps not all, of the observations can also be explained by mechanisms that do not require the apes to possess a concept of self or a theory regarding others' minds. Indeed, there is evidence that the minds of chimpanzees, our closest relation, may work in very different ways to those of humans. In a series of very careful and painstaking psychological studies, Povinelli found that chimpanzees seem to use lower-level rule-based approaches rather than a ToM when interacting with other apes or researchers[97]. The following is just one example from a detailed sequence of experiments. Chimpanzees, like us, will follow an experimenter's gaze (for survival in the wild, it could be important to do this if, for example, another animal has seen a predator). Moreover, chimpanzees beg for treats from an experimenter sitting facing them but tend not to do so from one facing away. These two observations seem rational to us and seem to suggest that, like us, the apes understand that other individuals can see things and that they know that someone facing away from them cannot see them. However, an alternative

---

[94]   Byrne (1999) discusses the evidence for anticipation in primates, and the implications of an awareness of the future for welfare.
[95]   Boyd Group (2002); Mackta (2010).
[96]   Byrne (1995a), pp. 130–131.
[97]   Povinelli (2000).

hypothesis is that the apes' behaviour is based on rules such as follow another's gaze, and only beg from an individual whose front is towards you. To test this possibility, Povinelli presented individual chimpanzees with two experimenters, both sitting with their backs to the ape, but with one experimenter looking over her shoulder towards the ape. If the apes possessed a ToM, then one would expect them to preferentially beg from the experimenter looking over her shoulder and clearly establishing eye contact. Surprisingly, the chimpanzees showed no preference between the two experimenters.

Other evidence that apes might possess a ToM comes from mirror studies. In these, after an animal has gained experience of mirrors, a spot is placed on the animal without its knowledge, in a position where it can only be seen by using the mirror. One way of ensuring that the animal does not know the position of the mark is to place the mark while the animal is under anaesthetic. If, on recovery, the animal responds to the mirror image by touching or paying other attention to the appropriate spot on its body, then one interpretation is that the animal recognises its image as an image of itself in the mirror, in other words it has a concept of self. In humans the ability to pass the mirror self-recognition (MSR) test emerges around 18–24 months of age, and MSR has been termed the gold standard of self-identity. For a while, only humans and great apes were shown to pass the test, other non-human primates tending to react to the image in the mirror as if it is another individual. This, combined with other evidence about their cognitive abilities, has been used to argue that apes should be a special case for moral consideration amongst non-human primates. However, apes and humans are no longer alone in passing the mirror mark test. Dolphins, elephants and, perhaps surprisingly, magpies have all passed the test, or at least a few individuals of these species have been shown to do so[98]. While MSR evidence is compelling, there is ongoing debate about what it actually means to pass the test, and some argue that it does not provide conclusive evidence regarding self-identity[99]. Other possible interpretations include that the animal might be using the information from the mirror as simple feedback to guide its movements without a concept of self. Moreover, even if responding to an image in a mirror in a way similar to how we do does indicate self-awareness, this can no longer be used as sufficient evidence to separate apes alone as a taxonomic group that requires special protection when used in research.

What then about monkeys – do they have a concept of self that might justify special treatment? Some monkeys do use deceptive techniques to obtain resources from other individuals, a type of behaviour that could indicate ToM, but this behaviour seems to be learnt by trial and error[100] rather than being the result of

---

[98]   Reiss and Marino (2001); Sarko *et al.* (2002); Plotnik *et al.* (2006); Prior *et al.* (2008).

[99]   Boyd Group (2002); de Waal (2008); Heyes (2008).

[100]   Byrne (1996). Byrne also argues that apes may possess a deeper depth of understanding both in the social sphere and in more technological areas, but see also Povinelli (2000), pp. 294–308, who suggests that chimpanzees either do not understand or only weakly understand the causal structure of problems and that their experience of objects and events may be very different from that of humans.

deliberate planning as may be the case with some great apes. Rhesus macaques are able to observe where another individual is looking, and to use that information to decide whether to steal a grape from that individual or not[101]. However, again, the simplest explanation is that this behaviour is based on rules rather than an understanding of another individual's point of view. Macaques and marmosets also consistently fail the mirror mark test for self-awareness, although in one recent study[102] two macaques fitted with head implants did use mirrors as visual aids while grooming themselves. The authors hypothesised that previous failures in this species may have been because of an innate gaze aversion resulting from the aggressive nature of a stare amongst macaques, and that in this study the head implants were such a dominant feature that they might have overcome the monkeys' inhibitions at looking at faces. It is, however, only one study and more evidence is needed before we can conclude that monkeys have a ToM or a concept of self.

Despite the lack of firm evidence on whether monkeys are capable of thinking about themselves or others in an abstract way, the Boyd Group's view was that monkeys should be accorded special moral status on the basis of the general richness of their social lives and their mental abilities. These characteristics, it was felt, meant that the restrictions of captivity and research would render them more likely to suffer than other animals. However, intelligence and social complexity have been studied in greater detail in primates than in many other animals, partly because of their close relatedness to humans. There also tend to be biases in the types of studies carried out on different species. Primatologists have tended to concentrate on areas of similarity with humans, while ornithologists have, until very recently, focused on behaviour and ecology rather than cognitive issues such as tool use and social behaviour[103]. It is also important to remember that the significance of different modalities of social communication varies with species and that we naturally tend to understand and empathise with species that use similar modes to ourselves. For example, much mouse communication is olfactory, a modality that is less important for humans and thus more difficult for us to understand, although progress is being made[104]. Olfaction is also important in primate communication, particularly for groups such as the marmosets and lemurs, but primates also make more obvious use of vision, touch and sound, and these modalities are easier for researchers to document and understand. The problem with the Boyd Group's conclusion is that there are no ultimately compelling data that primates actually do suffer more than other animals used in research. Moreover, as the next section demonstrates, primate cognitive abilities may not be as unique as has previously been thought.

---

[101]   Flombaum and Santos (2005).
[102]   Rajala *et al.* (2010).
[103]   Emery (2006).
[104]   Beynon and Hurst (2003).

4.4.3 What about the birds?

Many people tend to be more concerned about the use of primates in research than the use of birds, perhaps partly because birds are mistakenly perceived as being more primitive. Birds last shared a common ancestor with mammals and humans 310 million years ago[105] and there is a mistaken tendency to view animals that are more distantly related to us as being more primitive. The epithet 'birdbrain' is used disparagingly to signify a lack of cognitive abilities, but science is providing increasing evidence that birds are no less cognitively able than mammals. In terms of their anatomy, birds have brains that are roughly the same size in relation to their body weight as those of equivalently sized mammals[106]. It is true that there are differences in avian and mammalian brain structures, and that birds lack the columnar organisation of neocortex[107], but it is now recognised that areas of the avian brain previously thought to have derived from so-called primitive structures are in fact sophisticated processing regions that are homologous to areas such as the mammalian neocortex[108].

Studies of avian behaviour are also revealing that the cognitive capabilities of some birds are much more impressive than had previously been thought. Pigeons, for example, are capable of discriminating between categories of images such as trees and bushes and even between paintings by different artists[109], but the corvids (crows, magpies, jays, etc.) and parrots seem to be particularly able[110].

Corvids show remarkable convergence in their cognitive abilities with primates, to the extent that some of their social, spatial, memory and other skills have been compared to those of great apes. A number of birds have been documented as using and making tools, both in the wild and in the laboratory. Of these, the Caledonian crow is probably the most sophisticated, as at least one individual, Betty, is able to use unnatural tools, bending a piece of wire to hook out a reward from a tube. Not only this, but some of these crows are able to solve tasks that require a sequence of tools to be retrieved in order to finally obtain a food reward, in some cases on the first trial. Scrub jays appear to be capable of episodic memory, or something like it (episodic memory is the ability to remember an episode at a particular place and time), an ability that had previously been thought to exist only in humans[111], and we have already discussed the fact that magpies (corvid species) may be able to pass the MSR test.

---

[105]   Dawkins (2004).

[106]   Brain–body weight ratios for birds and mammals overlap considerably (Butler, 2008). Some bird brains are much smaller than would be predicted from their body weight; migratory birds, for example, have relatively small brains, but the crow brain is equivalent in relative size to that of the chimpanzee (Emery, 2006).

[107]   Butler (2008).

[108]   Macphail (2001), p. 416; Jarvis *et al*. (2005).

[109]   Watanabe *et al*. (1995).

[110]   Emery (2004). Emery (2006) argues that many corvids and parrots have been under similar social and environmental selection pressures to those of primates, which has resulted in similar cognitive solutions.

[111]   Güntürkün and Durstewitz (2001); Emery (2004, 2006); Wimpenny *et al*. (2009). See also Butler (2008). Videos of tool making and use can be viewed at http://users.ox.ac.uk/~kgroup/tools/movies.shtml, accessed 29 March 2013. There is recent evidence that cockatoos also show remarkable abilities to arrive at solutions to complex sequential problems (Auersperg *et al*., 2013).

A number of studies have purported to show that some apes are able to use signs or symbols in rudimentary forms of language to communicate with humans[112]. Similarly, at least three parrots have been taught to use speech in a meaningful way[113]. Irene Pepperburg spent many years teaching Alex, an African grey parrot, to speak, not just parroting words but to use speech in context. Her studies indicated that Alex could use the appropriate adjective ('colour', 'shape', 'material') to distinguish between two objects by answering questions such as 'What colour/shape is X?', 'What object is colour Y?', with an accuracy of about 80%. In other studies Alex was able to use concepts such as 'same' and 'different' to indicate what feature distinguished between two items, could indicate what attribute was possessed by a particular item in a collection of different objects, and even showed some understanding of the concept of number. Sadly, Alex is now dead, but it seems that he was not just a one-off; Irene has trained two other parrots to match the early stages of Alex's abilities. Suggestions that apes were capable of language have since proved extremely controversial[114], and studies such as these need to be both designed and interpreted carefully. Nonetheless, it seems that these birds had abilities that in this respect (the ability to use a word or thing to refer to objects or properties of objects) were not that different to those of the apes.

There is even research indicating that birds may possess a ToM. Western scrub jays that have hidden food in caches behave as if they perceive a difference between jays that watched where they hid food and those that were not present at the time[115], which is reasonable as jays that see the food being hidden may try to steal it. Moreover, scrub jays behave differently towards other birds depending on whether they themselves have previously pilfered food from other individuals. It seems that they change their view of the likely behaviour of other birds on the basis of their own past behaviour.

It is true that the data for some of these avian studies, such as the ones on language, have been collected from relatively few animals, but the same was true for many of the early ape studies, and the accumulation of evidence for avian cognition has been impressive. In 2001, only one of a list of cognitive functions that in humans require some form of consciousness[116] had been found in birds (Box. 4.1, function 8)[117], but by 2008 only two remained to be demonstrated in birds (self-recognition in a mirror and use of a complex syntactical language) and, following Prior's publication of mirror self-recognition by magpies, birds have now demonstrated all the functions on this list that have been demonstrated by non-human primates[118].

[112] Byrne (1995b), pp. 164–174 provides a good summary of these ape language experiments.
[113] Pepperburg's studies on Alex are reviewed in Griffin (2001), pp. 182–183. See also publications at http://www.alexfoundation.org/support_research.html, accessed 24 April 2013.
[114] See, for example, Povinelli (2000), pp. 337–338.
[115] Clayton and Dickinson (1998); Emery (2004).
[116] Roth (2001).
[117] Butler (2008).
[118] Prior *et al.* (2008).

> **Box 4.1   Some cognitive functions that according to Roth (2001) in humans require at least some states or levels of consciousness**
>
> 1. Imitation in the sense of task structure or task principle learning and tool learning
> 2. Taking the perspective of the other in deception and counter-deception
> 3. Anticipation of future events (e.g. the preparation of tools in advance)
> 4. Comprehension of underlying mechanisms, for example in the use of tools
> 5. Knowledge attribution/theory of mind
> 6. Self-recognition in a mirror
> 7. Teaching
> 8. Understanding and use of simple syntactical language (up to three-word sentences)
> 9. Use of complex and tactical language

It is not just certain clever bird species that are making the cognitive abilities of apes seem less exceptional. Gaze-following has not only been demonstrated in apes, but also in ravens and goats. Monkeys, when given mirrors as enrichment in captivity, will often tilt (i.e. use) them to see areas that would otherwise be invisible from their enclosures, and pigs can also learn to use mirror reflections to locate food[119]. Dogs perform better than chimpanzees in experiments in which they are required to follow a human's gesture to locate a food reward[120]. Those making the case for special cognitive abilities in primates sometimes refer to their complex social lives, the suggestion being that the need to keep track of a complex web of relationships both requires a high cognitive ability and may lead to a greater understanding of other individuals and the interactions between them[121]. However, not all primates live in complex social groupings, and other species such as elephants, whales, dolphins and some carnivores also form long-lasting social groups in which there can be complex relationships. While certain primates may exceed other animals in the number of different types of relationships in which they may be involved, primates are not unique in terms of many specific types of complex social interaction. Shettleworth points out that hyenas also form coalitions and engage in reconciliation behaviour. Similarly, birds and fish are also capable of learning about a stranger from its interaction with a known individual (social transitive inference)[122]. Indeed it has been suggested that many of the complex behavioural phenomena that have been extensively studied in primates, such as individual recognition, cooperation (including

---

[119]   Broom *et al.* (2009); Shettleworth (2010), p. 440.

[120]   Hare *et al.* (2002). However, see also Bräuer *et al.* (2006), who suggest that chimpanzees are adapted to understand causal cues, while dogs' ability to understand human signs is probably a result of domestication. Further, van Rooijen (2009) argues that the ability to understand pointing is seen in a wide variety of species and is not likely to be a good indicator of the ability to have insight into the meaning or intention of the pointer.

[121]   Boyd Group (2002), p. 26.

[122]   Shettleworth (2010), chapter 12 reviews similarities in social intelligence between primates and other orders.

cooperative hunting) and social learning, are also to be found in various fish species[123]. Sticklebacks have trusted partners when dealing/approaching predators (cooperation) and female guppies that observe another female being receptive towards a less preferred male become more receptive themselves (information gathering about relationships). Triggerfish have apparently learnt from others new methods of attacking their prey (social learning), and groupers solicit and lead moray eels to cooperatively hunt fish in coral (cooperative hunting). As Shettleworth notes, 'the research on social knowledge in nonprimates sampled here seems consistent with the conclusion that primate-like social cognition is not unique to primates'.

4.5 An Evolutionary Approach to Assessing Suffering

In this chapter I have argued that the decisions we make as to which species to protect and how we should treat those species that we do protect are often made on the basis of limited or controversial scientific evidence. Also, that attempting to place species on a scale in terms of their *general* ability to suffer is extremely difficult, but that there may be *particular* reasons for choosing a species relating to its likelihood of suffering in particular research projects (see also below). I have also suggested that the differences in protection given to species are often aimed at meeting public opinion rather than being based on biological differences. Some may be surprised that as a primatologist and animal welfare scientist I have played down the differences between primates and other species used in research. However, my aim has not been to lessen the ethical issues relating to the use of primates in research, rather to argue that if our aim is to reduce the amount of suffering of animals used in research, it may not always be best to assume that using a mouse involves less suffering than a primate. The risk is that we give special status to some species at the expense of other species that may, in a particular study, experience as much suffering or perhaps even more. Of course there are some very valid and good reasons for not using primates. For some species these may relate to conservation (all primate species are listed in CITES Appendices I or II) or to concerns about transport times (many macaques are imported from overseas breeding sites, and in some cases travel times can be as long as 60 hours). However, if, as in a case that I came across, there is a choice between using six primates in a study that results in no clinical signs (the test relied on post-mortem findings) versus an alternative test that requires 100 mice of which a proportion will experience ill effects, is it really correct to choose the mice, a choice that will result in more pain to more animals? Different people will have different views about this, but it seems to me that if we consider only the issue of the deliberate harm caused to the animals by the procedures, in this case it would have been better to use the primates[124].

---

[123]   Bshary *et al.* (2002).
[124]   In practice, other issues such as sourcing, transport and housing (see Chapter 5) and public concerns have to be taken into account.

So, where there is a real choice between species, rather than attempting to develop some scale of suffering so that an animal lower on the scale is always used in preference to one higher up, a much more rigorous approach is to evaluate the impact of the proposed study on the candidate species[125]. In the process, it is necessary to judge the impact of the proposed procedures and associated housing and husbandry against what we know about the animals' biology and how the animal is adapted to respond to the environment[126]. The analysis should identify strongly motivated behaviours that are likely to result in frustration if they cannot be expressed, and if standard conditions do not allow such behaviour consider whether measures can be taken that would permit them. Such behaviours may be either internally or externally stimulated. The motivation of sows to prepare a nest site, by rooting and digging as they approach parturition, is largely internal and probably driven by hormonal changes. However, a later stage of nest building involves the collection and arranging of nesting materials such as straw, and these behaviours are externally driven, being triggered by the presence of suitable material[127]. It is therefore important to consider what conditions might trigger strong motivations to perform behaviours.

Feelings and emotions are drivers of behaviour, and different species are adapted to behave in different ways that maximise their chances of surviving to reproduce. Negative subjective states can be seen as a mechanism that warns the animal about its predicament and triggers appropriate corrective behaviour[128]. Conversely, positive subjective states reward appropriate behaviour. It follows that certain emotions or feelings might be either more or less valuable for some species than others. Some feelings such as pain and fear, and possibly thirst and hunger, are common in their function, and are likely to be experienced by all sentient species. Others may not. One example might be grief, a feeling that we humans experience on the loss or departure of a loved one. However, humans have evolved to live in close social communities with long-term family ties in which the expression of emotions is an important part of the social dynamic and which can serve an adaptive purpose. The grief or panic of a mother who temporarily loses her infant would alert other group members who then might be able to assist in finding the child. It is relevant that we are a strongly K-selected species – we produce few offspring and care for them over a very long period. We know that we feel grief on the loss of a loved one, and chimpanzees also show signs of grief. Although somewhat less K-selected, the cow bellowing for its weaned calf may be feeling something similar, although for a rather shorter time, At the other end of the spectrum are species that are r-selected – they

---

[125] Smith and Boyd (1991) provide a list of points they recommend should be taken into account when making species choice (table 5.5, p. 91). These include capacity to experience harm, tolerance to human presence, knowledge of its natural history, special housing or husbandry requirements and its suitability for the study in question.
[126] Barnard and Hurst (1996). See also Chapter 5.
[127] Jensen (1993). See also Würbel (2009).
[128] Barnard and Hurst (1996). See also van der Harst and Spruijt (2007) who discuss pleasure as the 'currency' within the brain that guides decisions.

produce large numbers of offspring, but provide minimal care for them, relying on numbers to compensate for the high offspring mortality. In r-selected species such as salmon, which produce many eggs and leave them to their fate, there is simply no point in an emotion like grief. Other drives and emotions may, however, be more relevant. Herd-living animals like cattle and schooling fish are strongly motivated to remain close to other individuals, particularly when threatened. When making species choices we need to take into account how drives and behaviours such as these might be restricted or frustrated. We need to consider how restrictions on social behaviour might impact on animals of a particular species, to think about, for example, the impact of early weaning, and to assess how burrowing species such as the gerbil might cope in an experiment that requires solid floor housing. These examples are relatively simple, but to do the job properly requires a deep understanding of the species that might be used, and is much more complicated than relying on some naive idea about species hierarchy. In the UK, the regulatory regime theoretically already requires these thought processes, as investigators making applications to carry out research must indicate why the species chosen for the study is the best model to use, and indicate how any of the experimental or special husbandry provisions might adversely impact on the animals. The question is there in the application form, but in my experience investigators do not always consider properly the evolved behaviour and needs of the animals that they use. The use of an evolutionary analysis and, where there are alternative species that could be used, a comparison of the likely harm that includes the adaptations of the two species would go a long way towards ensuring that the experiment causes the least pain, suffering, distress or lasting harm.

The Harm–Benefit Judgement

5

It is increasingly accepted that decisions as to whether research should go ahead require an assessment of the harm likely to be caused to the animals and of the benefits likely to accrue from the research. In this chapter I discuss how these decisions can be made, who should be involved and the criteria that are used to make the decision.

5.1 Introduction

As research on animals can result in suffering, it is essential at the planning stage to weigh the likely benefits against the likely harms to the animals in order to decide whether the research is justified, whether it can be improved so as to reduce the harms or increase the benefits, or whether the harms are so great that no benefits could justify their infliction. My experience has been that when ethics committees or other authorities consider these matters, the judgement is usually in favour of the research continuing (researchers will normally have already given thought to ethical issues so it is unlikely that they would propose unjustifiable research), but there is often some tweaking aimed at reducing the risks of suffering or increasing the benefits. Nonetheless, there are, of course, instances where the committee or authority decides that research should not be permitted or should be stopped. During my career I have come across several cases where I felt that harms were such that the research was not justified. In one, a researcher (from the developed West) was surveying the numbers and distribution of small tropical mammals, and while doing so kept trapped wild animals in capture bags for some considerable time so that some died of stress or perhaps of hunger or thirst. When remonstrated

The Welfare of Animals Used in Research: Practice and Ethics, First Edition. Robert C. Hubrecht.
© 2014 Universities Federation for Animal Welfare. Published 2014 by John Wiley & Sons, Ltd.

with, he could not see the problem, as he saw the animals as unimportant, akin to vermin, and to be used as he wished. Another case involved a sloth, which researchers were operating on without analgesia or anaesthesia. I cannot recall the precise aim of the research, but it was to gain fundamental knowledge about the animal, rather than for an applied purpose. Both these examples are from many years ago when attitudes to animals and research were very different from those today. Nonetheless, they illustrate how important it is that research proposals are carefully considered and scrutinised by a group with relevant expertise so that sound decisions are made about whether projects should go ahead and, if so, what refinements may be needed.

Scientific review of proposed research projects has occurred for many years as part of the normal process of achieving good scientific quality and value for money, but the idea that there is a need to judge the benefits of the research against potential animal suffering is rather more recent. The idea of licensing research was introduced in the UK in 1876, but an explicit assessment of harms and benefits was only required in 1986, when the UK Animals, Scientific Procedures Act stated that:

> *In determining whether and on what terms to grant a project licence the Secretary of State shall weigh the likely adverse effects on the animals concerned against the benefit likely to accrue as a result of the programme to be specified in the licence.*

By 2007, 16 of 20 Federation of European Laboratory Animal Science Associations (FELASA) member countries surveyed had mandated ethical review prior to carrying out research on animals even though general ethical review at that time was not a European legal requirement[1]; and from 2013 the European Community mandated harm–benefit assessment under Article 38d of European Directive 2010/63/EU. Turkey and Iran also instituted legal requirements for ethical review between 2004 and 2006[2]. In the USA, although there is still no legal requirement for harm–benefit assessment, Institutional Animal Care and Use Committees (IACUCs) are reported to be increasingly carrying out this function (see Chapter 2). Not all researchers have welcomed the introduction of harm–benefit judgement, seeing it as a potential infringement of academic freedom that could block important advances in medicine and fundamental research[3], but the necessity for some process of ethical review of animal research is increasingly accepted. Usually, the major ethical issues to consider are any likely animal harms, whether the harms and benefits are in line with those predicted at the start of the project and whether improvements can be made to the animals' welfare or the quality of the research; however, biological research can raise other issues of public concern, such as potential damage to the environment or biosafety risks, which may also be the subject of ethical review.

---

[1]   Article 12 of EU Directive 86/609 did, however, require justification for experiments causing severe and prolonged pain.

[2]   Smith *et al.* (2007); Izmirli *et al.* (2010).

[3]   Dresser (1989) describes how, at that time, the idea of balancing the benefits and quality of research against harm to the animals was a controversial idea in the USA; see also Rollin (2006).

Ethical review is much more than just a yes/no decision about the ethical acceptability of a research project[4]. It is an opportunity to review research plans, the facilities and the care of the animals so that improvements can be made, if needed. It can also be a forum for raising other issues of public concern, such as potential damage to the environment or biosafety risks. Formal reviews at the outset of a project and on a regular (i.e. annual) basis are necessary to ensure that the research is justified and should continue, but ethical review is a process that should continue throughout the project. To help achieve this the UK, in 1998, introduced the concept of retrospective review, both during the project and after its completion. (From 2013, retrospective assessment for some projects is also required in Article 39 of the 2010 European Directive, but it is up to the competent body that authorises research in each member nation to determine if and when this should be carried out.) Effective continuous review requires that all parties involved – researchers, care staff and those with oversight – should be alert for opportunities to improve the benefits and for ways to reduce animal suffering and maximise the animals' welfare[5].

Harm–benefit judgements are, for most people, a utilitarian decision process, although some consider that no harm can ever be justified and some consider that some types or degree of harm can never be justified. The term 'cost–benefit analysis' has been used to describe this weighing up of likely benefits against likely harms, but the term can be misleading. I have come across cases (fewer these days) where scientists, unfamiliar with the ethical issues, mistakenly think that cost refers to the financial costs of the research. Another potentially misleading word is 'analysis', which suggests a rather exact and mathematical process. Indeed, one UK Chief Inspector produced a series of equations linking harms and benefits to questions in the licence application form, probably to try to convince prospective licensees that ethical review could be both rational and scientific, as the idea of external ethical review of research projects was still rather new. The trouble is that harms and benefits often cannot be measured on the same scale, for example if the harm is nausea and the benefit is alleviation of pain, it is not possible to add up benefits, subtract the sum of the harms and produce a numerical ethical score. The process is not therefore a mathematical analysis, but an assessment or judgement where the decisions are easier at the extremes (i.e. where the benefits are likely to be large and the harms likely to be small or vice versa). So, if harms and benefits are like apples and oranges, you might wonder if a process of weighing one against

---

[4] The same FELASA report (Smith *et al.*, 2007) noted that for longer-term projects there is a need for ongoing ethical evaluation. See also RSPCA & LASA (2012) report on guiding principles for ethical review. Within the USA, Public Health Service (PHS) policy requires review of PHS-funded research at least once every 3 years. US Animal Welfare Regulations require annual review, but only to ensure that the research has not changed from the original protocol (National Institutes of Health, 2002).

[5] A UK Chief Inspector described the regulation of animal experiments in the UK as 'a process rather than an event. It generally begins before a formal application is made and continues throughout the duration of the programmes of work, rather than being applied only at the time that authorities are granted or refused' (Animal Procedures Committee, 1997, p. 50).

the other stands scrutiny, but judgements like these are part of everyday life. When buying a house, you might have to weigh up the location of the house with respect to schools, shops and your work against factors such as its size, aspect, attractiveness and cost. Decisions such as these, where the outcome is important, can be difficult (so much so, that they sometimes feel painful), but they have to be made.

5.2 Who Should Be Involved?

We discussed in Chapter 1 how it makes sense for ethical judgements to be made by a group of people rather than one or two. Numbers provide some protection against the vagaries of individuals, and the inclusion in the process of some people not directly involved in the research makes it more likely that decisions will reflect a wider societal view. On the other hand, increasing numbers bring risks of increased bureaucracy and expense and time penalties, so there is a law of decreasing returns. In fact, most bodies that carry out ethical review of research amongst their other functions (local ethical review bodies, IACUCs, national advisory committees, funders, etc.) do have a varied membership to provide balance[6], and additional input (e.g. from experts) may be sought in difficult cases. It is true that in the UK, a single individual, the Secretary of State, makes decisions regarding the licensing of research[7], but these are taken with advice from a local inspector and when necessary other inspectors and the national advisory committee[8] may also provide input.

I have suggested that a broad input should help produce some consistency to decisions, so a US study that seemed to show a lack of consistency between ethical review committees may seem worrying. The study compared decisions made by an internal and an external IACUC of the same proposals and found that 79% of the reviews differed between the two[9]. However, in most cases the second committee was more negative than the first, and later correspondence on the article suggested that this may have been because they were not given information about the investigator and institution and so lacked knowledge that was available to the internal IACUC about the skills and previous experience of the scientist and the quality of the institution.

5.2.1 Inclusion of members with specific expertise
Input from animal care and veterinary staff is valuable because of their animal welfare expertise and because it gives them a voice on proposed research. In many establishments, animal care staff routinely carry out scientific procedures, such as

[6] See Chapter 2. Local ethical review bodies and national regulators usually also have wide-ranging responsibilities to ensure the quality of science and the welfare of animals used. Harm–benefit analysis is only one, albeit high profile, component of this task.
[7] Except in exceptionally difficult cases, the decisions are made by Home Office officials who act as agents of the Secretary of State.
[8] Formerly, the Animal Procedures Committee, now replaced by the Animals in Science Committee.
[9] Plous and Herzog (2001), see also letters published in *Science* 294, 1831–1832.

dosing or sampling animals, and so can have a wealth of experience to offer on the most effective and humane techniques, signs of suffering, and the setting and implementation of humane endpoints[10]. Although some technical staff may need support to give them confidence during project review discussions, in my experience their inclusion has been extremely worthwhile.

It is also important to include people with the scientific expertise and confidence to ask searching questions about justification and procedures. Scientists who are active in an area within or close to the researcher's own field are well placed to judge the quality of the research, but distance can also provide a useful perspective. In one establishment where I acted as an external advisor, the chair of the Ethical Review Process (ERP) was a chemist. His questions were challenging, forced people to think hard about what they were doing and encouraged them to reconsider options that might reduce animal suffering or their use of animals. People with specific expertise, such as in a particular area of science, the biology or care of particular species, or more general animal welfare expertise also add value. Such persons might well have to be brought in from outside the establishment.

5.2.2 External persons

To provide an external, independent, input to the decision-making process it is generally accepted that there is merit in including people with no personal stake in the research or the institution, and this principle is reflected in some countries' regulations. In the UK, it is the Home Office that is the independent external body that carries out the harm–benefit analysis required by law. In addition, prior to 2013, UK guidance on establishments' internal ERP suggested that they consider including a lay person, independent of the establishment carrying out the research[11]. At the time of writing the guidance was being rewritten, and the ERP has been replaced by the Animal Welfare and Ethical Review Body (AWERB), but it still stated that 'We will expect you to take into account the views of people who do not have responsibilities under ASPA, as well as someone who is independent of your establishment'. Within the USA, USDA regulations and PHS policy specify that IACUCs must include a non-affiliated person. Additionally, PHS policy specifies that there must be one person whose primary concerns are in a non-scientific area, and that there may be occasions when one person is able to fit both roles. The Australian code goes further, as it specifies that an institution's Animal Ethics Committee must include an external person with a professional interest in animal welfare, such as a veterinarian from an animal welfare organisation, as well as a person independent of the institution who has never been involved in the use of animals in research or teaching.

So two types of disinterest are seen as adding value to the process: independence from the institution and independence from the scientific process. The value of

---

[10] Humane endpoints are pre-specified descriptions of points at which action will be taken to end or minimise an animal's suffering; see Chapter 6.
[11] Home Office (2000).

having someone independent from the institution is easy to understand. Persons who are distanced from the institution or the research help to ensure the integrity of the review process, as well as providing an external and independent perspective. For some organisations the idea that external people should have access to internal information is troubling, but experience shows that it can be done. Many have served as such external persons; I personally have served on ERP committees of a pharmaceutical company, a charitable research organisation, a university and a primate breeder and on advisory committees for various organisations, including the UK Ministry of Defence.

The need for a non-scientific external person is, perhaps, a little harder to understand, but non-scientific members have no scientific baggage and can bring a fresh perspective by asking important questions about aspects sometimes overlooked by those too close to the detail. They also bring a different perspective to discussions about the likely benefits of proposed research, the need for it, and may be well placed to fulfil a liaison role between the research community and the public. Sometimes the presence of non-scientists is resisted on the grounds that their lack of understanding is a handicap that may frustrate and delay proceedings[12], but if researchers cannot clearly describe the proposed benefits of the research and its implications for the animals then it is likely that they are not thinking clearly enough about these issues themselves.

The presence of both types of external person on committees addressing harm–benefit questions, although not always possible, is valuable. Apart from their independence, and the public reassurance this provides, external persons can also provide input on what is publically acceptable. Rollin describes a case in the USA, in 2004, of research at a veterinary hospital aimed at finding better ways of treating fractures to dog limbs. The research involved the deliberate breaking of young dogs' limbs, albeit with measures to reduce pain and distress, and the research had been approved by the IACUC. However, the general public, the student body and the mass media were horrified by the research[13]. An external lay person is well placed to advise on any likely public reaction to research that affects animal welfare or other issues of public concern. External persons are also well placed to question or comment on husbandry practices, the care of the animals, and what evidence has been used to inform these practices. For example, if an establishment keeps dogs, it is perfectly reasonable for the external person to ask questions about the opportunities they have for socialisation and about time spent outside their home enclosure. If rodents, the external person might ask whether the rodents are kept in cages with nesting material, and if not why not. Some external persons may have had experience working for more than one establishment, and so may be able to draw attention to improvements that have been tried elsewhere.

---

[12] Robb (1993) provides an account as an unaffiliated member and non-scientist working in the US system, at a time when the role was perhaps not as well accepted as today.
[13] Rollin (2006), pp. 301–302.

5.2.3 Training

Training on legislation, ethics and animal welfare can be very useful for those involved in harm–benefit decisions. Officials charged with this job, for example Home Office Inspectors in the UK, are trained and have a period of induction, but in the past specific training for those who carry out local ethical review was rare. In some cases training will not be necessary or appropriate, but institutions should offer training that includes induction, or a period of overlap with a preceding holder of the role. Some institutions may not feel competent to provide in-house training, but various organisations have seen the gap and produced training and training materials. In the UK, the RSPCA has for some years run one-day meetings, and produced resources for lay persons, which contain advice that would also be useful for those in other countries. In the USA, training materials are available from, amongst others, the American Association for Laboratory Animal Science (AALAS), the Animal Welfare Information Center (AWIC) and Scientists Center for Animal Welfare (SCAW). The COST Manual provides background and a checklist of issues that need to be taken into account in ethical review.[14]

5.3 Factors To Be Considered in Harm–Benefit Decisions

We now come to the issues that need to be considered when judging proposals at the initial stage and when monitoring during the project. Fundamental to the process is implementation of the 3Rs principles, first developed by Professor Bill Russell and Rex Burch: (1) that sentient animals should not be used if replacement methods or non-sentient animals are available; (2) if there are no alternatives, that the numbers of animals used should be reduced to the minimum required to achieve the objective; and (3) that the use and care of the animals used should be refined so as to reduce suffering to a minimum (these are discussed in detail in the next chapter). Implementation of the 3Rs is one of the necessary conditions for research to be ethically justifiable.

Before beginning any research that uses sentient animals, researchers should have considered whether there are non-sentient alternatives (e.g. tissue culture, computer models or animals considered to be non-sentient; see Chapters 4 and 6) that could be used for the entire project or for sections within it to achieve the same aims. This will normally require a literature survey. If, having carried out this preliminary research, the researcher considers that the use of animals is still necessary, then the extent of the literature survey and the justification for not using alternatives should be included in the evidence provided for the harm–benefit review. Searching for alternatives is not just a process carried out at the planning stage; it should be repeated throughout the research project as new developments in replacements are being made all the time.

---

[14] Lane and Jennings (2004); Smith and Jennings (2009); *Ethical Review of Animal Research: A Training Resource Developed for Members of Ethics Committees*, Ethical Review Course Manual–Poland (2006) available from RSPCA Science Group, http://preview.tinyurl.com/bn9jr3y, accessed 14 May 2013. See also IACUC.org, http://www.iacuc.org/, and Kalman *et al.* (2011).

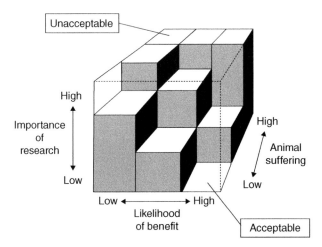

Figure 5.1   Bateson's cube, illustrating key aspects of the decision-making process when judging the acceptability of research. © Reed Business Information UK. All rights reserved. Distributed
by Tribune Media Services.

If there are no alternatives to the use of animals, then the next step is to weigh
up the benefits that will accrue from the research against the harms to the animals,
remembering that the former are dependent on the quality of the research.
The Bateson cube[15] models this decision process, the three dimensions of the cube
representing the importance of the research, the likelihood of benefit, and animal
suffering (Figure 5.1). Depending on the scores given to each of these factors the
result will fall either into the solid or transparent portion of the cube. If it falls into
the solid part of the cube, then the suggestion is that the research should not be
allowed. The cube has a refreshingly simple feel to it, but is sometimes misunder-
stood. It is not intended to generate answers to particular issues, but rather to
illustrate how rules, arrived at by consensus, can be used in harm–benefit decisions.
In the remainder of this section we will consider in more detail review of the benefits
of research, quality of science and harms to the animals.[16]

5.3.1 Research benefits
5.3.1.1 The purpose of research and judgements on need
As we saw in Chapter 2, legislation in some parts of the world restricts the purposes
for which animals can be used in research, reflecting societies' decisions as to what can
be justified. However, just because a particular piece of research falls within a generally permitted subject area does not necessarily mean that it is ethically justifiable.

---

[15]   Bateson (1986, 2011a).
[16]   There are a number of good discussions on harm–benefit assessment, which although focused on
the UK regulatory system, have broader applicability and are well worth reading. For example, see
Smith and Boyd (1991), chapter 13; Animal Procedures Committee (1997, 2003); Smith and Jennings
(2009).

Table 5.1 Categories of benefit as listed in various schemes that have been proposed to score the harms or benefits of research.

Authors	Benefit
Smith & Boyd (1991)	Economic value
	Educational value
	Other value
	Originality
	Timeliness
	Implications for other research
	Potential to lead to further benefits
Porter (1992)	Alleviation of substantial human or non-human pain
	Alleviation of moderate human or non-human pain or suffering
	Clear benefit to human or non-human health or welfare
	Some benefit to human or non-human health or welfare
	Fundamental research for the advancement of knowledge (no clear alleviation of pain or, benefit to human or animal health)
Stafleu et al. (1999)	Gains in health interest taking into account suffering, mortality and morbidity
	Gains in knowledge interest taking into account originality and whether the problem is worth solving
	Gains in economic interest taking into account effects on industry, national economy and human welfare

Most regulatory systems permit research on animals that is likely to result in benefits that include advancing human knowledge, for example in human and veterinary medicine, or result in protection of the environment (Table 5.1). It is also generally accepted that research on animals should not be carried out for trivial purposes. But what do we mean by trivial? What is important to one person may be far less so to another.

The pharmaceutical industry carries out research, at least partly, to make a profit, and the profit component of the benefit of animal research might seem an insufficient justification for research that causes animals harm. On the other hand, others might point to the tangible benefits to society in terms of employment, standards of living and, most importantly, the production of needed drugs that can arise from profitable research companies. Most seem to accept that balances have to be struck. In 1993, the then UK Home Office Chief Inspector stated that legitimate benefits could include economic benefits affecting profitability, employment, and conservation of natural resources. A later Chief Inspector made a finer distinction between cheaper healthcare, which would benefit many, and company profitability, which he did not consider an acceptable benefit[17].

---

[17] Animal Procedures Committee (1997), p. 56. Bateson (2005) considers that benefits, in addition to human and animal medicine, should include likely contributions to human understanding, education, the economy and the environment.

Most people accept that research aimed at helping to prevent or cure disease in humans or other animals can be justified in some circumstances, but even these benefits vary in their importance, and people differ as to where to draw the line. Is it acceptable, for example, to use animals in research or safety testing to produce a drug when a similar one already exists and where the benefits may be more economic than medical? If not, then are the potential benefits of producing a cheaper drug for poorer people being ignored? How does one assess the likely benefits that will result from a generic licence proposal drawn up by a contract research organisation to allow the regulatory testing of a wide range of future client's drugs of unknown value? What about using animals to solve problems that we humans have inflicted on ourselves? The exposé of the use of beagles in the UK in the 1970s to test a new, possibly safer, tobacco product caused much disquiet and subsequently the use of animals to test new tobacco (or alcohol) products was stopped. On the other hand, while some will consider smoking and drinking as self-inflicted problems, they have undoubtedly resulted in much disease and distress and research on health issues arising from them and treatment is still permitted. The problem is, of course, that it is often hard to draw lines; many medical conditions are to some extent self-inflicted and very few of us live a perfectly healthy lifestyle.

The use of animals in testing of cosmetics also raised widespread concern that this use was too trivial. However, some cosmetics are used by people to cover severe facial blemishes or burns. If cosmetics provide them with the self-confidence to be able to go out into public without embarrassment, then is their development trivial? The use of animals in the testing of household products is also controversial. On the one hand, the Boyd Group's statement that 'it is unacceptable to use animals in developing and testing new products that are widely perceived to be convenience products for which there is little potential need because similar non-medical products with adequate efficacy are already available' seems eminently reasonable as such use is clearly trivial. On the other hand, there is an obligation on manufacturers to ensure that products are safe, or if they are not that warnings or safety packaging are provided. Fortunately, whether cosmetics or household products are trivial or not, we are now at a stage where there is often no great need to test them. Very often, sufficient information on toxicity is already available for the constituents, and there may be little real need to develop new constituents. For this reason, finished product testing of cosmetics has been banned in Europe and cosmetic ingredient testing is also banned in the UK. A similar ban on the testing of finished household products is planned in the UK.[18]

Some people might want to make a distinction between applied and fundamental research. The benefits of applied research are easily understood as it can lead directly to solutions such as new or better cures for diseases. The scientist carrying out fundamental research, on the other hand, can seem idiosyncratic, the research

---

[18] Discussions on the ethics of cosmetic and finished product testing, and other ethical issues, are available at the Boyd Group website, http://www.boyd-group.demon.co.uk/.

benefiting, on the face of it, few other than the researcher and colleagues working in the same area. However, the benefits of fundamental research often lie a long way in the future, are hard to predict, and can lead to unexpected and important advances resulting in substantial long-term benefits[19]. Some may argue that there is a difference between fundamental research that has clear links to medicine and more esoteric less applied research. After all, while research that achieves a better understanding of, for example, physiology or parasites can lead to better medical treatments or a healthier population, it may seem harder to justify the use of animals in non-medical research. Nevertheless, non-medical fundamental research, for example into ecology and ethology, can also result in real benefits to humans, the animals or the environment in general, by providing us with a better understanding of how animals interact with each other, with us, and how they respond to changes in the environment. In fact, it has been suggested that with over 7 billion persons alive on our planet, environmental issues are now becoming so pressing that some of these issues may be more important subjects of research than human health and disease[20]. Still, it would be reasonable to argue that as the benefits become more speculative, the permitted harm to animals should also be less.

There are some areas of research where it may be easier to reach consensus that animal use is not justified. A UK committee provided examples of what it considered trivial research, including the use of genetic modification research aimed at the creation or duplication of favourite pets or for the development of animals as toys or fashion accessories[21], to which might be added research to find ways of improving the performance of sporting animals.

Even when the purpose of the research is not in itself trivial, the outcomes may be if, for example, the results turn out not to be helpful. Therefore, the assessment process must consider how much benefit the research is likely to bring. Again, making this judgement is tricky, but it seems reasonable to consider both the number of people or animals who would gain from the research as well as the scale of the benefit per individual. To some extent these aspects might be traded off against each other. So, for a given harm to the animals used, a benefit that affected only a small number of people would only be considered worthwhile if the benefits for those people were very large. However, the converse situation when the numbers of people affected are very large but the benefits are very small is less easy to justify. It does not seem right to sum trivial benefits, even when the number of people receiving them is large, as this could be used to justify virtually any amount of animal suffering in order to find cures for very unimportant conditions or merely to satisfy curiosity.

---

[19] For example, the link between hormones and breast cancer was made on the basis of earlier research into the role of hormones in the development of the breast.

[20] Cuthill (2007) makes the case for the benefits of fundamental research with respect to the study of animal behaviour, ecology and the environment. He also argues that human curiosity is such a fundamental part of our nature that satisfying our appetite for knowledge is a utilitarian benefit.

[21] Animal Procedures Committee (2001).

One approach to judging the importance of fundamental research, used by UK Home Office officials in the past, is to consider the number of researchers working in that field. While this provides an indication of the number of people who may be directly affected by (and perhaps benefit from) the research, my personal feeling is that it does not tell you much about the value of the research. Important developments in new areas of science are often the product of a single individual or research group, and it can take time for others to become involved. At an early stage it is not always possible to tell how important a research topic will prove to be, so the numbers of people working in a research area has no direct relationship to its future value or importance.

5.3.1.2 Scientific quality and model validity

Most research requires external funding and, for obvious reasons, medical and other funding bodies wish to back good-quality research that is likely to lead to real benefits. The standard system for assessing the quality of proposed academic research (as well as for assessing benefits) has been the use of peer review, in which a range of factors including the aims, proposed model, methodology and statistics, skills and experience of the researcher and their team, together with the support structures and facilities available at the host institution, such as husbandry facilities, skills of the animal care staff, and statistical expertise, are all taken into account. Such reviews are intensive, involving referee reports, external experts and experienced internal staff, and applicants may be given the opportunity to reply to referees' comments. In commercial pharmaceutical organisations the process is somewhat different but advice on research concepts might be sought from scientists with a track record in the area. In addition, there is, or should be, programme review at various levels with input from scientists not directly involved in the research. This process is not quite as independent or wide-ranging as the peer review process practised in academia, and research in pharmaceutical institutions might be improved by more openness in some cases. This idea is supported by the fact that when industrial cooperation does occur, opportunities to refine experimental procedures have been identified[22]. Finally, in those countries where there is appropriate legislation, the scientific quality of the proposed research is subjected to further review by the regulators. Often these will have had a scientific background.

Some may think that this self-governing process by scientists is a little too cosy, but it is hard to think of an alternative to peer review for making judgements on the scientific quality of research applications. Reassurance may also come from the fact that as funding has become harder, competition has also become greater, and only the very highest ranked proposals stand a chance of achieving funding. However, grading research is not easy, and even projects that are ranked highly may

---

[22] A UK Industry/Animal Welfare Initiative to minimise dog use in preclinical toxicology has suggested that, depending on the results of some further research, it may be possible to abandon the final 3- or 6-month study in regulatory toxicology by maximising data from earlier studies (Hasiwa *et al.*, 2011).

fail to achieve their objectives. This issue – the extent to which research leads to medical advances – came under scrutiny in the Weatherall report, which recommended that the major UK funding organisations should undertake a systematic review of the outcomes of all research using non-human primates that they had supported over the last decade. The review[23], chaired by Professor Bateson, duly reported 5 years later with some challenging findings. It turned out that in many cases it was very difficult to determine exactly how the research had impacted on developments in medicine. Indeed, in most cases there was little direct evidence available of actual benefit in the form of changes in clinical practice or new treatments, although the report acknowledged that for some of the research this might have been because too little time had elapsed to allow the development of practical changes. More concerning was that although citation rates (the number of times the publications describing the results were referred to in other scientific papers) were generally high in this sample, 4 of 14 studies on vision failed to produce any published papers, and two of these were assessed as having imposed a high welfare impact, albeit on a small number of animals.

Reasons for failure are many and not all are the fault of the researcher or are predictable. Research is always speculative, and as proposals can cover a research programme of up to 5 years, it is inevitable that some elements may fail and that the direction of the research can change over the funding period. The aims may have been too ambitious, or it may have been successful, of good quality and have resulted in published papers but, for some reason, the research was not cited. There may also be causes other than the science. As in many other areas of human activity, projects may be affected by staff changes, and halfway through a research programme, postdoctoral research assistants or fellows may be thinking about what they will do when the funding ceases and, given a good offer, may leave the project before its completion. This can have disastrous effects on a research project. However, we should not be too negative. Another way of looking at the Bateson report is that the majority of the studies produced successful research and that most had a low welfare impact. Moreover, practical outputs and quality of science do not always correlate. The report noted that the scientific impact of a subset of studies on the biological basis of vision was judged as high but considered low in terms of the short-term impact on medicine.

Model validity

An area of more general concern is the validity of some animal models in biomedical research. An animal model is the use of certain techniques on animals with certain characteristics (species, gender and age) to mimic some aspect of a disease, such as its signs[24], development or treatment. An important feature of animal models is that they should be repeatable and replicable, so that the same results should be obtained

---

[23] Bateson (2011b).
[24] Symptoms are voiced by patients; with animals one has to rely on signs.

when studies are repeated, either in the same laboratory or elsewhere[25]. Some models accurately mimic the causes and signs of disease and respond to the same treatment; these are called homologous models, but many, termed isomorphic models, have only a partial resemblance to the condition. Isomorphic models usually do have comparable symptomology and response to treatment, but the causes of the condition can be very different. At first sight, one might think that isomorphic models would be less useful than the more accurate mimic, the homologous model[26]; however, models, to be useful, do not need to be completely accurate representations of the thing they are modelling. Michael Festing uses an analogy to explain this. Maps can be made of paper, be two-dimensional, and to the untutored eye look nothing like the real world and yet, despite sharing virtually no features with the real thing, be extremely useful models of the real world. Like maps, most biomedical research animal models are isomorphic and are not exact models, but they may nevertheless adequately predict aspects of the condition, such as response to drug treatments.

It is clearly in the researchers' own interests to use reliable and predictive models, but how are those who review project proposals to judge this important factor? In my experience it is rather rare to find evidence on these matters in project proposals. Arguments that the model is well established, or has been previously used by the researcher should be treated with caution. That it has featured in a prior publication does not necessarily make a model valid, and a model that is valid for one purpose may not be for a slightly different one. Lack of evidence about model quality in research applications would be of less concern if all were well in terms of model validity. Unfortunately, it is becoming increasingly apparent that many of the models used in biomedical research have not been as successful as expected. Recent research has not led to as many useful treatments as pharmaceutical companies would have wished, and there has been a substantial drop in the number of new chemicals that have been approved by the FDA[27]. One study of ten big pharmaceutical companies found that 89% of compounds failed at some point during the drug development process[28]. Tellingly, most compounds fail once clinical testing begins in humans because of lack of efficacy and/or safety problems, suggesting that the animal models have not been as predictive as one might have hoped.

It is becoming apparent that model validity has been at least part of the problem[29]. Some areas, such as the development of compounds aimed at dealing with oncology

---

[25] Geyer and Markou (1995, 2002). Geyer and Markou (1995) list a bewildering number of types of validity in models that have been described in neuropsychopharmacology: predictive, construct, etiological, convergent and discriminant, and face validity. For those interested in this issue and definitions of these terms, the document is available at http://www.acnp.org/publications/psycho4generation.aspx, accessed 29 March 2013.

[26] In some cases bacteria may be more useful for research on a human disease than a primate. Russell and Burch (1959), pp. 80–84 refer to the error that animals used as models of human disease should be as closely related as possible to humans as the *high-fidelity fallacy*.

[27] Paul *et al.* (2010) cite a 50% drop in the last 5 years prior to 2009 compared with the previous 5 years.

[28] Kola and Landis (2004).

[29] Tricklebank and Garner (2012).

or disorders of the central nervous system (CNS), have proved to have particularly high rates of failure (95% and 92%, respectively), and Kola and Landis point out that these also happen to be areas of research where the models are not very predictive of the human pathophysiology. They give the example of oncological models in which a tumour cell line is injected into nude mice. In this model, the cell line has little relevance to the naturally occurring tumour, the nude mouse (unlike the human) is immunologically deficient, and the site of injection is different to where the tumour develops in the human. While models need not be homologous to be useful, perhaps this model lacks predictability because it shares too few features with the human condition.

Stroke models also have not been as effective as hoped. Systematic reviews of published results of interventions in animal stroke models overstate their efficacy in humans by around one-third, while compounds that appear promising in mice models of Alzheimer's disease have failed when the time came for clinical studies[30]. Similarly, anxiety models have had their problems, possibly because techniques such as the forced swim test, in which the behaviour (swimming, immobility, etc.) of rodents placed in a container of water from which they cannot escape is recorded over a set period, may have little relevance to human conditions of anxiety or depression. Moreover, when research aimed at finding anti-anxiety drugs (anxiolytics) depends on finding behavioural changes in rodents similar to those shown after they have been given benzodiazepines (a human anxiolytic), then useful anxiolytics that work in a different manner may not be detected[31].

Neuropathic pain[32] research is another example where there seems to be problems. Lack of negative publications and study selection bias[33] have resulted in overestimates of likely efficacy in clinical trials. One possible reason is that the animal model used to study analgesics for neuropathic pain usually involves a peripheral nerve injury model, but the causes of neuropathic pain frequently do not involve peripheral nerve injury. Moreover, the conventional measure of animal pain in nerve injury studies involves tests for hypersensitivity to stimuli. Unfortunately, spontaneous pain is the major symptom experienced by most patients with neuropathic pain, and this is not assessed by hypersensitivity studies. Other problems include poor correspondence between the animal models and the human patient population in terms of the incidence of the pain, the time course of the disease, and the age, gender and general health of the subjects.[34]

---

[30]   Sena *et al.* (2010); Zahs and Ashe (2010).

[31]   Work is being done to try to improve model selection, and the Laboratory Animal Science Association (LASA) together with the British Association of Pharmacology (BAP), the British Neuroscience Association (BNA) and the European Summer School on Whole Animal Pharmacology (ESSWAP) is producing guidelines that should help researchers planning appropriate behavioural studies aimed at CNS disorders. The planned publication date is end of 2013.

[32]   Neuropathic pain derives from injury to the nervous system rather than from some external cause of pain.

[33]   The problem of negative results is discussed later in the section on maximising output through dissemination of results.

[34]   Rice *et al.* (2008); Rice (2010).

Model validity can also be affected by husbandry, as animals are unlikely to be good models if kept in conditions that result in abnormal physiology or behaviour (see Chapter 7). For example, it has been suggested that because rodents are normally housed in cages with *ad libitum* access to food and with few opportunities for physical activity, they usually become obese over their lifespan. As obesity can affect health and a variety of physiological functions including the immune system, insulin resistance and predisposition to cancers, there is an obvious risk to the validity of research on these animals. In the words of the authors of one paper, 'Failure to recognize that many standard control rats and mice used in biomedical research are sedentary, obese, glucose intolerant, and on a trajectory to premature death may confound data interpretation and outcomes of human studies.'[35]

Other issues, identified by various authors as having resulted in failure of animal models, include poor experimental design and analysis, publication bias, inadequate preclinical trial and misinterpretation of the models[36]. This all sounds very negative, but I do not mean to suggest that all animal models are poor or that all research in these areas is poor. These challenges do, however, emphasise the need for funders and those charged with ethical review to carefully consider model validity.

5.3.1.3 Measures that can be taken to maximise benefits

When animals are used in research, there is a moral obligation to ensure that the maximum information and benefit is obtained from each animal used. Researchers should therefore be able to demonstrate how they will maximise the benefits and, if they have previously carried out research, show a track record. Some of the ways in which research benefits can be maximised are discussed in the following sections.

Maximising benefit through high-quality experimental design

We have discussed the importance of using valid models, but to maximise the benefits obtained from using animals in those models also requires good experimental design and appropriate statistical tests. This may seem self-evident but, unfortunately, reviews indicate that research data have frequently been invalidated by poor or inappropriate experimental design and statistical analysis[37]. Kilkenny and her co-authors, for example, reviewed a large sample of peer-reviewed biomedical publications and found serious flaws including, in the majority of the papers, a lack of evidence that the authors had used randomisation and blinding for animal selection (87% and 86% of the papers, respectively). Such failures would result in more incorrect positive results, which in turn may lead to further investigations with the associated financial, time and animal welfare costs. The solutions are obvious. Scientists need to know when to seek expert statistical advice and must be properly trained in the subject (see Chapter 6).

---

[35]   Martin *et al.* (2010).
[36]   Jucker (2010); van der Worp *et al.* (2010).
[37]   For example, Festing *et al.* (2002); Kilkenny *et al.* (2009); van der Worp *et al.* (2010).

Funding and ethical review bodies need to satisfy themselves about research quality, which includes the proposed design and analysis. Applicants therefore need to be able to demonstrate competence, or should be able to show that they have access to advice and used it. On publication, it is important that authors clearly demonstrate the quality of their design and analysis by providing sufficient details but, regrettably, reviews suggest that papers are often published with insufficient information to allow replication or to judge whether the experimental design is adequate[38]. In an attempt to address these failings, the ARRIVE guidelines provide recommendations on information that should be provided in publications with respect to regulation, ethics, design and analysis of research, and which should improve the quality of published work on animals. The US National Institute of Neurological Disorders and Stroke convened a workshop that has also produced recommendations with the same aim[39].

Maximising the information obtained per animal

A particularly useful way of maximising output is to piggyback research, by collecting additional data from animals that are already being used in a project. For example, welfare data might be collected to test the effectiveness of various anaesthetic or analgesic regimens on animals that are being used in disease or pain research. Piggybacking can be a very efficient use of animals and allows research that would otherwise be ethically difficult to justify, but it is always necessary to take into account any extra harm that might be caused to the animals.

Maximising outcomes through dissemination of results

Successful research should be disseminated widely so that maximum use can be made of data that have been won at a cost to animals and so that research is not duplicated resulting in unnecessary use of animals (bearing in mind that replication of research to check results and ensure consistency within and between laboratories is a fundamental part of the scientific process). Funding bodies are aware of these needs and some therefore include questions about plans for dissemination in their application forms. Most academic scientists need little persuasion to publish their research in learned journals or at conferences as this boosts their status and advances their careers. These traditional publication routes ensure quality but there can also be benefit in publishing in the more popular press or at venues aimed at target groups, the general public or non-specialists. All this costs money: travel, conference fees, and open-access publication can be very expensive. It is therefore best if research applications include financial provision for this at the outset as this increases the likelihood of good dissemination of results.

---

[38] Landis *et al.* (2012).
[39] The ARRIVE (Animal Research: Reporting In Vivo Experiments) guidelines, Kilkenny *et al.* (2010). See Landis *et al.* (2012) for the US National Institute of Neurological Disorders and Stroke workshop report.

Not all research is successful and studies that result in discarded hypotheses are often not published. However, the Bateson report rightly emphasised the moral obligation to publish research, even if results are negative, to prevent unnecessary repetition. The report described a vaccine development study that was terminated early because the concept did not work and on ethical grounds to avoid using more primates. Although the report's authors would have liked to have published the data they collected, they did not achieve this, probably because it is not easy to persuade a journal to accept a paper reporting a failed study and with an incomplete dataset. Publication of negative results is necessary to avoid bias in favour of positive results; nonetheless, it has been estimated that 50% of laboratory animal research never gets published, and the figure may be much less from for-profit organisations[40]. Journals have often been perceived to be reluctant to publish negative results, but the reduced cost of online publishing has provided new opportunities and there are now journals that do publish otherwise 'unpublishable' results. While researchers may never be as motivated to spend time publishing negative data, at least it is becoming easier to publish accounts that might help others to avoid fruitless line of research and prevent unnecessary animal suffering.

In some research areas publication may be problematic. For example, in defence research, the findings, methods and sometimes even the research activity might need to be kept confidential. With this in mind, I was surprised to find, when I served on a committee overseeing the welfare of animals used in British defence research, that much of the research could be published. It turned out that approximately 75–80% of the biomedical work was published in open peer-reviewed scientific journals and oral papers had also been presented at scientific meetings[41]. Pharmaceutical research is another area where companies may wish to keep commercially valuable information to themselves, and because they may work on the same diseases as other pharmaceutical companies, animal protection groups have expressed concerns about the risks of research duplication. The BUAV made this point in evidence presented to a House of Lords Select Committee, in which it argued that a number of influential persons and organisations, namely the European Commission, the OECD, the APC and a former Chief Inspector and a Home Office Minister, have all accepted that duplication either was, or could be, an issue[42]. On the other hand, I have spoken with representatives of pharmaceutical companies who argue that true duplication is likely to be very rare. My own view is that nobody knows whether unnecessary duplication occurs, and that without full publication of results by pharmaceutical companies (which is very unlikely) we may never know.

---

[40] ter Riet *et al.* (2012).
[41] AWAC Fifth Report 2001 and AWAC Sixth Report 2002. The latter is available at the UK Government Web Archive, http://preview.tinyurl.com/carhae5, accessed 13 May 2013.
[42] House of Lords (2002), p. 81.

Data obtained from animals that have suffered in order to provide it are valuable, and are sometimes needed to answer or clarify questions after the initial research is completed. Data should therefore be archived to avoid unnecessary animal use – a reasonable period might be at least 10 years. Ideally, the data should also be made available to other researchers, and to encourage this some research funders ask applicants whether and how this will be done. Sharing of commercially sensitive data is not always possible, but even in the world of the pharmaceutical industry some information sharing has proved possible. A UK joint industry/animal welfare initiative aimed at minimising dog use in preclinical toxicology has established a central database containing toxicological information on non-active ingredients used in drug formulations, such as vehicles, solvents and preservatives. The database is available free to participating companies, and can be accessed for a fee by others[43]. Ensuring that safety and efficacy test results are widely accepted is another way of reducing wastage, and to encourage this process the 2010 European Directive requires member states to accept regulatory testing data generated in other member states, where this is appropriate[44].

Maximising applied outputs

Publication is an important output of research but it is also important to consider how practical outputs can be maximised. In these days of financial uncertainty and stringency, universities and other research establishments are much more aware than they used to be of the need to exploit research. Nonetheless, as Bateson's report on academic non-human primate research showed, even successful published research with the potential for practical developments can be disappointing in terms of its actual impact in the world. It is possible that funding bodies and government departments could do more to maximise the returns of publically funded research, and the Bateson report has suggested that one solution might be a more joined-up approach by funders in which high-level horizon scanning for problems is integrated with a process that reviews research outputs and encourages their take-up and development by scientists and engineers[45].

5.3.2 Research harms

Having considered quality of science and likely research benefits as one side of ethical review, we now turn to the other side of the equation – the process of assessing, and where possible minimising, harms caused to the animals. Again, this is a process that should be continued throughout the research project. For convenience, in the remainder of this chapter I will concentrate on providing examples of the sorts of harms that can occur, and in the next chapter discuss options to eliminate

---

[43] Hasiwa *et al.* (2011).

[44] European Directive 2010/63/EU, Article 46 requires that 'Each Member State shall accept data from other Member States that are generated by procedures recognised by the legislation of the Union, unless further procedures need to be carried out regarding that data for the protection of public health, safety or the environment'.

[45] Bateson (2011b), p. 15.

or ameliorate harms by implementing the 3Rs. In practice, those involved in ethical review of research should combine these two aspects into an iterative discussion in which the harms are identified, ameliorated if possible, and judgements made regarding ethical acceptability.

Despite all the advances in animal welfare science, assessing likely harms in research proposals is often difficult. We have already seen that making judgements about welfare is not always easy even when the animal is in front of you. At the outset of a project decisions have to be made on projected outcomes, sometimes with rather limited information. In the UK system, research plans are described both when applying for funding and when applying to the Home Office for licences to carry out research that might cause suffering. In my experience, the level of detail on harms in funding proposals has often been rather low and much less than would be expected in a licence application. I think that there are several likely reasons for this. In the past, funding application forms often did not focus on potential harms, and even if relevant questions are asked, research proposals can cover five or more years during which time plans may change. It can also be hard for the applicant, at the funding application stage, to predict all the procedures that may be used on the animals. Given these difficulties in making harm–benefit judgements, funders may need to require further information throughout the course of the research.

Another obstacle to assessment can be a tendency to overemphasise likely benefits and underplay possible harms. Such a tendency, while wrong, may be quite unconscious and result from at least four factors.

1. Researchers, given the reduction in whole animal biology teaching, may not be expert in animal welfare, welfare assessment or natural history of the animal that they are studying.
2. The reason why researchers wish to carry out the research is because they believe in the benefits. The harms that occur are an unpleasant consideration and a potential barrier to their research, so there may be an understandable unconscious tendency to underestimate them.
3. Hardening and reduction in the sympathy towards animals of those involved in the research may also occur over time. That this can happen, even amongst those charged with championing animals, seems to be born out from surveys of veterinary students. One study found that students in later years of their course, particularly males, scored lower in terms of their empathy for animals, while in both genders there was a reduction in beliefs regarding the sentience of dogs, cats and cows. Another found that there was a tendency for students later in the course to indicate that they would be less likely to treat for pain[46]. If the students' views changed simply as a result of desensitisation, rather than because of increased knowledge, then there is a problem. Desensitisation can be accentuated by jargon used to distance the researcher and other staff from the reality of procedures. Examples include talking about animals as 'subjects' rather than

---

[46]   Hellyer *et al.* (1999); Paul and Podberscek (2000).

stating what they are (e.g. 'mice'), using euphemisms such as 'sacrificed' for 'euthanased' or 'killed', and the awful term 'bioreactor' for animals genetically modified to produce products such as pharmaceuticals in their milk. Evasions such as these are not helpful for developing a sense of responsibility and empathy for the animals.

4. Research projects may not clearly describe all the harms. This can be because the applicant has not identified the harms, or because the applicant's description is poor so that it is hard to disentangle what will actually happen to the animals, or because the harms fall outside of the scope of the regulation.

5.3.2.1 Direct harms

The most obvious harms to animals used in research are so-called direct harms, that is, harms that come from the experimental protocols and that are an unavoidable consequence of the research[47]. As these harms differ according to the type of research, they are extraordinarily diverse but examples include the effect of test substances, the method of dosing, surgery, body fluid sampling, harms resulting from genetic alteration, deliberate injury, food and water restriction, restraint, exposure to known stressors such as pain, fatigue, white noise, single or solitary housing, and so on.

For the purposes of ethical review, a description of a procedure (e.g. that an animal will be subjected to an operation to ligate a sciatic nerve, or that animals will be subjected to a 15% weight loss in order to motivate them to work for rewards) tells us little about what the procedure will be like for the animal. Descriptions of harms therefore need to be more than just a list of what will be done to the animals, and should also attempt to describe what the implications will be for the animal's experience during the research[48]. For the sciatic nerve ligation example, an attempt should be made to describe the severity and duration of pain likely to be experienced by the animals, as well as any consequences for locomotion, ability to perform strongly motivated behaviours, etc. To do this properly is not easy and requires an understanding of the animal's biology and evolutionary adaptations. Breed/strain and previous experience can all impact on how an animal experiences a particular harm, so to do a good job of providing an assessment of the quality and extent of likely harms requires some considerable expertise mixed with a dash of common sense and empathy.

5.3.2.2 Contingent harms

Direct harms are not the only type that needs to be taken into account. Contingent harms are those that occur as an unintended consequence of using animals (e.g. harms associated with transporting and keeping animals), that are not necessary for

---

[47]   Russell and Burch (1959) made a distinction between direct and contingent harms. See also next section on contingent harms and Chapter 6.
[48]   Animal Procedures Committee (2003).

the success of the research and which, in principle, could be avoided or ameliorated. It is easy to overlook or ignore contingent harms, partly because they are not caused deliberately and perhaps also because some harms, such as those due to husbandry, are not specific to laboratory animals and may thus be seen as just being the way things are. Nonetheless, harms arising from acquiring, transporting and keeping animals in captivity can be substantial and, if the experimental procedures are mild or non-existent, may be worse than those resulting from the research. For this reason, even studies that are not expected to result in direct harms should be subject to some form of ethical review to ensure that the benefits of the research justify the contingent harms. Some of the more common causes of contingent harm are listed below (measures to ameliorate them are considered in the next chapter).

Disease

Infectious disease has a great potential impact on colonies and research. Disease can result in suffering and, in laboratory animals, may require euthanasia of individuals or, if there has been a health breakdown, sometimes of the entire colony. Good modern housing and husbandry systems are designed to prevent infections entering or spreading within colonies, although health breakdowns can happen and some disease organisms are extremely hard to eradicate. Transfer of animals between institutions, either when supplied from a breeding colony or when institutions collaborate on a research project, always carries a risk of disease transfer, and the increased use of genetically modified or genetically altered animals and associated large number of movements of animals between facilities has increased this risk.

Pain

In some cases pain is the subject of the research, in which case it may be a direct harm. More often, where it occurs, it is an unwanted side effect. Pain following surgery is perhaps the most obvious unintended cause of pain for animals used in research, but contingent pain may also be caused by the use of common techniques such as tattooing, clipping and tagging, dosing and sampling, and some forms of restraint.

The increased used of anaesthesia and various analgesia regimens for procedures have undoubtedly been important in reducing contingent suffering due to painful procedures, but are all animals that should be receiving pain relief actually getting it? The answer, at least in the past, seems to have been no. A survey on the use of analgesics for laboratory rodents found that in the years 2000–2001 postoperative pain relief was provided in only 10% of surveyed papers, although this increased to 20% in the years 2005–2006[49]. Even when analgesia is provided, it may not always be effective as our ability to assess pain and knowledge of appropriate drug regimens to control pain are still limited for many commonly used species. Signs of pain are not always obvious[50], may be misinterpreted, vary between species and

---

[49]   Richardson and Flecknell (2005).
[50]   Stokes *et al.* (2009).

their detection requires specialist staff training (see Chapter 6). Moreover, the effectiveness, time-course and toxicity of analgesic and anaesthetic compounds varies between different species and with route of administration, so analgesics may be given without a clear understanding of their effectiveness or appropriate dosages. Some may argue that pain relief either would or could invalidate their studies. There will be some cases where this is true, but in others the basis for concern may be more speculative and should be challenged to establish the strength of the case or whether alternative approaches might permit pain relief. Those who argue that their research would be affected should also consider the likely effects that pain-related stress might have on their research (see Chapter 7).

Transport

Very often animals used in research are not bred at the facility where the research is carried out. They may be purchased from a specialist breeder, or sometimes the researcher requires a strain newly developed at another research institute. Breeding animals are also sometimes transported. There are several publications that review transport as a potential source of contingent suffering, some of which also give guidance on best practice to reduce the risk of stress and suffering[51]. Transport stressors may include disruption of social groups, unfamiliar housing and stimuli, changes in husbandry and handlers, changes in temperature and humidity, exposure to noise, vibration and movement, all of which may cause physiological changes and adverse mental states such as fear and frustration.

There has been much research on transport as a stressor for farm animals, which has shown large differences between species in their responses to travel[52]. Unfortunately, there is rather little information for non-farm species used in research. Pharmaceutical companies have, however, collected data showing that rodents can be exposed to large changes in temperature during journeys, and plasma corticosterone levels remain high for at least 16 days after transport while blood pressure and heart rate are decreased. Many Old-World primates used in research are bred in their indigenous countries, and so may experience long travel times of up to 58 hours. Recently transported animals can show disrupted behaviour for months, suggesting that there are animal welfare problems which not only impact the animals but may also affect research carried out on them as the adaptation times given to animals between transport and use are usually a matter of weeks rather than months.[53]

Housing and husbandry

Housing and husbandry that does not meet the animals' needs can also result in contingent harms, including adverse physiological and psychological states, abnormal behaviour and injury. Although many research and breeding establishments do

[51] For example, Swallow et al. (2005); White et al. (2010).
[52] For example, Knowles (1999); Friend (2000); Broom (2003, 2008); Stewart et al. (2003).
[53] Honess et al. (2004); Swallow et al. (2005); Syversen et al. (2008); Arts et al. (2011).

now provide enriched housing (objects, structures, materials or social environments that are beneficial for welfare, see Chapter 6), experimental requirements some-times preclude the provision of an adequately enriched environment. There are also national and institutional differences in enrichment take-up, so that some animals used in research are still unnecessarily kept in barren un-enriched housing. Even when efforts are made to enrich the environment, not all options are well validated and the enrichment may not meet all the animal's needs. Individual animals vary, and so do their needs, so even well-designed enriched housing may not meet the needs of all the animals at all times and at all life stages.

Husbandry procedures such as cleaning and rehousing can disrupt the animals' environment by destroying nests or scent marks. Stress can also result from splitting or joining animal groups with the resultant disruption of social bonds, relationships and networks. Finally, farm animals used in research may be subject to standard husbandry procedures such as castration, tail docking, beak trimming, and early weaning which, despite their routine nature, undoubtedly cause harm to the animals.

Capture and restraint

Capture and restraint can be very stressful for animals; indeed, restraint is some-times used as a stressor in research to study the effects of stress. Yet animals are routinely captured and restrained for health checks, weighing and experimental procedures. Domestication tends to reduce arousal and therefore the stress of han-dling, but even laboratory strains of mice that have been subject to many genera-tions of domestication still show stress responses when captured and held using the tail[54]. In wild animals, capture is likely to be even more stressful. One should not conclude, however, that all handling results in stress. Dogs, for example, which have had previous good experiences of socialisation with humans may benefit from handling, but even in this species some individuals remain nervous.

Marking and identification

Animals used in research need to be identified, particularly when housed in social groups. A great variety of identification techniques have been developed, ranging from visual identification, marking with dyes or marker pens, genetic markers in faeces, coloured plastic beads in faeces, hair clipping, ear tags, injection of or tagging with microchips (PIT tags), tattooing, ear-notching and toe clipping[55]. All have their advan-tages and disadva ntages, some last longer than others, and some are much more likely to cause suffering than others. Visual identification using the natural marking of ani-mals is likely to be harmless, but many other methods of identification are not. Marking an animal that will be released back into the wild can affect its survivability after release. For example, flipper bands used to mark penguins have been known to cause the birds up to 28% increased mortality and to increase energy expenditure

---

[54]   Hurst and West (2010).
[55]   Dahlborn *et al.* (2013).

by 24% during swimming and 11% overall[56]. Tattoos, microchips, ear, tail or toe clipping, or attachment of tags are all invasive to a greater or lesser extent, and are likely to cause some pain, techniques that involve amputation of a digit or part of the tail particularly so, as this involves cutting through cartilage, bone and nerve fibres[57]. Toe clipping may affect mobility and agility, and wild animals, even if marked using non-invasive techniques, may suffer other harm such as an increased risk of predation. Collars and tags can become snagged, trapping the animal, and may also result in direct injuries, such as sores, or obstruction of vital functions. Non-permanent techniques such as marking with dye can be less invasive[58], but in longer studies the animals may need to be recaptured and remarked, which can also be a stressor[59].

Euthanasia

Most animals that are no longer needed in research are euthanased. Rehoming is sometimes possible for animals when the procedures have been sufficiently mild, but usually animals have to be killed as there is a need to study their tissues or carry out post-mortem examinations. Euthanasia (from the Greek *euthanatos*, good death) is aimed at ensuring that further suffering or distress is minimised but it may be hard to avoid all stress and psychological distress. Animals that are to be euthanased need to be captured and restrained, and are usually transported to a separate room, all of which can cause stress. More worrying is the fact that while the euthanasia techniques are often the most humane options available, some do not result in pain-free loss of consciousness, and some have worryingly high failure rates in achieving a rapid death. Rodents are often killed with CO_2 but, at the concentrations needed to kill, the gas is strongly aversive for rats and likely to result in pain for at least 10–15 seconds, and similar concerns have been raised about chemical methods of killing fish and other aquatics[60]. Cervical dislocation, another commonly used technique for rodents, was found in one study to fail to result in immediate death for 21% of the animals, although whether these animals remained conscious for a period was not clear[61].

Harms caused to animals that are not the subjects of experiments

Animals not directly involved in the research may also be harmed. Depending on the methods used, the production of genetically modified animals can involve a range of techniques, including superovulation of young donor females to obtain

[56] Culik *et al.* (1993). However, the incidence of problems depends on factors such as the band and penguin maturity at the time of banding.
[57] Tail tipping and ear notching both lead to a transient (1 hour) elevation in body temperature, motor activity and heart rate, and tail tipping can result in hypersensitivity to pain (Zhuo, 1998; Cinelli *et al.*, 2007).
[58] Even procedures as apparently benign as marking with a black felt tip should not be thought of as being neutral. Burn *et al.* (2008) found that marked rats showed increased chromodacryorrhoea responses to handling. Paradoxically, other behaviours indicated a reduced stress response.
[59] Wells and Mazlan, Survey of the welfare impact of different identification methods, in Hawkins *et al.* (2013).
[60] Hawkins *et al.* (2006); Animal Procedures Committee (2009b). Killing rats by gradually increasing CO2 concentration is probably more humane as it induces unconsciousness at a concentration below that which causes pain. CO2 at 10% induces fear behaviour in mice (Ziemann et al. 2009).
[61] Carbone *et al.* (2012).

eggs, vasectomy of males for using to induce pseudopregnancy in females, surgical embryo transfer and genotyping. All of these result in some harm to the animals involved[62]. Breeding animals, as well as animals that are bred and not used, may suffer harms relating to supply, housing and husbandry. In some cases, breeding stock may be captured from the wild and these wild animals will suffer any stresses resulting from their trapping, and may be less able to cope with captive conditions than subsequent generations. Early weaning can be another issue. Weaning is a term used by those who manage animal colonies to describe the process of removing young from their mothers or carers, but the time at which this is done is often considerably earlier than would be normal in the wild. Moreover, even if weaning is done at the correct age, it is very sudden, whereas natural weaning usually occurs over a period. Concerns about these issues have been raised particularly with respect to non-human primates[63], but the issues are also relevant to other species.

It is not just the parents of the subject animals that may be affected. Capturing animals can disturb and disrupt wild groups, and the subsequent turmoil can, depending on the species, lead to aggression and death. If animals are trapped while breeding, then dependent animals may die. During trapping, the wrong individuals or animals of the wrong species can be caught. These harms probably occur only infrequently but can be substantial for the animals involved.

5.3.3 Categorisation and scoring of harms and benefits

In this section (5.3) we have considered some of the factors that need to be taken into account when judging the ethical acceptability of research that involves the use of animals. We have seen that researchers have to demonstrate need, show that the research will be done to a high standard, and have identified and show how they will minimise any harms to the animals. This is a lot of information for those tasked with making the judgement on the balance of harms and benefits, so to help with the process various guidance and scoring schemes have been developed. Fenwick[64] describes the approaches that various individuals and countries have used in developing severity classification systems. These classification systems are not just used for ethical review, monitoring and setting limits to the degree of suffering permitted in a project, but may also be used for purposes such as allocation of inspection resources, education of those who carry out the research about the principles of the 3Rs, and for public accountability.

5.3.3.1 Grading of harms

Under EU legislation, research procedures have been categorised using a system based on a prediction of the worst potential outcome for any animal involved, but which also takes into account any measures to alleviate harms (Table 5.2). The basis of this ranking system is a three-grade ranking of severity (mild, moderate and

---

[62]   Joint Working Group on Refinement (2003c); Lane and Jennings (2004).

[63]   For example, Maestripieri *et al.* (2008); Joint Working Group on Refinement (2009); Prescott *et al.* (2012).

[64]   Fenwick *et al.* (2011).

Table 5.2 Descriptions of severity limits of protocols, as described in European Directive 2010/63/EU.

Severity category	Examples take from Annex VIII, section 3*
Non-recovery Procedures which are performed entirely under general anaesthesia from which the animal shall not recover consciousness shall be classified as 'non-recovery'	
Mild Procedures on animals as a result of which the animals are likely to experience short-term mild pain, suffering or distress, as well as procedures with no significant impairment of the well-being or general condition of the animals, shall be classified as 'mild'	Superficial procedures, e.g. ear and tail biopsies, non-surgical subcutaneous implantation of mini-pumps and transponders Induction of tumours, or spontaneous tumours, that cause no detectable clinical adverse effects (e.g. small, subcutaneous, non-invasive nodules) Studies involving short-term deprivation of social partners, short-term solitary caging of adult rats or mice of sociable strains
Moderate Procedures on animals as a result of which the animals are likely to experience short-term moderate pain, suffering or distress, or long-lasting mild pain, suffering or distress as well as procedures that are likely to cause moderate impairment of the well-being or general condition of the animals shall be classified as 'moderate'	Surgery under general anaesthesia and appropriate analgesia, associated with post-surgical pain, suffering or impairment of general condition Models of induction of tumours, or spontaneous tumours, that are expected to cause moderate pain or distress or moderate interference with normal behaviour Use of metabolic cages involving moderate restriction of movement over a prolonged period (up to 5 days) Withdrawal of food for 48 hours in adult rats
Severe Procedures on animals as a result of which the animals are likely to experience severe pain, suffering or distress, or long-lasting moderate pain, suffering or distress as well as procedures that are likely to cause severe impairment of the well-being or general condition of the animals shall be classified as 'severe'	Toxicity testing where death is the endpoint, or fatalities are to be expected and severe pathophysiological states are induced Vaccine potency testing characterised by persistent impairment of the animal's condition, progressive disease leading to death, associated with long-lasting moderate pain, distress or suffering Use of metabolic cages involving severe restriction of movement over a prolonged period Immobilisation stress to induce gastric ulcers or cardiac failure in rats

(Continued)

Table 5.2 (Cont'd)

Severity category	Examples take from Annex VIII, section 3*
Research may also fall below the lower limit needed for licensing, above the upper limit permitted, be carried out under terminal anaesthesia, and may, subject to safeguards, may be permitted above the upper limit in exceptional cases (see text)	

*More detail and further examples are provided in the source document.
Source: European Directive 2010/63/EU. © European Union, http://eur-lex.europa.eu/.

severe), but research can also be below the threshold of suffering for which permission is required, above the maximum permitted, or be carried out under terminal anaesthesia (non-recovery). In addition, the Directive allows member states to provisionally permit research that would be above the normal upper limit in cases where there are exceptional and scientifically justifiable reasons, and subject to certain safeguards that include informing the European Commission, which can then either authorise or revoke permission for the research. Another approach to severity scoring, and one that has been examined by a group working on retrospective assessment of suffering, is to separately score intensity and duration[65]. Although initially this might seem a sensible approach, the group concluded that the process was too burdensome in practice and that the multiplicity of codes required meant that small differences in judgements about duration of adverse effects would lead to too much variation in assigning the codes. Some have argued that categorising harms into mild, moderate and severe is too coarse a grading system. On the other hand, the greater the number of grades, the greater the likelihood of inconsistencies between individuals in assigning grades, and perhaps for this reason a three-point system or similar seems to be one that has found favour in several legal jurisdictions[66].

Deciding either prospectively or retrospectively what an animal is likely to experience, or has experienced, is not always easy. Various bodies have therefore produced guidance in an attempt to achieve consensus, on a-priori assessment of the severity of the procedures, or on judgements of the severity of the clinical signs caused by the procedures. In 1994, FELASA produced a table classifying clinical signs such as behaviour, appearance and weight loss into mild, moderate or

[65] LASA/APC (2008).
[66] New Zealand uses a 5-point scale (Mellor, 2004; Bayvel *et al.*, 2007) based on a non-numerical, ranked system of grades for each domain of welfare impact. If one takes the most extreme as being broadly equivalent to a UK severity level that would not normally be authorised, and lowest as being below that where regulation is required, then these grades are roughly consistent with those used in the UK and Europe. The Swiss use a system of grades 0–3 for retrospective reporting, where grade 0 is no effect and grades 1–3 are again broadly equivalent to mild, moderate and severe (LASA/APC, 2008). Australia, however, uses nine categories, see Fenwick *et al.* (2011).

substantial categories while guidance on the UK's 1986 Act in 2000 provided examples of the sorts of techniques that would fall into each category[67].

The introduction of the 2010 European Directive has stimulated re-examination of these issues, and Annex VIII of the Directive now provides a rather longer list of procedures and the categories into which they fall than was provided in previous UK guidance. The Annex also lists factors and criteria that need to be taken into account in assessing severity, including the type of manipulation and handling procedures; the nature and intensity of any pain, suffering, distress or lasting harm caused by (all elements of) the procedure; the duration, frequency and multiplicity of techniques employed; cumulative suffering within a procedure; prevention from expressing natural behaviour including restrictions on the housing; and the husbandry and care standards that the animal will experience under the procedure. The new European requirement for retrospective reporting of severity will also prove a useful check on the severity assessments made at the outset of studies. Addressing the EU requirements for prospective and retrospective severity assessment is a novel experience for many within the European Community and, to assist, an expert group has developed useful guidance and examples concentrating on the experience of the animal rather than what is being done to it. The guidance includes sections on indicators of severity, consistency, and also covers who should take part in the assessment and examples of tools to use in the assessment[68].

5.3.3.2 Cumulative suffering

A novel feature of the 2010 European Directive is the requirement that the assignment of severity of a research procedure should take into account the cumulative suffering experienced by the animals subjected to that procedure. This assessment, while undoubtedly reflecting public expectations, raises some interesting challenges, including how one should deal with multiple elements within a procedure. In a recent survey, animal care staff, welfare experts and scientists were provided with a list of elements that might make up a procedure and asked to provide a severity assessment. They were then asked how their view changed if a further stressful protocol was added. Unsurprisingly, the results showed that as elements were added there was a strong tendency to increase the assessed overall severity[69]. On the face of it this is reasonable as, if we add protocols, each of which causes some degree of suffering, the animals will have experienced more overall suffering.

The European Directive limits the assessment to the harms in the procedure, unless there is a plan to re-use, in which case the lifetime experience of the animal has to be taken into account. However, I think that harm–benefit judgements for all projects

[67]   FELASA Working Group on Pain and Distress (1994). A consensus working document on severity assessment for European Directive 2010/63/EU and one giving illustrated examples can be found at http://ec.europa.eu/environment/chemicals/lab_animals/interpretation_en.htm, accessed February 2013. UK ASPA 1986 Guidance (prior to January 2013) is archived at http://preview.tinyurl.com/c95fw5h, accessed 13 May 2013.
[68]   European Commission (2012). Examples of severity classification can be found in European Commission (2013).
[69]   Wolfensohn and Anderson (2012).

should take account of all the harms that an animal may experience, and that these should include stresses due to capture, transport, husbandry as well as the procedure itself, not least so that these harms can be thought about and reduced if possible. On the other hand, adding the harms in the way that the respondents were guided to do in the study above may not be the best way forward. It is hard to see how some minor stressors, such as repeated handling, could ever add up to a moderate or substantial level of harm. My personal experience is that I undergo a series of minor stressors each day but do not feel that they accumulate in any meaningful way. In other cases experiencing two stressors may well be worse than experiencing one. The experience could even be worse if a previous stressor accentuated the effects of a later one, which might happen if the animal learns to expect an unpleasant experience, or if the previous stressors damaged the animal's ability to cope in some way[70]. So, rather than being a mathematical process, the judgement needs to be based on a very clear understanding of all the adverse events that the animal will experience. To help those making these judgements, it can be very helpful if the researcher produces a flow diagram illustrating when and what will happen to the animals. As with any assessment, people may well give different weights to the various harms and their cumulative impact. Including more harms provides more opportunities for disagreement, so the development of consensus examples of cumulative severity rating will also be useful.

Finally, while assessing cumulative suffering is challenging, we should not be too frightened by it. It is not, after all, so different from the current process of welfare assessment, at least when that is done properly. Those charged with reviewing projects have always had to make assessments based on what animals will experience over a period of time, the only difference being that, in the past, some contingent harms, such as transport, were not explicitly included in the regulatory requirements for review.

5.3.3.3 Maximum suffering
In the UK, regulated procedures are not permitted if they may cause severe pain, suffering or distress that is long-lasting and cannot be alleviated, and European Directive 2010/63/EU has a similar provision, albeit with an escape clause[71].

---

[70] Parker and Maestripieri (2011) describe, for example, the physiological consequences of early abuse in non-human primates. Moreover, abused infants often make poor parents. Wong *et al.* (2013) found that rats that had previously experienced a gaseous anaesthetic found subsequent exposure to the gas more aversive.

[71] European Directive 2010/63/EU

> *Article 15(2). Subject to the use of the safeguard clause in Article 55(3), Member States shall ensure that a procedure is not performed if it involves severe pain, suffering or distress that is likely to be long-lasting and cannot be ameliorated.*
>
> *Article 55(3). Where, for exceptional and scientifically justifiable reasons, a Member State deems it necessary to allow the use of a procedure involving severe pain, suffering or distress that is likely to be long-lasting and cannot be ameliorated, as referred to in Article 15(2), it may adopt a provisional measure to allow such procedure....*

Provisions such as these are of obvious importance in limiting research to publically acceptable limits, but are also important as they require humane endpoints to be described so that animals are euthanased before they reach such extremes of suffering.

It seems reasonable that there must be an upper limit on the suffering that can be inflicted on animals, but there are practical difficulties in defining and reaching consensus on this upper limit. To implement the limit it is necessary to assess the animal's state of welfare, which, as we have seen, is to some extent a matter of judgement. It is also necessary to agree on what is meant by severe pain, suffering or distress, and how long is 'long-lasting' (hours, days, weeks or months). At the extremes it is relatively easy. I think that most people would agree that abdominal surgery without anaesthesia or postoperative analgesia is likely to result in severe pain, and that most people would reject a project that resulted in months of such pain. However, how important is duration? If the pain is truly severe, perhaps seconds, minutes or hours is too long. Again, the only way in which consensus is likely to be achieved is by judging cases against examples that have been previously established for the purpose of illustrating concepts like 'severe' and 'long-lasting'. The difficulty of the task should not be underestimated. Suffering can occur in many forms, and it is not easy to equate nausea or fear with pain, for example. It will therefore be necessary to produce examples of maximum suffering that cover a wide range of different types and harm.

5.3.3.4 Numbers of animals

When weighing up harms to animals, it seems logical to factor in the numbers of animals that will experience those harms. Most of us would consider that increasing the number of animals that experience suffering in a project makes things in some way worse and therefore requires more justification. Not everyone agrees with this position. Richard Ryder[72], for example, has argued that the pains of several individuals cannot be summed, as pain is the experience of the individual rather than the group. He makes the point that the utilitarian idea of aggregating pains and pleasures might lead to the conclusion that gang rape might be justified if the pleasures to the rapists outweighed the sufferings of the victim. Others[73] have pointed out that excluding numbers from ethical decisions leads to the dubious conclusion that, given a choice, there is no moral difference between saving one or a hundred individuals and that the view that numbers matter is consistent with how we generally consider harms that occur to humans. Serial murder is, for example, viewed as a more heinous crime than single murder and while an individual's accidental death might be considered a tragedy, mass famine and earthquakes are considered disasters.

So numbers are important, but in practice, rightly or wrongly, numbers are often considered in conjunction with other issues. In Chapter 4, we discussed

---

[72] Ryder (2009).
[73] Leuven and Višak (2013).

species bias and how some people consider it right to use a larger number of mice rather than a smaller number of primates in a study, even when the mice apparently suffered more than the primates. Such a decision might be justified if the primate experience of suffering were truly more than that of a mouse but for many procedures this presumption is uncertain. It is not just mice whose ethical interests can be downplayed. In some toxicological studies, fish are used in much greater numbers than would be expected, or perhaps even permitted, for other vertebrate species. One study, on the effects of a potential carcinogen, used an incredible 40 800 trout[74].

Numbers sometimes also have to be balanced with the suffering experienced by individual animals. One study[75] surveyed prospective researchers taking a training course as to whether it was better to cause less harm to many animals or cause more harm to fewer animals. When participants were asked whether it was more acceptable to carry out one trial that would cause considerable pain and stress but with no permanent effects on each of 20 mice or 20 of the same trials on a single mouse, 58% felt that it was better to cause less suffering to 20 mice rather than to cause a single mouse to suffer more. One year after the course, this proportion had increased to 64%. It would seem that these researchers increasingly tended towards the view that the agony of one counts for more than the discomfort of several others.

5.4 Decision Tools

We have seen how difficult it can be to reach decisions that weigh up all aspects of the likely benefits and likely harms. It is not surprising therefore that some have tried to develop more detailed formalised scoring systems to use as tools to aid and rationalise decisions[76]. Some of these have attempted to use arithmetical systems to guide or even proscribe decision-making. Porter, for example, developed a relatively simple scoring system that used scores of the aim of the experiment, the likelihood of success, the species of animal, the intensity and duration of pain, the duration of the experiment, number of animals and the quality of animal care. These scores were added, and if the total score were less than an arbitrary cut-off point, then the experiment was considered acceptable. Similarly, Stafleu and his co-workers produced an arithmetical system that rated the benefits of the research with respect to economic interest, knowledge and health. The benefits score was then multiplied by an assessment of the relevance of the research (based on the availability of a replacement to animal

---

[74]   Bailey *et al.* (2009).
[75]   Franco *et al.* (2010).
[76]   For example, Smith and Boyd (1991), pp. 138–182; Porter (1992); Stafleu *et al.* (1999); Mellor (2004). See also Dolan (1999), pp. 218–243 for a summary of some of these approaches, and Honess and Wolfensohn (2010).

use, the necessity, probability of success and quality of the science and research group) to produce an overall assessment of the value of the experiment. Weighed against this score were the harms to the animals based on the intensity and duration of harm, the psychological complexity of the animals and the numbers of animals involved. If the scores for human interest were greater than the harms to the animals, then the experiment was considered justifiable. A similar system, using visual analogue scales to score assessments (placement of a cross on a line where the left-hand side is low and the right-hand side is high) is described by Smith and Boyd. A FELASA working group has produced a useful summary of the essential issues that should be taken into account (Box. 5.1).

Mellor and Reid's scheme, which is used in New Zealand, took a different approach. In this, only the harms are scored, and these are categorised by their effect on the animal in five domains of potential animal welfare compromise:

1. water deprivation, food deprivation, malnutrition – *nutrition*;
2. environmental challenge – *environment*;
3. disease, injury, functional impairment – *health*;
4. behavioural or interactive restriction – *behaviour*;
5. anxiety, fear, pain, distress, thirst, hunger, boredom – *mental state/experience*.

In each domain the severity of animal welfare compromise is graded, with the help of examples, on a scale from no effect (0) through increasing intensity (A, B, C and X where the predicted endpoint is death). Taking the environment domain as an example, experiments that elicit responses beyond the physiological adaptive capacity of the animals, but where the effects are readily reversed by restoration of benign conditions, are classified as B. Although harms are scored in all five domains, priority is given to the mental state/experience of the animal, as the grade given to this domain counts as the overall grade for the study, except when it is 0 in which case the highest grade in any of the other domains is used. More recently, Honess and Wolfensohn produced a scheme that only looks at the harms but which included a temporal component, so allowing judgements that would be useful in assessment of cumulative suffering.

With the exception of Mellor and Reid's scheme, none of these aids to decision-making have yet become widely accepted, perhaps because, in the end, the scores and decisions are subjective. Human and animal benefits need to be weighed against intensity and duration of harm, and the numbers of animals harmed but, as argued earlier, none of these factors can be related in any strict mathematical sense. Nonetheless, these aids have been useful in promoting debate and in some cases in achieving some consensus about the valuation of harms and benefits. Moreover, even if one rejects the arithmetical approach to decision-making, formalised systems and explicit lists can be useful aide memoires of the issues involved.

Box 5.1   Outline scheme for the assessment of benefits and harms in scientific projects involving animals

Assessment of potential benefits of the project

How will the results add to existing scientific and/or clinical knowledge and how might they be used? What practical applications, if any, are envisaged at this stage? And what is the potential value of these insights and/or applications?

- Are the objectives of the project:
  - o original, in relation to previous or ongoing studies
  - o timely, in relation to other studies that might be done (what is the need to do this study now?)
  - o realistic, in that they are achievable with the time and other resources available?
- If there is an element of replication of previous work, how strong is the case for this, and what efforts have been made to avoid mere duplication?
- If this is ongoing work, how does the present proposal relate to what has gone before? What progress was made in previous studies, and what scientific or other benefits have resulted?
- What is the relevance of this project to other studies in this field of research and what might be the implications for other areas of research, if any?

Assessment of likelihood that the potential benefits will be achieved in practice

Is there a reasonable expectation that the potential benefits will be achieved in practice, given the:

- choice of animal model and scientific approach
- validity of experimental design (e.g. use of appropriate number of animals, appropriate use of controls) and whether and how this has been informed by statistical or other advice
- competence of researchers and other staff, including their training, supervision, experience and expertise
- appropriateness and quality of facilities
- researchers' plans for communicating and using and/or building on the findings of the project?

Assessment of the harms caused to animals and possibilities for reducing these, in terms of

- the need to use animals at all (what efforts have been made to seek suitable alternatives to the use of animals in regulated procedures? Has as much information as possible already been gained from *in vitro* or other *ex vivo* work?)
- optimisation of the numbers of animals that will be involved (neither too many nor too few to achieve a meaningful scientific result) and quality of experimental design – again, what advice has been sought?

- the severity of the potential harms in the proposed studies, considering all potential adverse effects, psychological as well as physical, and their duration, in relation to:
  - the species and strain of animal used
  - the effects of the procedures themselves
  - wider factors, such as the source of the animals (including, where relevant, their breeding conditions) and, where relevant, the conditions of transport to the laboratory; and arrangements for their husbandry and care, including provision of environmental enrichment
  - the fate of the animals at the end of the experiments – will they be used in another procedure, killed (by what method?) or re-homed or released?
  - how all of these factors will be influenced by the competence of researchers and other staff, and the quality of the facilities involved
- possibilities for refining the impact of the study on the animals so as to cause less harm to the animals while achieving a valid scientific outcome, for example by
  - using a different species or strain
  - obtaining animals from a different source
  - adapting or enriching animal housing and care
  - modifying the techniques involved
  - enhancing the monitoring of the animals and implementing humane endpoints
  - better use of anaesthesia and analgesia and/or provision of other special care.

Source: Smith *et al.* (2007) with permission from SAGE Publications Ltd.

5.5 The Requirement for Ethical Review

We have seen that harm–benefit judgement, despite its complexities, difficulties and some resistance, is becoming increasingly accepted as an essential component of the research process. Indeed, I suggest that a requirement to weigh up the harms and benefits should be included as a standard component in any nation's regulatory controls on animal experiments. The experience of those who have already adopted systems for ethical review shows that it helps to ensure higher standards of animal welfare, shares the responsibility of deciding what is acceptable, and provides some reassurance to the public about the conduct of the research. Of course, there will always be disagreements and debate about what is acceptable and views about ethical acceptability are not constant. Just because certain research activities have previously been considered acceptable, they may not be in the future. In some cases technology will have moved on, allowing better techniques to be used, but views as to what is acceptable also evolve, generally in the direction of being kinder to animals. There will also be regional differences in attitudes to animals and to research, with the inevitable result that some things will be permitted under some jurisdictions that would not be allowed under others. This should be seen as an opportunity to educate and persuade so as to achieve higher welfare standards worldwide.

Finally, I would like to emphasise the point already made in this chapter that once research is approved, there should be a continuous process of review throughout its period of authorisation, in which opportunities are sought not only to improve the quality of the research and maximise the benefits, but also to find alternatives to the use of sentient animals, reduce individual suffering and reduce the numbers used, in other words to fully implement the 3Rs, which are the subject of the next chapter. This continuous ethical review should be the responsibility of those involved in the research, the regulators of the research and those who care for the animals.

Improving the Welfare of Animals Used in Research: The 3Rs

In this chapter I describe how the 3Rs originated, what they are and, using examples, describe how they can be implemented so as to minimise the harms caused to animals used in research. Unfortunately, addressing one of the 3Rs can sometimes conflict with others, so implementing them is not always straightforward. The chapter concludes with sections on sources of information for training and other resources.

6.1 Introduction: The Origins of the 3Rs

Having considered in the last chapter how research can cause harm to animals, we now turn to ways of avoiding or ameliorating these harms, and this involves applying the 3Rs (replacement, reduction and refinement). The originators of the 3Rs, William Russell and Rex Burch, defined the 3Rs as follows:

> **Replacement:** Means the substitution for conscious living higher animals of insentient material.
>
> **Reduction:** Means reduction in the number of animals used to obtain information of given amount and precision.
>
> **Refinement:** Means any decrease in the severity of inhumane procedures applied to those animals which still have to be used.

Broadly speaking these definitions hold today, although some authors have adapted the wording either to clarify and extend the principles or to provide definitions that are relevant to a particular legal jurisdiction. The National Centre for the Replacement, Refinement and Reduction of Animals in Research (NC3Rs), for

The Welfare of Animals Used in Research: Practice and Ethics, First Edition. Robert C. Hubrecht.
© 2014 Universities Federation for Animal Welfare. Published 2014 by John Wiley & Sons, Ltd.

example, uses a definition of replacement that takes into account the UK definition of a protected animal, and refinement has been redefined to make it explicit that it includes housing and husbandry (see section 6.4).

The 3Rs have been astonishingly successful in improving the welfare of animals used in research. They have changed the attitudes of researchers and legislators and are now broadly accepted as the fundamental ethical framework within which animal research should be conducted. In this chapter, I shall outline various ways in which the 3Rs can be implemented, and point to where more detailed advice and information can be found by those carrying out research on animals or involved in ethical review, but the formulation of the 3Rs, their years in the wilderness, and subsequent rapid uptake are, I think, an interesting story. If you wish to skip the history you can move directly to section 6.2.

The 3Rs came about following a decision in the early 1940s by the Universities Federation for Animal Welfare (UFAW) to concentrate efforts on improving laboratory animal welfare[1], but the 3Rs did not emerge immediately. UFAW began by funding studies on the effects of husbandry on the behaviour of animals used in research and produced the first ever handbook on laboratory animal care[2] with the aim of improving animal welfare and reducing animal use by finding ways of controlling and reducing experimental variation. These activities were groundbreaking in their scientific approach to animal welfare; nonetheless, it became apparent that a fundamental analysis of animal use was necessary to identify where problems lay and their solutions. Consequently, on the advice of the founder of UFAW, Major Charles Hume, and the renowned zoologist Peter Medawar, Dr William Russell and Rex Burch were employed to produce a systematic review of laboratory techniques and their ethical implications (Figures 6.1 and 6.2).

William Russell was a remarkable man. He was a polymath, described by Major Hume, in a 1959 lecture in Washington to the Animal Care Panel (later to become the American Association of Laboratory Animal Science), as 'a brilliant young zoologist who happens also to be a psychologist and a classical scholar'. He was also exceptionally energetic. According to his executor Cleo Paskal, Russell, while writing up his DPhil thesis, an activity that would normally occupy most of a student's time, simultaneously wrote a libretto for an opera, a play and a science fiction short story. His assistant Rex Burch was a skilled microbiologist, who collected much of the information necessary for the project. His major task, a difficult one that required an ability to inspire trust, was to interview scientists about their work using animals and to explore ways in which it could be made more humane.

In 1959, after extensive research and consultation with the UK scientific community, Russell and Burch published *The Principles of Humane Experimental Technique*[3], in which they set out why we should be concerned about the use of

---

[1] Kirk (2009) provides some early history on the development of the 3Rs. See also Russell and Burch (1959); Russell (2005); Balls (2007, 2009); Paskal (2009).

[2] The handbook (Hubrecht and Kirkwood, 2010) is now in its eighth edition.

[3] Russell and Burch (1959).

Figure 6.1 Bill Russell as a young man. With permission from Cleo Paskal.

Figure 6.2 Rex Burch and Bill Russell during the Sheringham workshop (1995). With permission from Michael Balls.

animals in research[4], and how the humane use of animals in science could be promoted without prejudicing scientific or medical aims. To make their intentions clear, Russell and Burch emphasised that the words 'humane' and 'inhumane' did not imply any ethical or personal criticism of the persons carrying out the research, instead referring only to the effects on the animals involved.

The Principles is a book that was well ahead of its time. It deals with complex issues such as consciousness, which types of animals might be conscious, the sorts of inhumanity that might be caused to animals and anticipates many of the issues that have been the subjects of later welfare research and debate. In it, Russell and Burch argued that we should be concerned about feelings such as fear, conflict and need in addition to the more obvious harm of pain. They suggested that the welfare states of animals might be described along a spectrum extending from complete well-being to acute distress, and linked these affective states to neurological and endocrine systems and physiological stress states. They reviewed the type and purpose of research carried out at that time, and described how the harms produced could be divided up into direct and contingent harms (see Chapter 5). They then graded the inhumanity of various procedures, from the relatively small discomfort caused by an injection to the much greater harm resulting from disease models and identified what could be done to minimise these harms under the headings of replacement, reduction and refinement.

Throughout their book, Russell and Burch emphasised that advances in science can help to find alternative ways of doing things in a more humane way, and also that humane research usually results in better results, as chronically stressed animals are unlikely to produce good data. One might think that scientists would have found this a strong selling point, but in fact the publication of *The Principles* had rather little impact at the time[5]. Quite why this was so is unclear, but perhaps the ideas were simply too far ahead of their time, and society and the scientific establishment were unready to accept any need for change. Looking back on the issue, Rex Burch recounted how the influential Littlewood Committee, set up in 1963 to consider the need for changes in the UK legislation on animals used in research, was negative about the possibilities for finding adequate substitutes for animal tests, and that perhaps this contributed to the general failure to grasp what *The Principles* had to offer. Whatever the reason, although UFAW continued to fund research, hold symposia and publish guidance on the housing and husbandry of animals used in research, progress in general understanding and implementation of reduction, refinement and replacement was at first slow. Then, in the 1970s and 1980s more organisations started to promote aspects of the 3Rs (the term was first used in print in the 1970s), and as a result of this increased activity and the associated increased concern for animal welfare in general, the 3Rs at last began to develop momentum.[6]

---

[4]   The reason being because of the possibility, or in their view the likelihood, that some animals might be conscious (see Chapter 4).
[5]   Balls (2009).
[6]   Burch (1995); Burch and Balls (2009); Stephens (2009). FRAME, the Fund for the Replacement of Animals in Medical Experiments, has done much to promote the 3Rs and deserves particular credit for helping to bring Russell and Burch's work out of obscurity.

By the 1990s the 3Rs were well established, although rather confusingly they sometimes were, and still are, referred to as 'Alternatives', a term that, as Russell pointed out, wrongly suggests that the objective is only replacement. The term is also unfortunate in that it suggests an equivalence between the animal-based research and its replacement when, in fact, replacements can achieve their goal while providing quite different information[7]. In 1992, UFAW reprinted *The Principles*[8], and by 1999 the Third World Congress on Alternatives and Animal Use in the Life Sciences attracted over 700 delegates from 35 countries. It was at about this time that I realised just how widespread and influential the concept had become. In 2001 I gave a presentation at a South American conference that was promoting the 3Rs in toxicology. I naively expected that the 3Rs would be a novel idea to many of the participants, but to my surprise found that the 3Rs were as well understood there as in many countries closer to their origin. I remember asking a toxicologist why his company was interested in the 3Rs, to which he responded 'for two reasons: first because it is ethically right, and second because it is good science and good business – our clients expect it'. I think that Russell and Burch would have approved; after all, what two better reasons could one have for uptake of the 3Rs?

Today, the 3Rs are incorporated into European legislation, are considered by the US Office of Laboratory Animal Welfare to be integral to an IACUC's consideration of research, and are promoted by numerous national and international professional and advisory organisations, one of which, the NC3Rs[9] in the UK, is the first governmental body to include the concept in its name. We have come a long way from those early discussions in the UFAW offices. In the next sections I will describe each of the 3Rs in the order in which they should be considered when planning research, provide examples of how they can be implemented and of how challenges to their implementation can be addressed.

6.2 Replacement

For Russell and Burch, replacement meant finding alternatives to conscious higher living vertebrates. They were fully aware of the difficulties involved in drawing the line between vertebrates and other animals, remarking that restricting concerns to vertebrates might seem arbitrary when comparing say a lamprey with an octopus. However, they were aware that extending their argument to a wider range of animals would have been much more controversial. Their decision was therefore a

---

[7] Russell (2005); Richmond (2010). Balls (2009) traced a letter from Rex Burch suggesting the use of the word 'alternatives' in introductory letters to heads of department for his fact-finding mission, but Bill Russell deleted the word and subsequently wrote that he was pleased that the term was falling into disuse (Russell, 2005). It is interesting to note that Richmond points out that the concepts that underpin the 3Rs can be traced back to medical publications in the 1800s.

[8] *The Principles* is now available online at http://altweb.jhsph.edu/pubs/books/humane_exp/het-toc, accessed 22 April 2013.

[9] http://www.nc3rs.org.uk, accessed 9 May 2013.

pragmatic one, which allowed them to make a case that could be reasonably defended. One might have hoped that we would be in a better position today, but sadly, as we saw in Chapter 4, we have not made much progress in developing the necessary evidence base to support these decisions.

6.2.1 Absolute and relative replacement

Russell and Burch made a distinction between what they called absolute and relative replacement. Absolute replacements are those where an alternative has been developed so that animals are no longer required at all, whilst relative replacement means that animals may still be required but in different types of procedures that cause, or are thought to cause, no distress. Examples of absolute replacement include *in silico* and *in vitro* methods as well as *in vivo* studies on organisms not thought to be sentient, such as the use of bacteria in the Ames test for carcinogenicity, nematodes in ageing research or social amoebae for emesis research, epilepsy and bipolar disorder. Other options are to use immature animals such as fish larvae that are thought not to possess the neurophysiological equipment to be sentient; the use of human tissue cultures (such as human keratinocytes in studies of skin irritation); and the use of existing knowledge through, for example, data mining techniques[10]. Examples of relative replacement include research in which animals are used in non-recovery procedures under terminal anaesthesia[11] or the use of animal material in tissue cultures (e.g. rodent central nervous system slices for the study of neurodegenerative diseases). Replacement can also be partial, as in some toxicity tests when animal testing only becomes necessary if pre-screening using non-animal techniques delivers a negative result. A further type of replacement, albeit with a sentient alternative, is sometimes possible if human volunteers can be used rather than animals.

Much of the effort to find replacements has been in the fields of vaccine testing and toxicology. One example is the almost total replacement of the use of rabbits in pyrogenicity tests (in which rabbits are injected with the test substance to find out if it causes fever). These tests used to be required by regulatory authorities to ensure that products intended for administration directly into the body (such as vaccines) were free of fever-producing contaminants. However, for over 30 years, the US Food and Drug Administration (FDA) has accepted the *Limulus* amoebocyte lysate (LAL) test for bacterial endotoxins, which involves using horseshoe crab blood *in vitro* as a replacement for many types of rabbit tests for pyrogens[12]. Other examples of replacements include the use of *in vitro* serological methods to assess the potency of vaccines for diphtheria, tetanus and whooping cough instead of live animal potency tests that in some cases end in lethality[13]. Replacements have also been found for tests

---

[10] Balls *et al.* (2012) provides a review of options that can be used in drug development.
[11] Some would consider the use of terminal studies, as an alternative to studies that involve conscious animals, as refinement. Russell and Burch, however, categorised them as replacement, provided that no suffering occurs.
[12] FDA (2012) Guidance for Industry. Pyrogen and Endotoxins Testing: Questions and Answers, http://preview.tinyurl.com/amu63zq, accessed 10 May 2013.
[13] Draayer (2011); McFarland *et al.* (2011); Stickings *et al.* (2011).

of products for eye irritation and skin corrosion (see Chapter 7). There have also been notable successes in areas other than regulatory toxicological testing. One innovation that led to a large reduction in animal use, albeit in routine testing rather than research, has been the replacement of animals (mice, rabbits and later frogs[14]) for human pregnancy testing with chemical tests that directly measure the presence of human chorionic gonadotropin (hCG) in blood or urine.

The challenge of finding a replacement is affected by the complexity of the system being modelled. Computer models of the relationships between molecular structure and activity of candidate drugs (see Chapter 7) have been successfully used for some time to provide early indications of efficacy or toxicology without the use of animals, and the tests described above indicate how some tissue responses have been modelled. Replicating the essential features of an entire organ is much harder, but progress has been made. One pharmaceutical company has developed a computer-controlled mechanical and chemical model of the human upper digestive tract, resulting in the virtual elimination of their need to use dogs in drug formulation studies to ensure that drugs reach their target areas at appropriate concentrations[15]. An *in vitro* lung on a chip shows promise in modelling drug toxicity that produces pulmonary oedema[16], while *in vitro* systems modelling multiple organ systems such as liver, fat, lung and other tissues have been developed and can be used to study organ-specific cytotoxicity or even to model a human with a tumour[17]. Mathematical and computer modelling have also been useful in modelling complex systems such as neurological organs, and even behavioural and ecological systems.

Another approach to avoid using animals in research is to reconsider the *need* for the research[18]. Regulatory requirements can be changed to eliminate inessential tests. That some tests with little added value have been carried out is illustrated by the recent elimination by the US regulators of the 1-year chronic toxicity test in a non-rodent species, usually a dog, as it was found to provide no more significant extra information than a 90-day study[19]. While applauding the decision, one cannot help wondering for how long this unnecessary extra testing had been carried out and at what cost to the animals involved.

Likewise, international harmonising of regulations can reduce animal use, by avoiding the need for different tests to satisfy different regulators. Researchers may also be able to avoid animal use by reconsidering their scientific questions and the strategies they plan to use to answer them. The importance of carrying out thorough

---

[14]   These tests involved injecting the animals with the woman's urine or serum. The ovaries of mice or rabbits were then dissected out to examine for follicular changes. Male and female frogs produce gametes after injection with a pregnant woman's urine and could be re-used.

[15]   Model of a human digestive tract, http://preview.tinyurl.com/cc3rz2l, (AstraZeneca, Our 3Rs Commitment, 2009), accessed 10 May 2013.

[16]   Huh *et al.* (2012).

[17]   Balls *et al.* (2012).

[18]   Richmond (2010).

[19]   Hasiwa *et al.* (2011).

literature searches to determine whether questions can be answered using existing data and knowledge cannot be overemphasised.

6.2.2 Challenges to implementing replacement options

Despite the advances that have been made in finding replacements to the use of sentient animals in research, much remains to be done. A survey of Canadian regulators and scientists found that an important obstacle to progress was thought to be a lack of appropriate replacement options[20], and although there are various helpful databases (see section 6.7 on sources of information on the 3Rs), finding suitable replacement options can be difficult. It is relatively easy to implement replacement options for simple well-understood biological systems, but developing replacements for more complex systems is harder, although researchers are working on solutions. One example of an attempt to address a problem of this type comes from recent work aimed at finding an alternative to the use of animals in emetic research. Nausea and vomiting (emesis) is a particularly unpleasant side effect of some cancer treatments, and the avoidance of emesis can also be a problem in drug development. Emesis is, however, a complex process that involves multiple signal inputs to the brain and the coordination of a number of various motor outputs (Figure 6.3), which makes finding replacements challenging. Despite this complexity there are encouraging indications that lower organisms, such as the social amoeba *Dictyostelium discoideum*, might be used as a partial replacement for animal use in the early screening of novel drug candidates for emetic action[21]. The research carried out to date has shown that when the amoebae are tested with a range of substances that produce varying degrees of emesis in humans they show a strong behavioural response to a number of bitter-tasting compounds that cause taste aversion and emesis in humans and/or laboratory animals.

The scientific challenges of finding a suitable alternative are not the only barriers to the greater uptake of replacements. The regulatory authorities that control the testing and acceptance of drugs and biological products are often perceived as slow to accept changes to tests that refine or replace animal use. That it can often be hard to overcome the hurdles and inertia of the regulatory system is supported by a report highlighting the need for harmonisation and standardisation of the procedures for validating and achieving acceptance of alternatives to animals for potency testing of vaccines[22]. The cost of developing replacements is another problem. Where replacement options result in better science and are ultimately cheaper than the use of animals, there should be a strong incentive to develop and use them, but unfortunately the financial advantage, if any, is not always known at the outset. Another difficulty is that institutions and scientists may not see the development of replacements as a priority, or even as part of their job. The sourcing of human tissue to use as a replacement can also be a sensitive issue, as illustrated by the public

---

[20] Fenwick and Fraser (2005).
[21] Robery *et al.* (2011).
[22] Fenwick and Fraser (2005); Casey *et al.* (2011).

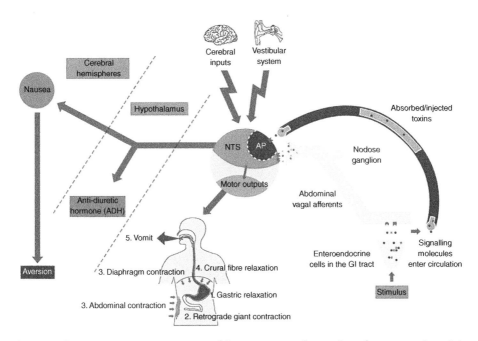

Figure 6.3 Vomiting is a consequence of the integration of a number of input signals and the coordination of a number of motor outputs within the brainstem. Inputs from the gastrointestinal tract via abdominal vagal afferents, the circulation via the area postrema (AP), the vestibular system (motion sickness) and the higher cerebral regions (fear-induced and anticipatory vomiting) are all integrated by the nucleus tractus solitarius (NTS) within the dorsal brainstem. The subsequent autonomic and somatic motor outputs arise from nuclei in both the dorsal (e.g. dorsal vagal motor nucleus) and ventral (e.g. nucleus ambiguus, pre-sympathetic, Botzinger complex) brainstem and occur in sequential order. (1) Gastric relaxation and (2) retrograde giant contraction via the vagus; (3 and 4) contraction of the anterior abdominal muscles via spinal motorneurons and the diaphragm via the phrenic nerve; and finally (5) opening of the mouth to allow oral expulsion. A major difference between retching and vomiting is the relaxation of the crural diaphragm during the latter to facilitate evacuation of gastric contents. Further outputs include increased antidiuretic hormone (ADH, vasopressin) release from the posterior pituitary and the induction of the sensation of nausea, presumably via the cerebral hemispheres. Reproduced from Holmes, A.M., Rudd, J.A., Tattersall, F.D., Aziz, Q. & Andrews, P.L.R. (2009) Opportunities for the replacement of animals in the study of nausea and vomiting. *British Journal of Pharmacology* 157, 865–880, with permission from John Wiley & Sons.

outrage in the 1990s following revelations that a pathologist at Alder Hey Hospital in the UK had removed and kept organs without patients' or parents' consent. Sensitivities about the sourcing of human tissue can lead to ethical concerns that act as a barrier to implementation.

Ideally, at some time in the future there will be no need to subject animals to research that causes them suffering or harm, but how do we measure progress towards this goal? The UK has for many years recorded statistics of animal use.

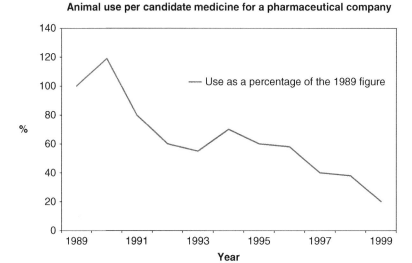

Figure 6.4 Numbers of animals used per new candidate medicine going into development at a pharmaceutical company expressed as a percentage of the 1989 figure. Redrawn from Samuels, G. (2000) Medicines: tried and tested in animals? In *Medicines: Tried And Tested – Or An Unknown Risk?* Medicines for Health Series, with permission from Pfizer.

Between 1945 and the mid 1970s the number of procedures on animals steadily increased, but then use decreased until the late 1990s. The reduction was largely due to the introduction of *in-vitro* techniques in high-throughput screening[23] combined with sophisticated statistical methods to access, mine and analyse data, both of which reduced the need for animal research in the early stages of drug development. This is illustrated by a publication in 2000, which showed that modern technologies in one pharmaceutical company had led to an 80% reduction in the numbers of animals used per candidate drug over 10 years (Figure 6.4)[24].

However, since the late 1990s there has been a steady rise in the numbers of procedures involving the use of animals in the UK, and one reason for this has been the development, and rapid uptake, of genetic modification (GM) techniques. These now account for 44% of the total, the vast majority of the procedures being on mice (92%). While some argue that the UK statistics have overemphasised the increase (see Chapter 1), there is no doubt that the increased use of GM techniques has substantially contributed to the increased use of animals. So, on the one hand, animal use can decrease as replacements are found for existing techniques but, on the other, numbers may increase with the development of new research techniques or as needs for new research emerge. It follows that if we wish to assess the success

---

[23]   High-throughput screening uses robot systems to carry out assays on hundreds of thousands of compounds a day, testing for specific biological activity.
[24]   Samuels (2000).

of replacement, crude statistics of the numbers of animals used in research are likely to be a very poor measure, yet those who campaign against animal experimentation hail every reduction in the numbers of animals as a victory and any increase as a reverse. The truth of the matter is that advances in replacement and reduction have to be judged in a more thoughtful way by assessing progress in specific areas of animal use or in the attainment of specific information. This is obviously a much harder job, and also harder to explain to the public.

6.3 Reduction

If it is not possible to use non-sentient alternatives to animals, the next priority is to consider how to use as few animals as possible, while still achieving the goals of the research. Various strategies can be adopted to achieve reduction, but essentially they come down to good planning and good experimental technique.

6.3.1 Using existing data

Before starting to use animals, researchers should consider how information already available to them might influence their plans and so contribute to replacement and reduction (they overlap somewhat). Bioinformatics is a rapidly developing area in which advanced computer techniques are used to mine large databases of biological information, such as those developed through genomics, and the information gained has potential to reduce some animal use. Re-analysis of existing data might also, sometimes, obviate the need for some research. Systematic review and meta-analysis, in which the results of several previous experiments are combined, are well established in clinical studies, but it is likely that more could be done to use these techniques to maximise the effectiveness of animal research. Experiments should therefore be designed so that they will be suitable for further analyses[25]. Another way of maximising the information obtained from a particular study is to make use of datasets and of samples or tissues obtained from animals post mortem. A European network of primate breeders, EUPRIM-net, has done just this by funding a biobank to make available and promote the use of tissue, blood, serum and genetic samples.

6.3.2 Choice of technique

Having studied the literature and thought about the background to the proposed research and the hypothesis, the next consideration is how to address the research question. Choices of approach or technique made at this stage can affect the number of animals that are needed. Here, some recent advances in technology can help researchers use fewer animals. It used to be necessary to maintain a small colony of

[25] Hooijmans et al. (2010) provides a checklist for researchers to facilitate systematic review. See also http://alttox.org/ttrc/emerging-technologies/-omics/ and http://www.cnprc.ucdavis.edu/animals/reduction.aspx, both accessed 10 May 2013.

the animals when a particular strain was needed for future use. Today, lines can be 'frozen down' by cryopreservation of embryos or sperm, dramatically reducing the numbers of genetically altered rodents that need to be maintained in breeding establishments. Other technologies such as imaging have reduced the numbers of animals required for drug safety evaluation. Instead of studies in which histology is carried out after euthanasia so that each animal is used once, imaging allows data to be collected from the same animals over an extended period of time. Animals can therefore be used as their own controls, individuals can be selected or deselected from the trial where appropriate, and intra-subject variation can be minimised[26]. In cancer research, animals may need to be euthanased at regular intervals so that post-mortem examinations can be carried out to monitor the progress of internal tumour development. However, imaging can be used to reduce animal use as tumour growth can be monitored on a regular basis in the same individuals and without any need for euthanasia. The ability to study several biomarkers at the same time may also reduce the numbers of experiments and animals needed[27]. Another animal efficient technique for collecting data is the use of remote telemetry with implanted or external sensors to record physiological measures such as body temperature, heart rate, blood pressure and activity that are unbiased by capture, handling or restraint. Like imaging, telemetry can be carried out over extended periods, which reduces the number of animals needed as data can be collected repeatedly from each animal and animals can serve as their own controls. Indeed, telemetry can reduce animal use by 60–70% in single studies and by more than 90% when the same animals are used to test a variety of treatments, allowing individual variation to be blocked out in analyses[28]. Nonetheless, the technique has downsides as well as benefits, which need to need to be taken into account (see section 6.4.2.9 on telemetry refinement).

For those involved in regulatory toxicity testing, animal use can be reduced by developing validated alternative tests that are accepted by the regulatory authorities. For example, the fixed dose procedure (FDP), which depends on detection of signs of toxicity at one of four fixed dose levels and which avoids using death as an endpoint[29], has largely replaced the LD_{50} for oral toxicity testing, so resulting in fewer animals used and reduced suffering. Likewise, a fixed concentration procedure with similar benefits is being developed that may be an alternative to the median lethal concentration (LC_{50}) test for acute inhalation toxicity testing[30]. Sometimes, comparison of methods between pharmaceutical companies can identify those which use fewest animals, as has been achieved with the use of non-human primates in chronic toxicology studies of monoclonal antibodies[31].

[26]   Wang and Yan (2008).
[27]   Cressey (2011b).
[28]   Studies reviewed in Kramer and Kinter (2003); see also Kramer and Hachtman (2010), p. 459.
[29]   Botham (2004) describes the LD_{50}.
[30]   Stallard et al. (2011).
[31]   Chapman et al. (2011).

6.3.2.1 Planning project strategy and sample size

According to a recent review, one of the biggest opportunities to minimise animal usage in regulatory general toxicity testing lies in the decisions made on sampling method, the nature of the toxicokinetic profile needed and the analysis of samples. For example, using dried blood spots that only require up to 100 µL of blood, as opposed to plasma samples that require up to 500 µL, allows serial sampling from individual mice and a substantial reduction in the number of animals needed[32]. Reduction in animal use can also sometimes be achieved by using one control group for several experimental groups, or by screening several drugs at the same time in the same animal (cassette dosing). Decisions on matters such as these have the potential to save thousands of animals per annum per company, which shows the importance of good planning for reducing animal use.

Research questions should always be clearly defined. It is also important to consider whether changes to the ordering of the various elements of a project might prevent wastage of animals (as well as finance and time). Pilot studies can be useful in the planning process to help identify decision points at which review might either lead to redirection or termination of unprofitable avenues of research. Figure 6.5 shows a flow diagram that can be used to help guide this process[33].

The right species must be chosen to answer the question, and for this it helps enormously if the researcher understands how the natural behaviour and physiology of the various candidate species might interact with the experimental paradigm. Most procedures on mice and rats have been carried out during daylight, which suits us humans, but this is the animal's inactive period and disturbance at this time affects both their behaviour and physiology. Imagine being asked to perform a cognitive or balancing task after being woken up in the middle of the night. Reversed lighting schedules are one way to address this problem. It is also important to consider the animals' sensory capabilities. One researcher examining the ability of hens to discriminate between various visual stimuli found that they were very bad at the task. However, the reason was not a cognitive problem, but simply that the visual acuity of the hens was poor at the distance required in the test. Changing the distance at which the objects were presented resulted in success. Species-specific behaviour matters as well. If one hopes to train an animal to respond to a stimulus in a particular way, it helps if the design of the study takes into account how evolution might have linked the stimulus and response. It is easy, for example, to train rats to associate taste of a food item with sickness (induced by X-rays or lithium chloride) but they do not readily form an association between an auditory or visual stimulus and later onset of sickness. Likewise, it is easy for rats to form an association between an electric shock and an audio/visual stimulus, but not with a taste[34]. The most likely reason for these predispositions is that evolution has favoured the animals' ability to make connections between a novel food and

---

[32] Sparrow *et al.* (2011).
[33] Gaines Das *et al.* (2009); Richmond (2010).
[34] Garcia and Koelling (1996).

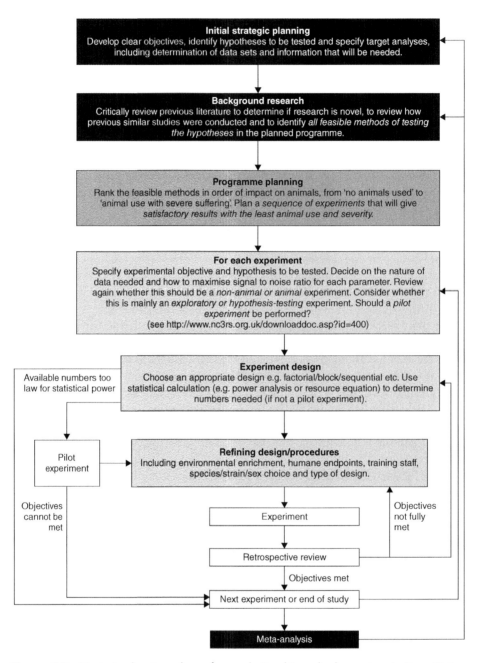

Figure 6.5 Strategic planning scheme for conducting biomedical experiments. From Gaines Das, R., Fry, D., Preziosi, R. & Hudson, M. (2009) Planning for reduction. ATLA, *Alternatives to Laboratory Animals* 37, 27–32, with permission from FRAME.

poisoning, and thus to avoid items that prove dangerous, but it has not favoured such an association with light because light will rarely, if ever, result in sickness.

Individual variation is the raw material of evolution, and some scientists make a career out of studying it. However, for others this variation is a nuisance, as the greater the variation between individual animals in a study, the more animals are likely to be needed. Some sources of variation can be parcelled up in the design of experiments and their analyses (block designs) so as to mitigate this problem, but Russell and Burch also drew attention to the need to 'make serious attempts to reduce unwanted variation at its source by controlling variation between individual animals, through control of the factors that determine it'. Variation is not just a product of genetics, but also of the various experiential influences on the animal (early physical/social experiences, disease, etc.), which is why most animals used in research are purpose-bred. Inbred strains provide another means of reducing variation, and Michael Festing has worked hard to persuade toxicologists to use several strains of inbred rodents in toxicological experiments rather than one outbred strain[35]. Using inbred strains allows for genetic variation to be parcelled out in blocks and allows one to untangle how results might vary with genotype. Sadly, as yet, these ideas have not been widely adopted.

The issue of how many animals to use in a study is critical to reduction. It is, of course, essential that the sample size is sufficient to answer the question, but too large a sample size is not just a waste of animals, it also increases the risk of obtaining statistically significant findings that have little biological significance. Russell and Burch described how in the early days of animal experimentation, the link between the size of the sample and the accuracy of the research was not understood, which led to a 'rat race' between researchers in their use of increased sample size and considerable animal wastage[36]. Fortunately, statistical techniques have improved. Pilot studies combined with power analysis provide a way of deciding how big the sample should be to provide a reasonable chance of detecting a defined level of effect at a particular probability. Many years ago, the ethologist and animal welfare scientist Arnold Chamove published a worked example showing different methods of sample size estimation for a proposed study on improving housing of cotton-top tamarins (a small South American monkey)[37]. His best guess as an experienced researcher, without carrying out any statistical calculations, was that 15 monkeys would have been needed. However, when he carried out a power analysis he showed that only nine were necessary. Most funding bodies now expect applicants to justify the number of animals they propose to use and to provide information from power analyses, if possible, when making their case for support.

It seems unnecessary to state that good experimental design is an essential prerequisite for research, and particularly for research on animals, but papers have been published with poor experimental designs that ignore the opportunities for reduction

---

[35] For example, Festing *et al.* (2002); Festing (2010, 2011).
[36] Russell and Burch (1959), p. 111.
[37] Chamove (1996) available from AWIC, http://preview.tinyurl.com/dydkben, accessed 9 May 2013.

offered by statistical approaches such as the use of factorial and sequential designs[38]. Analysing several variables in a single study using a factorial design uses animals much more efficiently than testing the factors separately. Sequential analysis is a technique in which the number of subjects to be used is determined by the sample observations as they are collected and which can result in the use of fewer animals with no loss of statistical power. In Chamove's worked example he went on to show that if he used sequential analysis he could further reduce the sample to only two tamarins, a remarkable reduction of 78% while achieving the same result and at the same level of significance. Substantial reductions using sequential analysis have been achieved in practice. In a recent study of cerebral ischaemia a reduction in animal use of 35% was achieved by using sequential analysis[39]. Sequential analysis may not be appropriate for all studies but its potential is sometimes ignored in biomedical research.

6.3.2.2 Challenges to implementing reduction

Reducing animal use to a minimum requires that scientists have sufficient statistical expertise and access to statistical advice. Many years ago, when I was a zoologist working in a psychology department, I was struck by the variety and complexity of statistical techniques psychologists used, many of which were new to me. Biologists' statistical training has greatly improved since then, but basic errors are still sometimes made, such as using repeated *t*-tests instead of ANOVA (analysis of variance) or using the individual, instead of the cage or group, as the experimental unit. Part of the problem may be inadequate statistical education and a reluctance to adopt modern methods, but difficulties in communication between statisticians and biologists can also be an issue. Some independent-minded researchers may feel that it is not necessary or too time-consuming to consult a statistician. However, Gaines Das[40] provides a thought-provoking quote from the biologist and statistician Ronald Fisher, who said: 'To consult a statistician after an experiment is finished is often merely to conduct a post mortem examination. He can perhaps say what the experiment died of'. No researcher wants to waste time, money or animals. Statisticians are the experts and, whether the experiment is simple or complex, researchers should check with a statistician that they are using the most efficient experimental designs and analyses. Fortunately, many institutions now provide dedicated statistical support, and require statistical input to the ethical review process. Research training has vastly improved and various organisations provide good statistical training and support materials[41].

---

[38]   Chapter 5 provides refererences for poor experimental design and analysis. Shaw (2004) also points to the limited use of factorial designs in the pharmaceutical industry at the time of publication.
[39]   Waterton (2000).
[40]   Gaines Das (2004).
[41]   There are many very detailed statistical textbooks and guides. In addition there are publications particularly useful to those involved in animal research, e.g. *ILAR Journal* 43(4), 2002, which is devoted to experimental design. See also Festing *et al.* (2002), Festing (2010, 2011) and Martin and Bateson (2007). FRAME provides training courses, and advice is also available on various websites, e.g. AALAS, CCAC, NC3Rs.

6.4 Refinement

Scientists still sometimes misunderstand refinement, thinking that it is about improving the quality of a scientific technique or experimental design. Refinement in the 3Rs sense is about the animals' experience of the research and trying to prevent or ameliorate any harm to them. Refinement is the R of last resort where all other efforts to avoid or reduce the use of animals has failed or reached a limit. This does not mean that refinement is the least important of the 3Rs or that it should only be considered at a late stage. The moral obligation to minimise suffering of the animals that are used is a heavy one.

Russell and Burch defined refinement in terms of reduction of the severity of inhumane procedures. This is a potential source of confusion as, in the UK, the term 'procedure' has come to refer to the experimental techniques that are carried out on the animals, and Article 3 of European Directive 2010/63/EU defines a procedure as:

> *any use, invasive or non-invasive, of an animal for experimental or other scientific purposes, with known or unknown outcome, or educational purposes, which may cause the animal a level of pain, suffering, distress or lasting harm equivalent to, or higher than, that caused by the introduction of a needle in accordance with good veterinary practice.*

However, Russell and Burch certainly did not mean to restrict refinement to experimental techniques performed on the animals. They were well aware that husbandry has an impact on experimental outcomes and could have a great impact on the welfare of animals used in research. In fact, they explicitly included the care of the animals in their description of the sources of inhumanity, when they wrote 'we cannot even in principle separate husbandry from the conduct of the experiment itself'[42]. However, to make clear to a modern audience the breadth of the issues that need to be considered under the heading of refinement, a revised definition has been proposed as follows:

> *Any approach which avoids or minimises the actual or potential pain, distress and other adverse effects experienced at any time during the life of the animals involved, and which enhances their wellbeing.*[43]

When considering how to refine a study we need to think about all the animals that may be directly or indirectly affected by the research. For a typical laboratory study this will include, in addition to the subjects of the research, the breeding animals that produced them and any stock animals that, for one reason or another, do not end up being used in research (e.g. if breeding resulted in more animals than required, or because their gender, age, etc. renders them unsuitable). In field studies there can be even broader ramifications. Capture and release of subject animals

---

[42] Russell and Burch (1959), pp. 54–55 and 64.
[43] Buchanan-Smith *et al.* (2005).

may perturb relationships within their social groups, and may even affect animals that are not directly socially linked (e.g. if an animal is released into another animal's territory, the resident animal may end up ejected and, like ripples from a pebble thrown in a pond, there can be knock-on consequences to neighbouring animals).

Opportunities to refine research need to be sought from the planning stage, through when animals first become part of the research process, and on to the end of the research or the animal's death. Lessons need to be learnt and disseminated as widely as possible and then acted on in future research.

6.4.1 Refining direct harms

In Chapter 5, we saw how Russell and Burch categorised suffering into direct harms resulting from the procedures, and unwanted and unnecessary contingent harms (e.g. arising from the way the animals are kept), that almost always damage the research as well as the animals. Refinement of direct harms poses special challenges, as it tends to be specific to the particular study or type of research. Nonetheless, some general points can be made and there are some commonly used techniques for which refinement affects many animals.

It is the individual that we are concerned about when considering welfare, and so measures to refine research procedures normally operate at the level of the individual animal rather than at group or even animal-room levels[44]. We need to know how animals are faring so that necessary measures can be taken to alleviate suffering. Animals therefore need to be checked by skilled personnel on a daily basis at the very least, and more frequently where the nature of the procedures demands this, while bearing in mind that the disturbance caused by viewing the animals can be a stressor (see Chapter 3, and section 6.4.1.1 on humane endpoints).

Some procedures inevitably result in suffering or stress but refinements should always be sought. It may be possible to use another less stressful experimental approach that provides the same information (or, ideally, still more information). The Morris water maze, for example, is used in studies of cognition in rodents. Animals placed in the maze (a tank of opaque water) have to swim until they discover the position of a hidden platform under the surface, and they are then tested on their ability to remember the location of the platform. Mice do not swim well, so one researcher has modified his apparatus so that the mice only have to paddle rather than swim. Mice also have to overcome their natural urge to remain close to the walls of the apparatus to climb on to a platform in the middle of the maze. The researcher therefore provided, instead of a platform, a number of tunnels in the side of the apparatus, one of which (the target) allowed escape from the water. Another example demonstrates how refinements are often unique to each experimental protocol. In epilepsy research, the use of chemicals to induce chronic seizures in rats has often resulted in high mortality rates (10–50%). However, the technique has been refined by using a muscle relaxant to reduce the intensity of seizures, by reducing the dose of

---

[44]   Individual checking of animals may not be possible, for example, in studies where large numbers of small fish are housed together. In such circumstances it may be necessary to make decisions at the group level.

the agent used to induce chronic seizures, and by treatment to terminate the seizures once data has been collected. The refined procedure remained a severe one, but it resulted in a lower level of seizure activity and a mortality rate of less than 2%. The researchers noted that while the refined technique did increase variability it could well be a more useful model of the human condition than the traditional technique.[45]

6.4.1.1 Refinements with broad potential

Although many refinements are specific to the research procedure, some may be applicable across many areas of research. Similarly, some housing refinements can be broadly applied for particular laboratory animals[46]. Refinements of these sorts can improve the welfare of many animals and are therefore particularly valuable.

Blood sampling

Most of us at some time in our lives have given blood, or had samples taken for diagnoses, and the procedure is, at most, a minor discomfort as long as one does not have an excessive fear of needles or blood. Animals, however, are not always as cooperative. They may struggle or move, and so require restraint, which increases stress and fear. Most samples are taken from superficial veins by needle or by cannula, but more invasive techniques have also been used such as cardiac puncture or decapitation. In the case of small rodents, blood may be taken by tail snipping and subsequent 'milking' of the wound. It is perhaps worth mentioning here that the use of retro-orbital sinus sampling is probably not a good idea as it requires considerable skill to avoid damaging the animal, is likely to have a considerable adverse welfare impact and, as the sample contains other fluids as well as blood, may not be suitable for some studies[47].

Adverse effects of blood sampling can include pain, stress, bruising and, more rarely, infections or thrombosis. Too large or frequent samples can also have health effects. Choice of the most appropriate and refined technique must therefore be considered at an early stage. There are many options and factors to consider, and detailed guidance is available on paper and online[48]. Examples of blood sampling refinements include automated sampling techniques and indwelling catheters to reduce the stress of repeated capture. However, these techniques can introduce other harms, such as tissue irritation, septic thrombophlebitis or generalised sepsis, or the need to house the animal singly with a tether. Depending on the study and the species, it may be possible to use positive reinforcement training, which reduces the stress associated with the sampling, for both animal and researcher (there is more on this in section 6.4.2.5). It is also worth considering alternatives to blood sampling.

---

[45]   Hawkins *et al.* (2012).

[46]   Rennie and Buchanan-Smith (2006), for example, describe refinements of husbandry and procedures commonly carried out on non-human primates.

[47]   Richmond (2010).

[48]   For guidance available, both generally and for individual species, see Joint Working Group on Refinement (1993), Holmberg and Pelletier (2009), Hubrecht and Kirkwood (2010), Dontas *et al.* (2011), and the NC3Rs blood sampling microsite. Holmberg and Pelletier (2009) describe an automated sampling device, showing the equipment and tether.

Some substances of interest, or their metabolites, are excreted in urine, faeces or saliva, so it may be possible to obtain the necessary information via these routes.

Dosing

Dosing is a common technique, but the methods, and hence options for refinement, vary. Both the active substance and the substance used as its vehicle have physico-chemical properties (viscosity, pH, etc.) that have to be taken into account when planning the best route and method for administration. If dosing is by injection, formulations should be chosen so as to minimise tissue damage and the substances injected should at least be at room temperature and ideally at body temperature. Adjuvants may be used to stimulate the immune system when injecting antigens to produce antibodies, and some cause chronic inflammation, tissue damage and other unpleasant side effects. Choice of adjuvant (if any), together with method of dosing and training of the operator, are all important to limit harm. Dosing by injection, like blood sampling, often involves capture and restraint, so training animals to cooperate by positive reinforcement techniques can ameliorate stress. Again, detailed advice is available on general refinements relating to dosing and on ways of refining specific procedures[49].

Humane endpoints

Some years ago I found myself speaking to a commercial toxicologist about his customers' requirements for tests on animals, and it was clear that there was considerable variation in their expectations. While his company did its best to educate customers that refinement could often be achieved by terminating research on a particular animal before extreme suffering occurred, some still required what he called 'blood on the floor', in other words mortality as an endpoint. His choice of words was shocking, and was meant to be, as it emphasised how much suffering could be avoided by clearly specifying in advance limits to animal suffering in a particular research programme (i.e. humane endpoints). The problem is not just confined to toxicology. Even as late as 2008, a review of research carried out on Huntington's disease models concluded that 'Our results indicate that the use of death as an endpoint is still relatively common, despite strong advice against this, and international efforts to introduce humane endpoints'[50].

But what is a humane endpoint? A number of definitions have been offered, but some do not cover all aspects. The OECD's definition, for example, limits the scope of the subject to euthanasia or to changing a treatment:

> *A humane endpoint can be defined as the earliest indicator in an animal experiment of severe pain, severe distress, suffering, or impending death… These adverse conditions, once identified should be minimised or eliminated, either by humanely killing the animal or, in long-term studies by (temporary) termination of exposure, or by reduction of the test substance dose.*[51]

[49]   For general advice see Joint Working Group on Refinement (2001); Dontas *et al.* (2011); Turner *et al.* (2011). Advice for different species groups is provided in Hubrecht and Kirkwood (2010).
[50]   Olsson *et al.* (2008).
[51]   Organisation for Economic Cooperation and Development (OECD, 2000).

The Canadian Council for Animal Care provides a broader definition, which makes clear that humane endpoints can also be used to determine when medical treatment or other types of support will be administered:

> *Humane endpoint refers to the point at which an experimental animal's pain and/or distress is terminated, minimized or decreased. It includes actions such as humane killing, terminating a painful procedure and giving treatment to relieve pain and/or distress.*[52]

Humane endpoints are used to terminate research on an animal at a point where sufficient information has been gained to predict a later scientific endpoint (such as death in toxicity testing, see below). Humane endpoints can also be used to provide an upper limit to the suffering experienced by the animals when the suffering is so great that continued use of that animal can no longer be justified. Such endpoints may be decided in the context of the harms and benefits of the research in question, but can also be set by legislation when the level of suffering breaches the limit on suffering for any research. Finally, humane endpoints can be used as a trigger to make decisions about the direction of research either for the individual animal or for the study as a whole.[53]

The proper implementation of humane endpoints requires careful planning and skilled staff able to assess the animals' welfare and to identify the endpoint criteria. During and after the research there should be reflection and feedback so that lessons can be learnt, incorporated into current and future research, and disseminated to other researchers. Choice of appropriate endpoint criteria is best done using a team approach, involving the researcher, the animal care staff, and those involved in ethical review. Pilot studies are useful in establishing the endpoint criteria, including when and how often the animals will need to be assessed.

Depending on the type of research and the nature of the challenge to the animal, it may be possible to predict death on the basis of changes in body temperature and weight, blood and other body fluid indices, heart rate, appetite and behaviour[54]. Choice of measures depends on the effects of the treatment or procedure. In cancer research, for example, the maximum tumour size that is considered acceptable is a mean diameter not normally exceeding 1.2 cm in mice or 2.5 cm in rats, with somewhat larger dimensions specified when the research is aimed at treating the cancers[55]. Measuring surface tumours is relatively straightforward, but determining endpoints when there are internal tumours may require imaging, or other indicators such as biomarkers or numbers of circulating cancer cells. Score sheets (Table 6.1) can help with consistency and the identification of trends[56]. Ideally, humane endpoints should be validated, and for regulatory testing the validation

---

[52] Canadian Council for Animal Care, http://preview.tinyurl.com/cf4rpwc, accessed 10 May 2013. See also the NC3Rs website.

[53] Morton (1999).

[54] Hendriksen and Morton (1999); *ILAR Journal* 41(2), 2000, 'Humane Endpoints for Animals used in Biomedical Research and Testing'; Morton and Hau (2010), p. 560; Hendriksen *et al.* (2011). See also Appendix to NRC (National Research Council) (2008).

[55] Workman *et al.* (2010).

[56] Morton (1999, 2000).

Table 6.1 Example of a score sheet for assessing the welfare of rats used in streptozotocin-induced diabetes research.

Rat No. 3		Date on study: 5/9/99		Issue No. 8978		
On-study weight: 214		**Pre-on-study weight: 219**				
Date	10 May	11 May	12 May	13 May	14 May	14 May
Day	0	1	2	3	4	4
Time	8:40	9:00	8:50	8:55	8:05	11:00
From a distance						
Fed mash/jelly/grain	Y	Y	Y	Y	Y	Y
Inactive		–	–	–	+	+
Isolated		–	–	–	–	–
Walking on tiptoe		–	–	–	+	+
Hunched posture		–	–	–	+	+
Pinched face		–	–	–	+	+
Starey coat		–	–	–	+	+/–
Type of breathing*		N	N	N	120L	70L
Negative red light response	nd#	nd	nd	nd	+	+
On handling						
Not inquisitive and alert		–	–	–	+	+
Not eating		–	–	–	+/–	+
Not drinking		–	–	–	?	+
Vocalisation on gentle palpation		–	–	–	–	–
Volume water drunk (average of rats in cage) (mL)		50	113	133	140 av	0
Body weight (g)	204	209	203	192	170	168
% change from pre-starved weight	7	5	7	12	22	
Body temperature (°C)		37.5	37.4	37.6	32.4	34.7
Pale or sunken eyes		–	–	–	+	+
Dehydration		–	–	–	+	+
Distended abdomen/ swollen		–	–	–	+/–	+/–
Diarrhoea† +0 to 3 (+m or +b)		–	–	–	–	–
Cage wet		–	+/–	+	+	–
Condition score‡ 4 to 1		4	4	3	2+	2
Saline given s/c, volume/ sites?		–	–	–	2 mL/×2	–

Table 6.1 (*Cont'd*)

Rat No. 3	Date on study: 5/9/99			Issue No. 8978	
On-study weight: 214	**Pre-on-study weight: 219**				
Blood sugar level	nd	nd	nd	nd	nd
Nothing abnormal detected	–	–	–	–	–
Other	Day 0 streptozotocin @65 mg/kg				Animal killed
Signature	KL	KL	KL	KL	

Special husbandry requirements

Feed irradiated diet and adapt animals to it 2–3 days before diabetes induction
Animals should be cleaned out twice daily
Two bottles of UV water should be provided for each cage and filled twice daily
Deprivation of water overnight may be sufficient to cause death by dehydration

Humane endpoints and actions

1. Any animal showing signs of coma within the first 24–48 hours will be killed
2. Any animals weighing less than the starting weight after 7 days will be killed, or losing more than 20% than start weight at any time will be killed
3. Any animal showing tiptoe or slow ponderous gait will be killed
4. Inform veterinarian and principal investigator if more than one clinical sign occurs

Scientific measures

Tissues to be kept: kidney into a mixture of 10% formaldehyde in saline

---

* Breathing: R, rapid; S, shallow; L, laboured; N, normal.
[†]0, normal; 1, loose faeces on floor; 2, pools of faeces on floor; 3, running out on handling; +m, mucus; +b, blood.
[‡]Refer to condition chart: 4, normal to 1, emaciated.
[#]nd, not determined.
Source: reproduced from Morton (2000), with permission from Oxford University Press and David Morton.

needs to be accepted by the regulatory authority[57]. Identification of humane endpoints is easier when the same methods are used repeatedly, as in regulatory toxicological testing, but all who plan experiments on animals should consider whether it is possible to set or further refine endpoints.

Implementing humane endpoints is not always easy. Practical problems can include cost (detecting when endpoints are reached requires staffing for close vigilance), subjectivity in judgements about when the endpoint is reached, and lack of agreement on harmonisation for regulatory tests[58]. Researchers may be inclined to

[57] Hendriksen *et al.* (1999).
[58] Hendriksen *et al.* (2011).

concentrate too much on signs specific to the research model (such as on tumour size in cancer studies) whereas other signs can sometimes provide better predictive information. In some research models, humane endpoints have to be set at a point where animal suffering is already high. In others, humane endpoints may be, or seem to be, incompatible with the aims of the research. One example is experimental autoimmune encephalomyelitis (EAE), which is a mouse model for multiple sclerosis. The model results in paralysis of the hindlimbs and, as in the human disease, involves a relapsing remitting course so that at one point the mice have extremely poor welfare and a few days later they will be almost normal. If early endpoints are set, they are likely to conflict with the research aim, because the aim of the research is to prolong remission or reduce the severity of relapses. In difficult cases such as these, it is important to reconsider the justification for the research, the method of animal assessment and whether there is a real need to set the endpoint at a late stage.

6.4.2 Refining contingent harms

Even though contingent harms (all those that may occur which are unrelated to the research protocols) are unintended, the researcher and research establishment still have a responsibility to try to prevent or mitigate them. The following sections indicate how some of the contingent harms described in Chapter 5 can be ameliorated.

6.4.2.1 Reducing disease

In long-term studies, animals may become aged and develop ailments that justify treatment, euthanasia or removal from the research depending on the ailment and research in question. Treatment is often the most humane solution, but in some cases could result in greater cumulative suffering or might interfere with the research. Careful clinical and ethical judgements about the best course of action are therefore necessary.[59]

Outbreaks of infectious disease can have severe welfare (and financial) consequences, and subclinical or latent infections may bias results and result in increased variation. A great deal of effort has therefore gone into preventing infectious disease and ensuring specified levels of health within experimental facilities. In the 1970s many facilities kept animals in open cages with fairly ineffective microbiological barriers between the animals and the world outside. Today, physical barriers, such as the room itself, protective clothing, plastic film isolators, filter top boxes, independently ventilated cages (IVCs), laminar flow hoods, isolation booths, and rigid cabinets, as well as systems to control and filter airflow and various sanitation and sterilisation systems, are all used to reduce the risk of infection (and to protect those working with animals from animal infections or allergens)[60]. Disease control measures include limiting access to animal rooms, health checks

---

[59] I am indebted to Maggie Lloyd, who drew attention to these issues in a presentation given to RSPCA Ethical Review Process Lay Members Forum, UK, 2011.
[60] Dennis (2010).

and maintenance of records, vaccines, a prophylactic regimen to control parasites and sometimes the use of sentinel animals (animals housed in the same unit as the experimental animals solely for testing to monitor health).

These measures have achieved a great deal, but more is known about disease control for some species of laboratory animals than others. Zebrafish are increasingly used in research but relatively little is known about the control of subclinical diseases, or about the basic biology of genetic diseases in this species[61]. Quarantine facilities and obtaining accurate health information from supplying institutions are essential elements in a strategy to reduce the risk of introducing disease. In barrier systems it may be necessary to re-derive breeding stock to eliminate infectious diseases[62]. Maintaining, or working towards better, health status is a complex veterinary responsibility that requires specialist expertise, as well as standardisation between facilities. Fortunately, there are a number of guidelines and recommendations available for those charged with this responsibility.[63]

6.4.2.2 Reducing or eliminating pain

It is reasonable to adopt the precautionary principle that if a procedure or condition causes pain to humans it should be assumed to cause pain to animals also, unless there is evidence to the contrary. Certainly, in the face of painful stimuli, animals show many of the signs that one would expect from our own experience of pain and (some species at least) will learn to self-dose with analgesics after operations likely to cause pain.

Pain control through anaesthesia, analgesia or euthanasia is usually the responsibility of the researcher or of the veterinary staff, but identification of signs of pain, stress or distress is also the responsibility of animal care staff. Species responses to commonly used drugs vary and so it should not be assumed that effective pain control has necessarily been achieved just because an analgesic compound has been administered. Choice of agent may be constrained by factors such as the duration and objectives of the research. For example, injectable anaesthetics can result in more physiological variation than gaseous anaesthetics, so their use might require larger sample size. Pain control is a highly complex issue and those responsible should ensure that it is regularly reviewed and that staff receive adequate and ongoing training on pain assessment and on the use and administration of effective drugs. Again, technical resources are available to those responsible for pain control[64].

---

[61] Kent *et al.* (2012); Spitsbergen *et al.* (2012).
[62] This is a procedure in which animals that are to be future breeders in a barrier housing system are surgically removed from a mother under aseptic conditions and placed with a surrogate mother of known health status.
[63] Dennis (2010); Hansen (2010); Hessler (2010); Hardy and Wolfensohn (2011). Guidelines on health monitoring for various species are available from the FELASA website, http://www.felasa.eu/, accessed 10 May 2013.
[64] For example, Flecknell (2009); Hellebrekers and Henenqvist (2010); Hedenqvist and Flecknell (2011). See also Animal Welfare Information Center (AWIC) website for analgesia resources, http://awic.nal.usda.gov/pain-and-distress/analgesia, accessed 10 May 2013.

6.4.2.3 Refining transport

Many animals used in research are transported at some stage in their lives, either from a breeder to an institution or between institutions. Refining transport to reduce stress caused to these animals requires action in three areas:

1. ensuring that the animals' needs are met as far as possible during transport;
2. effectively planning the transport so as to reduce potential stressors; and
3. minimising the transport period as far as possible.

The first two of these require little explanation, although their implementation is not always easy, and sometimes compromises have to be made. For example, for practical reasons or because of the risk of aggression, it may be necessary to transport animals in much smaller groups than those in which they would normally be kept, and single housing is sometimes necessary. For some species the duration of the transport time may be relatively less important than the stresses of loading and unloading but, in general, the shorter the transport time the better. Increased transport time brings greater risks of unforeseen events such as accidents, strikes or delays. Moreover, it is may be hard to control the environment during transport (temperature, vibration, noise, etc.) or to intervene in the event of aggression.

Before considering how to refine transport, it is worth thinking about whether there is a need to transport animals at all. Transport welfare issues may be avoided if it is possible to transport frozen embryos or gametes, or if it is practical to move the researcher to the animals. If transport cannot be avoided, then it is necessary to consider the species, ages, sexes and condition of the animals to be transported. Animals should generally be healthy and capable of caring for themselves. Shipping pregnant or unweaned animals is risky, and special precautions are required for immunocompromised animals or for strains or genetically altered animals whose phenotype otherwise affects their abilities to cope with the stressors of transport. Containers need to be well designed so as not to result in injury, and there should be provisions for feeding, watering and, if necessary, for checking the animals during transport. All transport requires careful planning, to meet the animals' needs and to satisfy national and international regulations. The International Air Transport Association (IATA) provides regularly updated information on the regulations pertaining to transport of live animals by air, and detailed guidance relating to all aspects of transport is available in guidance documents.[65]

6.4.2.4 Refining housing and husbandry

Ensuring that housing and husbandry are refined means providing an environment that meets the animals needs, whether these are related to nutrition, health, environmental factors such as temperature, humidity or noise, or the ability to satisfy

---

[65]   IATA, www.iata.co.uk/. For reviews of welfare issues relevant to transport, see Swallow *et al.* (2005) and White *et al.* (2010).

strongly motivated behaviours[66]. Handling, monitoring and welfare assessment are also part of husbandry so refinements can also be used to reduce the stress associated with these procedures. Ideally, of course, there would be no need for environmental enrichment, as all housing and husbandry would be designed to meet animals' needs. However, as we discussed in Chapters 3 and 4, concerns about health and standardisation resulted in traditional laboratory housing that was often stark, offered few opportunities for the animals to occupy themselves or to express species-specific behaviour, and resulted in abnormal behaviour and physiology. Housing and husbandry affects all animals directly or indirectly used in research and, what is more, affects them throughout their lives. Housing and husbandry refinement therefore has the potential to improve the lives of all animals used in research and can also ameliorate experimental harms.

Enrichment is the term used to describe the process of adding to, or changing, basic housing systems so as to improve animal welfare. Legislators and regulators have broadly accepted the value of effective enrichment, and the concept has been incorporated into European legislation (it is mentioned 15 times in European Directive 2010/63/EU as well as in various national codes of practice, e.g. US National Research Council documents[67]). Despite this widespread acceptance, enrichment is an unfortunate term in several ways. First, it is used by different groups of scientists in different ways, and in some cases used too loosely. Enrichment was originally used in neurobiological research, aimed at studying the effects of the environment on brain development, to describe social or physical changes to animal housing systems, with no expectation that the animals' lives were necessarily improved by the additions. Second, enrichment suggests that provisions required to meet animals' needs are optional extras, rather than standard baseline requirements. Third, animal care staff often use the word enrichment to describe any husbandry changes aimed at improving the animals' welfare, regardless of whether the changes are actually known to be beneficial, and some are not. Indeed, some so-called enrichment has proved harmful or stressful to the animals it was supposed to benefit, as shown by the provision of shelters for mice. Mice seem to like shelters, but males of some mouse strains respond with increases in testosterone, cortisol and aggression that may be related to territorial behaviour[68]. Some injudicious attempts to enrich enclosures can also result in injury to the animals if they become trapped in the enrichment or ingest or choke on items. Given these risks, Würbell and Garner's proposal[69] that we should use the following terms to distinguish between useful and non-useful additions to animal housing is, I think, a good one, as it points to the need to carry out risk assessments and to properly and scientifically assess proposed enrichment and other husbandry changes[70]:

---

[66] Advice on the husbandry of many species used in research can be found in Hubrecht and Kirkwood (2010).

[67] Most notably, *The Guide*, NRC (National Research Council) (2010).

[68] Haemisch and Gärtner (1994); Haemisch *et al.* (1994); Nevison *et al.* (1999).

[69] Würbel and Garner (2007).

[70] Bayne (2005); Nelson and Mandrell (2005).

- *pseudo-enrichment*, to describe enrichment that does not benefit the animal;
- *conditionally beneficial enrichment*, to describe enrichment that is beneficial for some animals or under some circumstances; and
- *beneficial enrichment*, to describe enrichment that is known to work, i.e. results in an animal with a better state of welfare.

The aim of enrichment is to provide animals with a complex environment that meets their needs, as determined by their evolution and selection, without adversely affecting experimental outcomes. More specifically, enrichment increases complexity and choice, and provides animals with a measure of control over their environments. Beneficially enriched environments stimulate normal rather than abnormal behaviour and can reduce animals' stress levels[71], so resulting in better research subjects (see Chapter 7). Beneficial enrichment also helps prevent or reduce adverse mental states and may induce more positive mental states. For example, mice housed in an enriched environment self-administered less anxiolytic than those housed in unenriched cages, while rats kept in enriched environments (with wooden blocks, cardboard tubes and various types of shelter) tended to be more 'optimistic' when tested than those kept in unenriched cages.[72]

Meeting animals' needs

The range of strategies to provide beneficial enrichment is wide, but includes modifications of the enclosure structure or surroundings, housing animals in harmonious social groups, or providing them with opportunities for social interaction. This may involve subdividing the enclosure, provision of objects or structures within the enclosure, provision of various opportunities for sensory stimulation, changes to feeding methods, and various strategies to occupy the animals' time[73] (Figures 6.6, 6.7 and 6.8). It is useful to take into account the way in which enrichment items can be used, the number of controllable aspects and their predictability[74]. Given the scope of these tasks it is obvious that enrichment provision needs regular review and proper resources.

Some aspects of husbandry refinement are now widely accepted. For example, nesting material is routinely provided for mice in the UK, and it is generally accepted that animals of naturally social species should be housed in socially harmonious groups unless there are good welfare or scientific reasons for not doing so. There is always an element of risk in social housing, and in some cases aggression can be so bad that direct physical contact is not possible, but usually the advantages of social housing (options to behave normally and so occupy the animal, reduction in abnormal behaviours, and in some cases improved health) outweigh the risks. Sometimes social housing is considered risky because of possible interference with the subject

---

[71] For example, Young (2003); Hubrecht (2010).
[72] Sherwin and Olsson (2004); Brydges *et al.* (2011).
[73] Young (2003); Garner (2005); Hubrecht (2010).
[74] Sambrook and Buchanan-Smith (1997).

Figure 6.6 Guinea pig housing showing provision of an enriched environment including social housing, shelters, hay and fresh food. The enclosure provides the space benefits of a pen but is suspended from a wall, which allows the technical staff easy access. With permission from Novo Nordisk A/S.

Figure 6.7 Rabbit housing showing provision of social housing, structures that provide shelter and which the rabbits can climb on (promoting healthy bone growth), hay and fresh food. The division between the two sections of the enclosure has a pop hole at each end, which prevents a dominant animal from preventing access to one side. Slides allow the pop holes to be closed off so that the animals can be easily caught and handled. The enclosure is suspended from a wall, which allows the technical staff easy access. With permission from Novo Nordisk A/S.

Figure 6.8 Marmoset housing that allows them to be housed in social groups and showing perches, ropes and other enrichment items. The clear plastic top of the enclosure allows the marmosets good visibility outside of their enclosure. With permission from Cambridge University.

animal or its instrumentation. Macaques, for example, were routinely singly housed for these reasons in many parts of the world. However, the risks of social housing are not always as great as they appear; and with appropriate husbandry it is now understood that social housing is usually possible even for adult male macaques with cranial implants[75].

Predictability is also important in animal care. Unpredictable environments are known stressors for animals; indeed just varying the times of husbandry activities can be stressful. Predictability can be increased by providing cues of forthcoming husbandry events, such as by using a bell to signal feeding time for macaques[76], or by enabling the animals to see what is happening outside their enclosures (Figure 6.9). On the other hand, some environments can be too predictable leading to negative affective states that in ourselves we might call boredom, so there needs to be some, but not too much, predictability. Many animals are motivated to

[75] DiVincenti Jr and Wyatt (2011).
[76] Gottlieb *et al.* (2013).

Figure 6.9 Older and newer types of dog enclosures. Vertical bars reduce vision out of enclosure for the dogs, while horizontal bars and a smaller room provide much better visibility for both the dogs and the staff. Other options to increase visibility include the use of clear plastic or glass panels.

explore new objects or structures in their environment and so programmes to vary or rotate enrichment items can be helpful to maintain the animal's interest. Care, however, needs to be taken not to misidentify a negative stress response to the change as a positive one reflecting an increase in interest.

Much husbandry refinement is about providing what animals want. We know that animals will work hard to obtain certain resources, indicating the strength of their motivation for them[77]. Some might not think it very important to meet animals' desires, but even if you don't think that it is ethically important to have a satisfied mouse, there are other reasons for providing refinements and resources. Some provisions are almost essential, such as housing social animals in compatible social groups and providing appropriate nesting material for mice. Nesting material not only allows mice to make a refuge but also allows them to control the temperature of their microenvironment. This is important as mice used in research are typically housed at ambient temperatures of 20–24°C, which is below their thermoneutral range. If as a result they suffer thermal stress, then welfare and experimental outcomes can both be affected[78].

Gerbil housing provides another good example of an essential need. Gerbils have a propensity to develop stereotyped digging[79] when kept in standard laboratory cages, but why this was so was not understood until research showed that gerbils not only need a safe place such as a nest box, but also a tunnel of a certain length leading to that refuge. The reason is probably related to the gerbils' use of burrows in the wild as an anti-predator strategy. Short burrows are dangerous so gerbils are strongly motivated to dig tunnels of sufficient length. In captivity, when the cage and substrate do not allow them to produce a tunnel of the right length, they just carry on digging. Providing gerbils with sufficient burrowing material in

---

[77] Young (2003); Hubrecht (2010).
[78] Gaskill *et al.* (2009, 2012).
[79] Stereotypies can be a sign that housing/husbandry conditions have been inadequate, see Chapter 3.

Figure 6.10 Housing for older macaques. Steps and structures are designed to make it easy for these older animals to access different areas of the enclosure. With permission from Moshe Bushmitz, BFC Israel, Bioculture Group.

research facilities would often be impractical but, fortunately, an artificial burrow together with an access tunnel seems to be an effective solution[80].

Husbandry refinement is not just about meeting animals' needs during the period when they are used in research. Good housing is needed throughout their lives. During the early stages of an animal's development complex physical and social environments are necessary to produce physiologically and behaviourally normal animals with the ability to cope with future stressors. Most animals used in research never become old, but there is some long-term toxicity testing and geriatric research where this does occur. Rather little research has been done on housing older animals[81], but it is obvious that their needs can differ from those of younger animals. Thought needs to be given to how or whether they can make use of structures or devices within the cage. Geriatric primates might, for example, require ramped ladders and appropriate placing of enrichment items such as platforms and branches (Figure 6.10).

[80]   Wiedenmayer (1997); Mason and Latham (2004).
[81]   But see Eisele (2001) on provision of beds for geriatric dogs.

Figure 6.11 Visual barriers at a macaque breeding centre allow macaques to modulate their social interactions and provide a sense of security as they can hide behind them. With permission from Moshe Bushmitz, BFC Israel, Bioculture Group.

Complex or enriched environments can allow animals greater control over social interactions with cage mates. An anecdote relating to polar bears from the zoo community makes the point (the principles of animal housing are common for whatever purpose an animal is kept). A zoo had noted that two polar bears were becoming aggressive towards each other and solved the problem by the simple expedient of adding a fallen tree as a visual barrier to the enclosure, allowing the bears to get out of view of each other when they wished to do so. In some ways it is surprising that this worked. Bears have a powerful sense of smell, and a good memory, so they must have been aware that the other animal was nearby even if they could not see it. Nonetheless, the ability to control the visual aspect of their social interaction seemed to be enough to allow them to continue to be housed together without excessive aggression. Visual barriers not only allow social control, they can also provide the animals with a refuge and possibly a sense of safety (Figure 6.11).

Validating enrichment
Some of the early attempts to enrich mouse enclosures may have just caused clutter rather than providing beneficial enrichment. As enrichments may be beneficial, neutral or harmful, it is very important to properly evaluate them before use. Nothing should be introduced into an animal's environment without careful consideration, including a risk analysis, and without involving both researchers and animal care staff in the decision.

Refinements should be based on knowledge of the biology of the species and on any studies that have validated enrichment. Commercial suppliers of enrichment items (chews, mouse houses, etc.) should be asked to provide evidence that the enrichment is beneficial. Where refinements are specific to the facility or type of research, an in-house study of their benefits may be necessary, and the job of doing this can fall on the animal care staff. This task should not be undertaken lightly, as designing and interpreting studies aimed at assessing the welfare value of enrichments can be very complex, and it may be advisable to seek professional advice[82].

Ensuring harmonised high standards throughout the animals' lives

Very often, animals used in biomedical research are sourced from other institutions such as specialist breeders. Animals may also be purchased from other research centres, sometimes when a particular species, strain or line is only available from those centres, as is often the case with genetically altered animals. Institutions may also farm out research to contract research organisations, or may fund research in academia. In some of these cases the supplier or other institution can be in another country. Even though responsibility for these animals and their care lies with the organisation where they are currently kept, and even though standards of care may be different in other countries, there are opportunities for contracting bodies to help spread good practice throughout the industry by clearly communicating their expectations about quality of housing and animal care[83]. They can achieve this by requesting information on housing and husbandry provision as well as health records or by making visits. Passports, such as have been proposed for genetically altered animals[84], that provide information for each animal and on its husbandry requirements are useful for communicating special needs.

Species-specific enrichment

Advice on enrichment for commonly used species can be found in various literature including research publications, laboratory animal handbooks and some codes of practice. To give an indicator of current practice some very brief summaries of suggested enrichment for some commonly used animals are provided below. However, those responsible for animal care should refer to specialist publications and codes[85],

---

[82]   FELASA Working Group on Standardization of Enrichment (2006); Hubrecht (2010).

[83]   See Chapter 2, section 2.8.3.1, Review by clients.

[84]   Animal Procedures Committee (2007), GA Passports: The Key to Consistent Animal Care. A report of the RSPCA GA Passport Working Group (GAPWG), available from www.rspca.org.uk.

[85]   For example, for mice, Olsson and Dahlborn (2002), see Gaskill et al. (2012, 2013a) for the importance of nesting material for mice; rats, Patterson-Kane (2004); dogs, Hubrecht and Buckwell (2004), Joint Working Group on Refinement (2004), Hubrecht et al. (in preparation); primates, NRC (National Research Council) (1998), Wolfensohn and Honess (2005), Joint Working Group on Refinement (2009). Reviews for these and other species and information on general husbandry can be found in Young (2003), Wolfensohn and Lloyd (2008) and Hubrecht and Kirkwood (2010). See also the Animal Welfare Institute refinement resources including databases at http://preview.tinyurl.com/cezjrwr.

as the value of enrichment can vary between breeds, strains, husbandry regimes and the nature of the research carried out on the animals.

Mice

Mice are motivated to work for access to complex cages with shelters and raised platforms and for running wheels, but the needs of mice differ according to their strain and phenotype. Nude mice, for example, need greater care with respect to temperature control, and have been reported to develop eye problems when exposed to certain types of cotton nesting material. It is, however, generally accepted that most mice benefit from the provision of solid floors with bedding material, and most importantly nesting material.

Rats

The needs of rats are probably less well understood than those of mice. Rats show preferences for larger cages, foraging devices, shelters and nesting material. They do build nests, although these are not usually as structured as those of mice. As a result, they benefit from solid floored cages. Other forms of enrichment to provide complexity can be beneficial in helping them to cope with the laboratory environment. Social contact and shelters should probably be considered minimum requirements, but knowledge is still limited and more research is needed to determine exactly how such resources should be provided.

Dogs

Although some research on enrichment has been carried out, there is a need for more detailed studies on dog housing. Pair housing seems to work well and dogs make good use of chews and platforms, as long as these are correctly designed and presented. Platforms provide complexity and improve the dogs' visibility outside their enclosures. Good sight lines seem important, as does subdivision of a sleeping area within the pen. Exercise periods provide an opportunity to explore novel environments and to socialise with other dogs and staff. Soft bedding, in a chew-resistant dog bed, can be useful for older dogs.

Cats

Cats benefit from the provision of a range of shelving at different heights. Structures in which the cats can hide have been shown to reduce stress, and there should be sufficient of these to avoid competition. Toys can be used to encourage pseudopredatory play.

For those carrying out research in New South Wales, housing/enrichment recommendations based on reviews of the literature have been produced by the Animal Research Review Panel: NSW Department of Primary Industries and Animal Research Review Panel, http://preview.tinyurl.com/d8l2cuv. These are useful resources for those in other jurisdictions.

Reviews of husbandry requirements for commonly used groups of animals for the European Council Revision of Appendix A to ETS 123 can also be found at http://preview.tinyurl.com/cw6b8lr.

All websites accessed 13 May 2013.

Figure 6.12 Long-tailed or crab-eating macaques swim in nature and will swim and play in water in captivity, when given the opportunity. Here a group tests the water. With permission from Moshe Bushmitz, BFC Israel, Bioculture Group.

Primates

Perhaps more efforts have been made to enrich primate housing than that of any other species used in research. Many primate species have been used as research subjects, but the most common are macaques (the long-tailed or crab-eating macaque *Macaca fascicularis* and the rhesus macaque *Macaca mulatta*) and marmosets and tamarins, particularly the common marmoset *Callithrix jacchus*. Different primate species have very different natural behaviours and therefore their husbandry needs and appropriate enrichment also vary. *Macaca fascicularis* habitat includes mangrove and riparian areas. They naturally swim, sometimes even under water, and some breeders and researchers have successfully provided swimming/paddling/water foraging opportunities for this species (Figure 6.12). *Callithrix jacchus* gnaws the bark of certain trees to obtain gum and, again, the provision of wooden items and artificial gum-dispensing devices to gnaw have proved successful in research establishments. Commonly used forms of enrichment include manipulable items, devices to stimulate foraging behaviour, structures such as perches, platforms, branches and ropes to give the animals more options about where to sit or how to move around their enclosures, and the use of mirrors for some species to improve vision out of their enclosure. Some establishments have successfully provided additional, external, exercise areas.

6.4.2.5 Refining handling through cooperation

Positive reinforcement techniques (PRT) are increasingly being used to train animals to cooperate with regular husbandry and also with various experimental procedures. An animal that chooses to cooperate is likely to experience no, or much less, fear or stress than one that is caught and handled using force. Many tasks, such as moving animals between enclosures, weighing, sampling or dosing, that in the past have required the animals to be captured from the home enclosure and then restrained for the procedure can be refined by the use of PRT. PRT has become common in primate management, as shown by a 2009 survey of nine NIH-funded facilities in the USA that found all of them were using training techniques, but those using other species should also consider the possibilities of PRT, as it works well with many species.

PRT uses a positive operant approach whereby the animal is progressively rewarded for behaviour that approaches the desired response, until eventually it will do it on command. The technique has been used, for example, to train marmosets to touch or hold a target object such as a spoon held at the front of an enclosure. Having achieved this, it is relatively straightforward to direct and train them to sit on a weighing scale placed within their home cage. It is then possible to obtain the marmosets' weights without entering or reaching into a cage. PRT can be used in a similar way to direct an animal as needed so as to separate it from its social group. Animals can also be trained to urinate on command, or to cooperate with saliva collection. Macaques and great apes have been trained to extend their arms for blood sampling and to cooperate with dosing and inspection.

PRT can help improve the relationship between caregivers or researchers and the animals that they use. It can also be used to train animals *not* to perform undesirable behaviours, such as aggressive behaviour during feeding times, and can thus reduce the risks of social housing. However, training animals does require some skill, so training is required for those training the animals as well as for the animals themselves. PRT can be very good, but it is not a universal panacea, as it can be difficult to train animals to do things to which they have a strong aversion.[86]

Rodents are used in such numbers that it may often be impracticable to use PRT. Nonetheless, handling methods can be refined so that the animals are less stressed by the procedure and more cooperative. If mice are picked up by the tail, subsequent attempts to handle them result in an aversive response and behaviours indicative of anxiety. If, on the other hand, they are encouraged to enter a small tube and picked up in this or in the cup of a hand they show little anxiety-related behaviour and are much more accepting of subsequent handling[87].

---

[86] Laule (2010); Sørensen (2010); Perlman *et al.* (2012). A training video and various courses for primate users is available from EUPRIM-net, http://www.euprim-net.eu/network/prt.htm. See also the website Common Marmoset Care at http://www.marmosetcare.com, the NC3Rs site on PRT at http://nc3rs.org.uk, and Prescott and Buchanan-Smith (2005). The US National Primate Research Center survey is reported in Perlman *et al.* (2012). All websites accessed 10 May 2013.

[87] Hurst and West (2010).

6.4.2.6 Refining marking and identification

Given that some marking methods have the potential to cause considerable harm to animals, researchers need to think hard about the reasons for, and the consequences of, their choice of technique. The aim should always be to use the least invasive method compatible with the research being carried and, as in all areas of animal use, users should endeavour to refine invasive marking techniques. This may involve using smaller tags or microchips, using smaller tattoos, and where appropriate using anaesthesia or analgesia.

If tissue is required for genotyping genetically modified animals, then taking the tissue sample can often be combined with the marking method. In these circumstances when marking small rodents, ear notching is generally preferred to toe clipping. Nonetheless, toe tipping/clipping may sometimes be necessary, as when very young mice (less than 14 days) need to be identified and genotyped because the modified phenotype will later develop conditions that seriously affect welfare[88]. At this age, ear notching the mice would cause extensive ear damage.

6.4.2.7 Refining euthanasia

Euthanasia should by definition involve little or no suffering, but when euthanasia is not carried out properly it can cause extreme suffering. Even if the correct procedures are followed, as we saw in Chapter 5, not all euthanasia methods (even those accepted by regulators) are as humane as we, or the animals, would desire. Those carrying out anaesthesia tend to prefer non-physical methods such as the use of chemicals[89]; however, while operator preferences are relevant, the most important principle is that the chosen technique should be the most humane for the animals. Researchers, and managers of research facilities, should keep an eye on progress on research in this area, and not assume that well-established techniques are necessarily the best.

Refining euthanasia includes ensuring that the correct equipment is available and well maintained. Staff training, both basic and ongoing, together with adequate supervision is extremely important to ensure that animal handling and the techniques themselves are carried out properly as otherwise poor practice can develop. For example, a review of one establishment found that it had a standard operating practice that permitted the floor of CO_2 killing chambers to be filled with mice so that they could not move or change posture, and cervical dislocation had been carried out on the floor of the room rather than on tables, practices likely to reduce the reliability and humaneness of the killing methods[90]. Advice on general principles and on the techniques appropriate for specific species may be obtained

---

[88] See Joint Working Group on Refinement (2003a) and two Norecopa position statements (2008 and 2010) that review the literature on toe clipping and other marking techniques for mice available at http://www.norecopa.no/sider/tekst.asp?side=22, accessed 10 May 2013.

[89] Wallace (2008).

[90] Animals Scientific Procedures Inspectorate (2010), http://www.homeoffice.gov.uk/publications/science-research-statistics/animals/wickham-laboratories, accessed 10 May 2013.

from specialist publications[91], and in some jurisdictions legislation or codes specify permissible techniques[92].

6.4.2.8 Refining the production of genetically altered animals

The techniques used to alter the genome of experimental animals, particularly mice but also of other species, vary but involve mutation (either spontaneous or induced) or transgenesis in which genetic material is transferred from one organism to another[93]. Genetic alterations can cause direct welfare harms. Sometimes these harms are predicted, but unexpected harms can also occur. As harms can be hidden, or may occur at a particular stage of development or when the animal is stressed or when housing or environmental conditions are changed, all new lines should be screened for welfare issues and measures should be taken to reduce harm or discomfort. Initial screening should take place when the animals are young, and then further screening should be undertaken, as appropriate, according to the expected consequences of the modification and its interaction with the experimental protocol. Because the phenotypes induced vary greatly, the problem of improving the welfare of genetically altered animals is often specific to the laboratories that use them. On the other hand, the procedures to create genetically altered animals are common to many laboratories, so there are opportunities to refine them by spreading good practice. With this aim in mind, various sources of information are available to help users refine their techniques[94].

6.4.2.9 Refining telemetry procedures

Radiotelemetry allows experiments to be designed that use fewer animals (see section 6.3). It can also be a significant refinement in its own right. Data can be collected from animals housed normally in their home cages with their social group. The animals do not have to be caught, restrained or connected to apparatus and are therefore unstressed. This is good for the animals and also for the researcher as the reduction in stress-linked variation between individuals enables sample size reduction and the data collected are more likely to reflect normal responses. In the days before telemetry I saw a marmoset being restrained so that electrodes could be placed on its chest and connected to a polygraph machine in order to record its heart rate. Unsurprisingly, its heart rate was extremely elevated. Other benefits of implanted telemetry devices are that the animal is not encumbered with a bulky external object, and that the cage mates cannot interfere with the transmitter.

---

[91] Hawkins *et al.* (2006); American Veterinary Medical Association (2007); Hellebrekers and Henenqvist (2010); Wolfensohn (2010).

[92] For example, UK Schedule 1 under ASPA (1986) and European Directive 2010/63/EU.

[93] A good short description of the various techniques can be found in Pintando and Roon (2011).

[94] NC3Rs, Genetic alteration, http://www.nc3rs.org.uk/category.asp?catID=68, accessed 10 May 2013, and the 'Full report of GA mouse welfare assessment working group' also available from the NC3Rs site. Also see Joint Working Group on Refinement (2003b); Wells *et al.* (2006); Thon *et al.* (2010). A website is available that encourages standardisation of the terms used to describe signs of poor mouse welfare, http://www.mousewelfareterms.org, accessed 10 May 2013.

Nonetheless, telemetry does result in harms. Surgery is necessary to replace batteries, which may result in pain and infection, and ethical decisions have to be made as to how many times this surgery can be repeated. The weight of the device can be another issue, although advances in technology have led to them becoming smaller and so less of an imposition on the animal. Given that telemetry allows data to be collected from the animal in its home cage, it is paradoxical that sometimes telemeterised animals have been housed singly, either because the transmitters broadcast on a single frequency and so would interfere if animals were kept together or because smaller cages are used to ensure sufficient proximity of the receiver to the signal. Fortunately, companies producing telemetry equipment are responding to the single channel problem by producing multiple frequency systems, and the problem of enclosure size can often be overcome by the use of multiple aerials. Repeat surgery remains a problem, but again there are options for avoiding this. Implants can be turned off using an external magnet to increase battery life, and technological advances are resulting in longer battery life. For those carrying out this work, there are publications that provide advice on the welfare problems and on how to refine telemetry techniques[95].

6.4.3 Challenges to implementing refinement
Generally, there is little resistance to implementing refinements when there is evidence that the refinement is of real benefit and the cost is not too great. Nonetheless, there are certain barriers to the more widespread implementation of refinement options and these are worth looking at to see how they can be addressed.

6.4.3.1 Concerns that refinement might change experimental outcomes
Experimental continuity is a common concern when proposals are made to refine husbandry. I once discussed with a US physiologist the type of housing that he provided for his rats, the fact that rats prefer to have nesting material, and the difficulties of providing this in the types of grid-floored cages that he was using. I asked him what would make him change to solid-bottomed cages, and his reply was that for scientific reasons he wished to maintain continuity of housing from the breeder through to his research. In this case the answer to the problem would be to get the breeder to change his housing system, and given the numbers of researchers that laboratory breeders supply, such an action might have a very great impact. More generally, those involved in ethical review need to challenge researchers to provide good scientific justification for inertia about introducing refinements rather than giving way to vague fears about introduction of experimental bias.

Researchers may be concerned that the introduction of enrichment to an established protocol will compromise comparison of results with previous data or, in the case of regulatory toxicology studies, that it might cause regulators to reject

---

[95]   Joint Working Group on Refinement (2003c); Kramer and Kinter (2003); Kramer and Hachtman (2010).

the study. However, it should not take long to establish new baseline data using the new method. With respect to regulatory acceptance, the important issue is that a case is made that the enrichment has no effect on the results of the test and that the regulators receive sufficient information about the enrichment (such as a certificate of analysis if it is likely to be ingested).

Other concerns with respect to refinement relate to the possibility that it might bias results, increase variation between animals, thus reducing test sensitivity and reliability, and decrease replicability between laboratories. We shall discuss these objections and the answers to them in more detail in Chapter 7, but in most cases one would expect that successful refinement should result in better science as well as better welfare.

6.4.3.2 Concerns about refinements relating to biosecurity

The development of various forms of barrier systems to control the spread of pathogens or allergens (see section 6.4.2.1) might be thought to be a barrier to effective enrichment. Barrier systems certainly create both physical and procedural difficulties when moving objects in and out through the barrier whilst maintaining its effectiveness, but experience shows that it is possible to provide shelters, bedding and nesting material for mice in individually ventilated cages. Similarly, primates housed even in the higher categories of containment systems can be group housed and provided with a range of enrichment opportunities such as platforms and perches and other enrichment objects that can be sterilised[96].

6.4.3.3 Issues relating to lack of generalisability of some refinement options

We have already discussed in this chapter how options for reducing suffering are very specific to the research in question. Russell and Burch recognised this in *The Principles* when they stated that refinement is 'so protean in its aspects, that it almost seems to require a separate solution in every single investigation and refinement might be regarded as an art or the ability to improvise'. This diversity means that each researcher has to work to consider how best to reduce suffering in a particular experiment and the work done may not be transferable to other areas of animal use. Nonetheless, it is important to publicise new and successful refinements, as even if they are used only by a small number of researchers they reduce animal suffering, and the research area might grow so increasing the refinement's value. Unfortunately, there is often little incentive for researchers to publish successes or failures in this area. It may be difficult to publish refinements in journals in the researcher's own field and publishing in the laboratory animal science literature often does not benefit their research career. Nevertheless, some research scientists and their animal care staff do publish papers on refinements and present on them at meetings, and such behaviour should be

---

[96] Dennis (2010).

encouraged. Awards and prizes are one way of doing this but funding bodies and employers probably possess the greatest leverage. Funding bodies can encourage the development and dissemination of refinements in the conditions of their grants, while employers of researchers can make these activities part of the culture of the institution. Learned societies that organise meetings on research that uses animals could also do more to encourage the presentation of novel refinements at their symposia.

6.4.3.4 Resistance to refinement due to lack of training or experience in implementing refinement

Perhaps because refining research procedures is so specific to the research in question, most researchers do not receive much education on refinement. Devising better research comes naturally to scientists, but thinking specifically about the causes of poor welfare and then devising refinements to deal with these is a skill for which, to a large extent, researchers have to train themselves. In the UK, modular courses are required to be taken by those who wish to hold a licence to carry out research, but these are very short. Improvements in this area must lie in better training and continuous professional development (CPD).

6.4.3.5 Concerns about refinement related to cost

Expense is sometimes given as a reason for not introducing a refinement. Obviously there are limits to what is affordable, but often the additional costs represent a rather small proportion of the overall costs of the research. Major funding bodies, at least in the UK, have emphasised the importance of refinement and have indicated that they would consider offering funding to enable establishments to achieve high standards[97].

6.4.3.6 Resistance to refinement resulting from the 'not invented here' syndrome

Globalisation, as we shall see in Chapter 7, presents both opportunities and challenges. One of the more frustrating aspects of working in the field of animal welfare science is seeing advances that are accepted in one country not taken up in others. International harmonisation of good practice has undoubtedly helped to improve standards in some countries, but can also act as a break on innovation. International companies can sometimes struggle to find consensus among their care staff at different localities as to what is best for animals. Even laboratory animal science journals published in different global regions can be seen as coming from a different culture and may therefore be disregarded, and language can also be a problem. Nonetheless, if proposals for refinement are based on good science, in time they should become generally accepted.

---

[97] See 'Responsibility in the Use of Animals in Bioscience Research: Expectations of the Major Research Council and Charitable Funding Bodies', pdf available from MRC, Wellcome, BBSRC and NC3Rs websites.

6.5 Conflicts Between the 3Rs

One might think, given the obvious value of the 3Rs for reducing animal suffering, that the ethics of implementation would be straightforward. In fact, for replacement, the ethical issues are usually fairly simple even if the practical issues are not. There is either a valid non-sentient replacement or there is not, and if there is then ethically the right thing to do is to use it, if necessary getting regulatory acceptance.

However, implementation of refinement or reduction can result in more complex ethical issues. People may differ, for example, in their views as to what counts as refinement and what does not. Some of those opposed to animal experimentation may even dispute whether refinement is the right thing to do, if they believe that efforts to refine husbandry could distract attention from the need for reduction or replacement[98]. A common problem is that of balancing the aims of reduction and refinement. For example, in one study the authors found that rats conditioned using an electrical shock as a stimulus were less variable in their response when a higher shock intensity was used[99]. As a consequence, the authors suggested that it would be preferable to use the higher shock in order to reduce the sample size needed. However, is it really better to use the smallest number of animals possible when they will suffer more? A similar issue arises with telemetry devices where, as we have seen, the potential for increased data collection from individual animals may reduce the numbers of animals needed, but the animals that are used have to undergo the costs of surgical implantation of the equipment. Another example is the use of animals in studies to assess the pathways involved in drug metabolism and the rate of metabolism of drugs. As long as a reasonable 'wash-out' time is allowed between the various test drugs, then the same animal can be used to evaluate a large number of drugs, but is it better to use one animal repeatedly in this way or to use more animals and limit the number of procedures on each?

Experimental design changes aimed at reducing numbers can also set refinement against reduction. One technique, which deserves more prominence amongst laboratory animal scientists than exists at the moment, is the use of unequal sample sizes. If a particular treatment will cause significant harm to the animals involved, then it is possible to reduce the number of animals in the treatment group by up to half, while retaining the power of the analysis by increasing the number of animals in the untreated (control) group[100]. Unfortunately, the downside is that the number of animals added to the untreated group has to increase by a greater number than that of the animals taken out of the treatment group. So, a decision has to be made as to whether to reduce the numbers of

[98] Olsson *et al.* (2012).
[99] Pietersen *et al.* (2006).
[100] For example, Suresh and Chandrashekara (2012). Standard statistical packages for calculating sample size and power can be used to plan unequal sample size designs, or seek statistical advice.

animals experiencing suffering at the cost of using more animals in the study, albeit ones that do not experience suffering as a result of the treatment[101]. The ethical issue becomes even more complicated when animals in both groups are expected to suffer or experience discomfort, but to different extents. If you believe that you know the relative levels of suffering that will be experienced by the two groups, then it is possible to calculate how such decisions on sample sizes should affect the total suffering experienced[102]. Unfortunately, as suffering is subjective, it is very hard to put precise figures on any relative difference in suffering experienced by the two groups of animals. It is also not clear that suffering can be treated in this mathematical way (i.e. numbers × suffering), so judging the right course of action remains subjective. The use of humane endpoints raises similar ethical dilemmas when a choice to limit the suffering of one or several animals results in animals being wasted or more animals being used in the long run.

The answers to these questions are not clear-cut – there is no right answer. Nonetheless, as we saw in Chapter 5, people tend to be most concerned about each animal's individual experience. This is understandable: if we put ourselves in the animals' place in the different scenarios, most of us would rather be one of a larger group experiencing lesser suffering than one of a smaller group experiencing greater suffering. If this view is correct, then in most cases where refinement and reduction clash, refinement should win over reduction.

6.6 Training

Proper implementation of the 3Rs requires appropriate training of all involved in the conduct, support, oversight or review of animal research. Even though a person may have a long and distinguished scientific career, that is not sufficient to guarantee competence in the skills necessary to carry out the research in the most humane way. Experimental techniques and methods of housing, handling and sourcing animals are constantly changing, and becoming more sophisticated. Regulations change, and so do attitudes as to what is deemed acceptable, so it is a good thing that the importance of training and CPD is recognised in the legislation of many countries[103].

---

[101]   Any ethical analysis of a plan to reduce suffering by reducing the treatment group and increasing the control group would of course need to take into account any contingent suffering (marking, handling, sham dosing, etc.) caused to the control group.

[102]   Sedcole (2006).

[103]   See, for example, US Animal Welfare Regulations (Animal Welfare Act as amended, USC, title 7, sections 2131–2156). The latest edition of *The Guide* places increased emphasis on training, which is the responsibility of the IACUC (NRC (National Research Council) 2010). For Canada, see the CCAC website. For Europe, European Directive 2010/63/EU, preamble and Article 23 also places responsibility for ensuring and documenting training with the institution.

There is no one-size-fits-all solution to education in laboratory animal science as many educational needs are specific. The requirements of researchers, veterinarians, technical staff and those involved in review are different, and even within these areas of responsibility training needs differ by institution or research group because of the diversity of types of research carried out, the species used, and types and sizes of research institution. Education is necessary to ensure that those involved with research are competent and aware of the legal framework, but education is also needed on ethical issues, an understanding and commitment to the 3Rs principles, knowledge of how to implement the 3Rs within the research programme, and where to find resources to support their implementation. Training is not a one-off process. Those involved in research or animal care require an initial period of supervision and throughout their career should be able to demonstrate that they have kept up to date with good practice.

Specific requirements for training and accreditation vary by nation although there are efforts at harmonisation. To meet these educational needs there are many national and international organisations that variously provide training, accredit training programmes, or CPD opportunities. Guidance on the goals of education and materials for education programmes is available from organisations such as the Office of Laboratory Animal Welfare (OLAW), AAALAC, the American College of Laboratory Animal Medicine (ACLAM), Canadian Council on Animal Care (CCAC), European College of Laboratory Animal Medicine (ECLAM), Federation of European Laboratory Animal Science Associations (FELASA), the Chinese Association for Laboratory Animal Sciences (CALAS), and the Japanese Association for Laboratory Animal Medicine (JALAM), and see section 6.7 on sources of information on the 3RS, whilst detail on the goals and practice of education and training can be found in specialist publications[104].

6.7 Sources of Information on the 3Rs

Implementing the 3Rs in a particular research programme requires research. Online databases such as NORINA and InterNICHE provide information on replacements for animals in teaching resources, while AltBib and ALtTox are useful sources for toxicology. The vast quantity of online information that is available can in itself be an obstacle to finding specific information, so it is helpful that guidance on search strategies is available from a search guide produced by the European Centre for the Validation of Alternative Methods (ECVAM). More general overviews on the 3Rs are available from learned journals, and governmental and non-governmental sources of information such as the Animal Welfare

---

[104] Links to some of these bodies can be found through the International Council for Laboratory Animal Science (ICLAS) website, http://iclas.org/committees/education-and-training-committee/education, accessed 10 May 2013. For specialist publications covering education and training, see for example Duffee *et al.* (2010); Howard and Nevalainen (2010); Howard *et al.* (2011), and *ILAR Journal* 48(2), 2007, 'Training and Adult Learning Strategies for the Care and Use of Laboratory Animals'.

Information Centre (AWIC), the American Association for Laboratory Animal Science (AALAS), the Norwegian Reference Centre for Laboratory Animal Science and Alternatives (NORECOPA), the Canadian Council on Animal Care (CCAC), the Center for Alternatives to Animal Testing (CAAT), the National Centre for the Replacement, Refinement and Reduction of Animals in Research (NC3Rs), the International Council for Laboratory Animal Science (ICLAS), the Federation of European Laboratory Animal Science Associations (FELASA), the Fund for the Replacement of Animals in Medical Experiments (FRAME), the Universities Federation for Animal Welfare (UFAW), the Royal Society for the Prevention of Cruelty to Animals (RSPCA) and many other sources[105].

---

[105]   ECVAM Search Guide (Roi and Grune, 2011), http://bookshop.europa.eu/en/the-ecvam-search-guide-pbLBNA24391/, accessed 10 May 2013. For sources of databases, etc., see appendix in Smith and Allen (2005). For support organisations by region of the world, see Vergara and Demers (2010).

Science and Animal Welfare: A Partnership

The recurring theme in this book, that improving animal welfare usually goes hand in hand with good science and often facilitates it, is a powerful argument for motivating change. In this chapter I give some examples from across the 3Rs of how improving animal welfare has resulted in better, cheaper or easier outcomes. The chapter ends with a brief look at the implications of current trends in research for the welfare of animals used in research.

7.1 Replacement Methods: Scientific and Other Advantages

7.1.1 Replacements in research

Pharmaceutical companies, in their research for new and improved drug therapies, make considerable use of computer-based techniques that do not directly involve the use of animals[1]. Examples include data mining, similarity searching, structure–activity relationship (SAR) and qualitative and quantitative structure–activity relationship (QSAR) models, pharmacophore models[2], and homology methods. These methods have been developed because the industry sees that they offer considerable

---

[1] Examples of alternatives widely used in pharmaceutical development are provided by Balls *et al.* (2012), see also Boyer (2009). Ekins *et al.* (2007a) reviews the development of *in silico* techniques and Ekins *et al.* (2007b) list a selection of target molecules for which *in silico* techniques have been used to discover molecules that will bind to those targets.

[2] Pharmacophores are three-dimensional and electronic descriptions of the binding of a ligand to target receptors. Once a model has been developed, it can be used to search a three-dimensional chemical database for potential candidate drugs, so-called pharmacophore-based virtual screening (Yang, 2010). These screens are widely used in the pharmacology industry to identify candidates for a variety of different targets.

The Welfare of Animals Used in Research: Practice and Ethics, First Edition. Robert C. Hubrecht.
© 2014 Universities Federation for Animal Welfare. Published 2014 by John Wiley & Sons, Ltd.

scientific and other advantages. There are obvious welfare benefits in avoiding animal use, but the tests also offer savings, in the cost of product development, in time
that would otherwise be taken up in lengthy and expensive animal tests and, if the
tests are more predictive, by eliminating leads that would otherwise waste time and
resources[3].

Taking QSAR models as an example, these *in silico* computer models are used to
predict the activity of molecules on the basis of prior knowledge within a particular
applicability domain (the physicochemical, structural or biological space, knowledge or information on which the training set for the model was derived). QSARs
are now the most frequently used model in drug design decisions where they are
used to screen candidate molecules for important drug characteristics such as
degree of effect, and absorption, distribution, metabolism, excretion and toxicity
(ADMET).

However, QSARs do have some fundamental limitations. They can only be used to
model the more basic types of toxicity (i.e. those that involve only one or two mechanisms of toxicity) and their capacity is limited when it comes to predicting more
complex organ-based toxicity that can occur through a multitude of mechanisms.
Other problems include the flexible structure of protein molecules, which makes
prediction difficult, and the quantity and quality of the data that may be available to
generate the model. Availability of data varies by area of research, with more being
available in the field of drug development than for chemical risk assessment. So there
are difficulties, but then, the *in vivo* studies that they replace also have limitations.
For example, a survey of multinational pharmaceutical company data analysed in
1999 found that there was human/animal concordance in organ toxicity of 71%
when data were collected from both rodent and non-rodent species, and for rodents
alone only 43% concordance[4]. Indeed, one of the reasons for developing *in silico*
QSAR techniques to predict the properties of compounds was recognition that the
data from *in vivo* experiments was often 'inadequate, inappropriate, or incomplete'[5].

In vitro techniques, like *in silico* tests, offer the ethical advantage of reduced
animal use, but also offer substantial other benefits. *In vitro* tests can be a much
quicker and cheaper option than using animals in the initial stages of drug discovery. *In vitro* high-throughput screening is, as its name suggests, fast and is likely to
become even faster. One author[6] has claimed that using advanced techniques it is
possible to screen 100 million compounds for relevant biological activity in about
10 hours, 1000 times more than would previously have been possible. In addition,
the new techniques use far less reagent (10-million fold less than conventional techniques) so reducing the research cost. Cell culture tests have also speeded up the
initial screening of new drugs for efficacy, one estimate[7] being that cell culture can

---

[3]   Cronin (2010).
[4]   Olson *et al.* (2000).
[5]   Cronin *et al.* (2003).
[6]   Agresti *et al.* (2010).
[7]   Festing and Wilkinson (2007); see also review by Balls *et al.* (2012).

be 10–20 times faster than using animals and can avoid the associated costs. Another advantage is that tests that use human material avoid the problem of differences between humans and non-human animals in their responses to compounds. Some may also allow very specific research questions to be answered. For example, the development of technology to create induced pluripotent stem cells (iPSCs) from adult somatic cells may lead to much more detailed tests than are available at present for drug, lifestyle and genetic interactions[8]. Many drugs fail at the clinical stage because of ADMET problems, so the routine use of high-throughput and less expensive *in vitro* tests has allowed companies to make earlier decisions, thus saving on development and other costs that can be in excess of a million dollars per day. It is obvious that such savings of time and money are a powerful incentive to adopt these replacements[9].

Early advances in replacement were relatively rapid because they were not subject to bureaucratic controls. Those involved in fundamental research were free to make their own assessment of the replacement's scientific, economic or practical value, as were individual companies for in-house alternatives required for internal safety or marketing decisions for products not subject to regulatory approval[10]. Unfortunately, when regulatory approval is necessary, progress can be slower. Nonetheless, between 1991 and 2011, the OECD adopted 18 *in-vitro* tests used variously to identify corrosive substances, skin irritants, severe eye irritants, skin photoxicity, skin penetration, and genotoxicity. The *European Pharmacopoeia* has also accepted a further nine methods relating to the potency and safety testing of vaccines and assessment of pyrogenicity (capacity to cause fever)[11]. The mills of the regulators may grind slowly but their approval is an indicator of the tests' scientific quality and value.

Assessing the value of replacements is not, however, always straightforward. Many replacements are not designed as one-to-one replacements for animal tests, but if the animal model does not itself give particularly good results it can be difficult to judge the relative scientific value of a one-to-one replacement. This seems to have been the case with the search for replacements for the Draize rabbit eye irritation test. Ideally, any replacement would be compared with human data but, for obvious ethical reasons, adequate human data on the tissue effects of irritant substances are rare. Replacements such as the EpiOcular Eye Irritation test and the Bovine Corneal and Opacity test have therefore been judged against data obtained from the existing Draize test. The problem is that while regulators accept the currently Draize test (with its current modifications to reduce the numbers of animals

[8] Balls *et al.* (2012).
[9] Tsaioun and Jacewicz (2009) review papers suggesting that drug development costs of tens of thousands of dollars per day in direct costs and over 1 million dollars per day in opportunity costs, pointing to the importance of early recognition of problems likely to lead to candidate drug withdrawal.
[10] Curren (2009).
[11] EC Joint Research Centre Document 'Alternative (Non-Animal) Methods for Cosmetics Testing: Current Status and Future Prospects – 2010' Frequently Asked Questions, http://preview.tinyurl.com/pswddnz, accessed 29 April 2013.

used and the pain and distress caused to the animals), it is a rather poor model of the human condition because of structural differences between rabbit and human eyes (the rabbit has a nictitating membrane, a thinner cornea and does not produce tears to the extent that humans do). Furthermore, the Draize test relies on subjective scoring of the results, which may explain why there is high inter-study variability[12]. It is obviously less than ideal that replacements might be rejected because they are judged against the gold standard of a test that has these limitations, but as there is no other standard against which to assess potential replacements, it is hard at present to see a way around these difficulties.

One issue that may make replacements more expensive in the future (an important factor in judging their utility) is the understandable desire to make *in vitro* alternatives closer replicates of the human system they are modelling. There has therefore been a trend towards more complex and sophisticated alternatives of the sort described as 'organ on a chip' and even possibly 'human on a chip'[13]. These are exciting developments but may come with higher costs and technical demands that reduce their suitability for general use[14].

7.1.2 Replacements in education

Animals are not just used in research but also in formal education, usually, but not always, at tertiary level. There are now a very large number of non-animal alternatives that include lifelike physical models on which techniques can be practised (examples include models of rats or portions of primates on which a student can practise dosing or bleeding techniques), and computer models that simulate many aspects of biology including physiological and behavioural responses[15]. Any educator considering the use of alternatives to animals will want to know whether students using them achieve the learning objectives for which animals have traditionally been used. Some skills, including animal handling, surgical and anaesthetic techniques, administration of drugs and humane killing, continue to require at least some animal use. On the other hand, other learning outcomes, such as knowledge of principles or facts, or the creation of data for analysis, are achievable using computer simulations, and replacements to animals in education result in as good or even better learning experiences for the students than when animals are used[16]. This may seem odd, as the animal is, after all, 'the real thing', but *in vitro* educational alternatives do have certain advantages. They allow students to work at their own pace without the fear of harming the animal; they allow them to repeat procedures if they go wrong without ethical concerns; and computer simulations

---

[12]   Balls, M. (2011) 'Toxicity Testing: The Need for New Maps for the Future' (editorial), *ATLA* 39, 417–418; Kaluzhny *et al.* (2011); Kolle *et al.* (2011).

[13]   Marx *et al.* (2012). For other examples see Chapter 6.

[14]   Balls *et al.* (2012).

[15]   For examples of replacements in education see, for example, the InterNICHE and NORINA websites.

[16]   Knight (2007); Patronek and Rauch (2007); Dewhurst (2008).

can provide feedback on performance and greater and easier integration of background information as the student collects their data.

7.2 Reduction: Scientific and Other Advantages

Finding documented examples of better science or of economic savings arising through the use of better experimental design or analysis to reduce the numbers of animals used is hard, probably because most researchers who adopt better methods do not have time to document advances over previously used less efficient methods. I have provided some published worked examples in Chapter 6, but the case often has to be made a priori on the grounds that a better understanding of experimental design helps one to formulate questions clearly and in a way in which they can be successfully answered.

Good experimental design should make research more reliable, so increasing the likelihood that the research will lead to valuable technical advances and that it will be accepted for publication. Success in both these areas can provide opportunities for further research, and improve the likelihood of future funding. Reduction in the numbers of animals used, through improved experimental design or techniques, such as the use of smaller blood sample volumes described in Chapter 6, will also result in savings in the cost of the animals and of their daily care. As animal costs can be a major component of the overall research budget, this is a substantial benefit for the researchers and for research funders. It is true that there may be times where better experimental design results in more animals being used but because poor-quality research wastes animals, affects the researcher's career prospects and the research may need to be redone, better design is still likely to be a win for both the animals and for the researchers. Poorly designed but published research may also lead other researchers up blind alleys so further wasting resources and animals.

We saw in Chapter 6 how new technologies can reduce the number of animals needed while also producing better-quality data. In addition to the immediate welfare and scientific benefits, the link between better welfare and better data can open up new avenues of research. Advances in imaging techniques such as computed tomography (CT), magnetic resonance imaging (MRI), positron emission tomography (PET) and biophotonic imaging[17] now allow scientists to accurately identify and track the progress of some diseases, and to study drug delivery mechanisms. Similarly, telemetry, besides reducing animal use, has been used to address a whole suite of research questions that simply would not have been possible using previous techniques. It is possible, for example, to study how heart rate and other physiological parameters change during social interactions or following changes in housing and husbandry. Moreover, as these data can often be collected from animals that

---

[17] Biophotonic imaging is a process in which small animals have genes inserted to make cells within them bioluminescent. The resultant light can then be imaged from outside the body, in some cases while animals are freely moving within an enclosure (Francis, 2005).

would have been kept in the colony for breeding or for other research purposes, more research can be done with little adverse impact on welfare.

7.3 Refinement: Scientific and Other Advantages

Attempts to refine housing, husbandry or research procedures have the potential to either adversely affect research or improve it, so each case has to be looked at on its own merits. As we have already seen, any proposal for refinement that did adversely affect research would not be a true refinement within the meaning of the 3Rs, as the net effect would be to waste animals, but it may be hard to establish whether this will be the case at the outset. Nonetheless, as we shall see in the following sections, often there is evidence or good reason to think that successful refinements have improved the quality of the science.

7.3.1 Refining pain control and better science

Pain control often involves the use of analgesics. In some cases researchers may have concerns that side effects could adversely impact on their research. Sometimes, these concerns are based on sound evidence suggesting an interaction with the compound being tested or that the analgesic might affect behaviour or other measures that are required for the research. In other cases, however, firm information is lacking, in which case those involved in reviewing the issue have to consider whether it is reasonable to withhold analgesia solely on the basis of a concern that an analgesic *might* have an effect on, say, metabolism or excretion of a compound.

It is true that in the absence of any known data on a drug interaction, there is a certain a-priori risk of drug interactions (when human patients take two drugs, the risk of interaction can be as high as 16%, increasing to 37% for four to seven medications, and up to 83% for those taking greater than eight medications[18]). Moreover, the metabolism of a drug in one pathway can inhibit metabolism in another pathway, resulting in the accumulation to toxic levels of drugs normally metabolised by that pathway[19]. Similar interactions are observed in animals used in efficacy, drug metabolism or toxicity testing, although the interactions are not always the same as those in humans because of differences in the ways in which drugs are metabolised[20].

On the other hand, uncontrolled pain is also very likely to impact on experimental results. Animals in pain are unlikely, for example, to perform well in behavioural trials. They may have impaired muscle activity or may guard themselves from pain and be apathetic. Additionally, they may, like humans, suffer from some degree of

---

[18]   These data, produced by Gaeta *et al.* (2002), come from elderly patients who because of their age are more likely to take multiple pharmaceuticals. It is possible that age predisposes to a higher rate of drug–drug interactions, but the highest risk factor is the frequency of taking multiple drugs.

[19]   Carey and Cleveland Clinic Foundation (2010).

[20]   Martignoni *et al.* (2006).

cognitive impairment. Pain can cause major changes in endocrine stress responses, resulting in initially raised levels of circulating glucocorticoids, and later in depressed levels if the pain is prolonged. In turn, these changes have a widespread influence on the body, affecting, among others, metabolism and immune responses, gut motility, urination behaviour and respiratory function and can lead to long-term changes in the CNS[21]. Such changes are likely to have profound effects on the quality of the data obtained from animals in pain.

Behavioural and physiological changes of this sort can be a real issue in reasonably common procedures, as demonstrated by a study in which midline surgical procedures were performed on rats to simulate various components of a laparotomy. The surgery resulted in significant reductions in food and water consumption and body weight, as well as locomotor activity. The effects were, however, significantly reduced if the rats were given analgesics. Similar results have been obtained for other surgical procedures, and the degree of reduction was related to the severity of the procedure so it is reasonable to infer that the changes were linked to pain. Studies such as these[22] suggest that in most cases effective analgesia should result in better science and should be given unless there is good evidence that it will interfere with experimental outcomes. If analgesia cannot be provided, it goes without saying that other options to reduce the intensity or duration of pain should be explored.

7.3.2 Refining pain research improves sensitivity

Research on pain and analgesics can involve the induction of pain followed by various types of stimulation tests that induce withdrawal of the stimulated part of the body, or other behavioural changes to assess the degree of pain[23]. The results may then be compared with those of animals to which putative analgesics have been administered. Unfortunately, these tests have not proved to be very effective for predicting the value of potential analgesics aimed at relieving neuropathic pain, as they frequently fail to screen out compounds that are not efficacious. One group of researchers[24] has suggested that a better approach might be to avoid these experimental tests and instead study naturally occurring behaviour. Pain produces changes in animals' behaviour just as it does in humans and these changes may better reflect the human response to the pain and analgesic. Additionally, ethologically relevant changes might be more consistent as natural selection will have acted on them to produce the best possible outcome for the rats. With these thoughts in mind, the researchers have been looking at the effects of pain and analgesics on the natural burrowing behaviour of the laboratory rat. Initial results are promising: the model seems to have better validity and to be more sensitive, as changes in

---

[21] See Chapter 3; also Richardson and Flecknell (2005).
[22] Flecknell and Liles (1991); Liles and Flecknell (1993).
[23] Abelson and Roughan (2011) describe a variety of tests, including thermal stimulation (e.g. tail flick test, Hargreaves paw withdrawal test and increasing temperature hot plate test) and mechanical stimulation (e.g. paw pressure and tail pinch tests or the use of Von Frey filaments).
[24] Andrews *et al.* (2011).

burrowing behaviour can occur before other clinical signs of pain are readily apparent. The model has advantages for the rats. If it is more sensitive, then less pain needs to be involved in the testing of analgesic agents. Moreover, the animals do not need to be handled or restrained in a test apparatus, and there is no need to induce pain by thermal or mechanical stimuli. The refined technique offers better welfare and all the indications are that the science will be better as well.

7.3.3 Refining surgical procedures saves animals, time and money

Reduction in the risk of infection in surgical procedures is another good example in which reduced animal suffering accompanies better research outcomes. It may come as a surprise to some that surgery on rodents has not normally been carried out under aseptic conditions, but some of those who carry out this surgery have assumed that rodents are peculiarly able to deal with postoperative infection. In fact, rodents are used as an experimental model of post-surgical infection and are at risk, like other animals, of infections that can result in death[25]. It follows that aseptic techniques should result in less wastage from animals dying or having to be euthanased, should reduce variation resulting from some animals in the sample becoming infected, and should certainly improve animal welfare. There is certainly some evidence that sterile technique can improve the outcomes. When non-sterile procedures were used to implant blood vessel catheters into rats, the survival rate over 25 days post surgery was only 50%, with 83% of the catheters becoming blocked and infected. In contrast, sterile techniques resulted in a survival rate of 100% with no complications, which may not be surprising as, in humans, intravascular catheter-associated bloodstream infections are common and most catheters implanted for more than 8 days contain a biofilm of microorganisms[26]. An anecdotal account also indicates that when GM mouse embryos were implanted, a change from using merely 'clean' to sterile instruments resulted in a procedure that had previously failed to result in successful pregnancies, becoming much more successful[27].

7.3.4 Refinement of macaque head implants saves time and costs

Macaques are very expensive and those used in neurological recording studies are often used over a number of years. Each one requires surgery and training to perform tasks during data collection, so it can take a great deal of time and effort before data can be collected. Unfortunately, infections can occur around head implants causing pain and malaise for the animals and which may result in the need for repeat surgery. The animal may have to be taken off study and, in some cases, euthanased. One group has attempted to refine these head implants by using tissue-friendly materials and MRI to produce a better fit. These refinements have reduced the rate of infections, resulting in better welfare for the macaques and considerable

---

[25] LASA (2010).

[26] Popp and Brennan (1981) cited in Ritskes-Hoitinga *et al.* (2006); Crump and Collignon (2000).

[27] Anecdotal account in Ritskes-Hoitinga *et al.* (2006).

cost, time and effort savings for the researchers[28]. Again, the development of these refined techniques has improved welfare and benefited the research.

7.3.5 Refining husbandry can reduce workload and costs

Husbandry refinements, in addition to improving the animals' welfare, have practical benefits for the scientific and animal care staff. One is the often reported but rarely measured increase in job satisfaction for animal care staff[29]. Another is that improved animal handling can make life a lot easier for the experimenter so improving the efficiency of the research. Sympathetically handled cattle, for example, not only have a reduced physiological stress response but are also easier to approach and handle[30]. Similarly, researchers have found that adopting positive reinforcement training (PRT) techniques results in much easier and faster sample collection from marmosets. For larger animals the results can be even better. One author estimated that PRT would result in savings of 40 hours after taking into account the time spent on training in a study of 16 minipigs weighed once a week for a year[31].

7.3.6 Husbandry refinement and data quality

We have seen how husbandry refinements such as enrichment can improve animal welfare, but concerns have been expressed that refining husbandry through the provision of enriched environments might adversely affect experimental results[32]. When these concerns are directed at enrichments that are not in fact beneficial (see Chapter 6), then there is no conflict. On both animal welfare and good science grounds they should not be provided. There may also be cases where there are good scientific reasons for thinking that a particular beneficial enrichment might compromise a study, in which case the enrichment would again not be justified, but what about when more general concerns about enrichment are expressed? Non-specific concerns that have been raised against enrichment (in general) are variously that that it might bias results, if implemented differently in different laboratories; decrease replicability between laboratories; or that providing a more complex and changeable environment might increase variation between animals thus reducing test sensitivity and reliability. These points are addressed in turn in the following sections.

7.3.6.1 Enrichment and bias

It is certainly true that animals in enriched environments can differ behaviourally and physiologically from those kept in un-enriched environments (this is hardly surprising as the goal of enrichment is to make things better for the animals). Various types of enrichment have been shown to result in improved memory,

---

[28] NC3Rs (2011), pp. 35–37.
[29] There are a number of publications that mention the work satisfaction that accompanies providing good housing for animals, e.g. Chang and Hart (2002).
[30] See work by Hemsworth and collaborators, e.g. Breuer *et al.* (2003).
[31] McKinley *et al.* (2003); Sørensen (2010).
[32] For example, Toth *et al.* (2011) set out concerns about possible effects of enrichment on experimental outcomes.

learning and behavioural plasticity, probably due to related changes in brain structures such as the cortex, hippocampus and thalamus. Enrichment can result in changes in neurogenesis, increased capillary development, changes in the formation and efficiency of synapses, changes in brain size and changes in neurotransmitter activity. Enrichment of rodent housing has also been shown to affect the senses, including eyesight and cortical representations of rats' sensory whiskers[33].

However, the real question to ask is, what are we comparing the animal kept in an enriched environment to? As Poole argued, in a seminal and rather provocatively titled paper 'Happy animals make good science'[34], if standard non-enriched environments are barren compared with an animal's natural environment and if the enrichment is beneficial, then enrichment will result in more normal animals and hence usually in better experimental models. Poole pointed out that most scientists working with animals take for granted that their subjects have normal blood pressure, heart rate, appetite, immunological competence and behaviour, etc., but that this assumption may not be valid because stressors that may occur (e.g. noise, light for nocturnal animals, inadequate environment, food or water deprivation, early weaning, handling and marking) can affect the subject animal's biology resulting in changes to heart rate, blood pressure, hormone and neurotransmitter levels, metabolism, growth rates, lactation and immunology and hence disease susceptibility. We also know that stress can induce behavioural changes which, depending on the animal's ability to cope with the stress, range from the development of more 'pessimistic' or cautious behaviour to much more obvious abnormal behaviour such as learned helplessness, repetitive stereotyped behaviour or self-injurious behaviour. Even an activity as routine as cage cleaning and changing can cause anxiety-like behaviours in mice many hours later[35]. Animals with stereotypies may show a general disinhibition of behaviour that can have a wide range of effects, including changes to extinction learning, home cage behaviour, responses to stimuli, response latencies, and switching between behaviours and if these responses are part of experimental paradigms, then the research will suffer[36]. Finally, as Healy and Tovée point out, if brain size is reduced when rodents are reared and raised in so-called standard non-enriched cages, then what are the implications for the validity of research on learning and memory carried out on these animals. Given such findings, it is not surprising that US and European guidance documents warn about the potential of stressors to either directly or indirectly impact on experimental data[37], rendering results obtained from these animals suspect. Beneficial housing and

---

[33] Reviews in Healy and Tovée (1999) and Hubrecht (2010).
[34] Poole (1997). His use of the word 'happy' was controversial as it implied that we know that animals have feelings such as happiness.
[35] Rasmussen *et al.* (2011). Castelhano-Carlos and Baumans (2009) discuss common husbandry techniques as potential stressors.
[36] Garner (2005).
[37] NRC (National Research Council) (2008), pp. 21–22 describes some of the effects that stress may have on experimental variables. See also FELASA Working Group on Pain and Distress (1994) and Garner (2005).

husbandry refinement that reduces stressors to levels that the animals can cope with will result in better experimental data.

It is not just enrichment that can reduce stress-related bias. We know that refining animal handling by adopting PRT can avoid or reduce the adverse affects of traditional non-cooperative methods of handling which, depending on the species and whether sedation is used, can include changes to haematological and serum chemistry values, an increased risk of stress-related abortions, fear-related behavioural changes and diarrhoea. PRT has even been proven to reduce the stress caused to cats when collecting blood from the jugular vein[38]. Other methods of refining handling techniques also result in changes that can be beneficial to research outcomes. It has been shown that petting dogs during and between taking blood samples from them reduces the increase in circulating cortisol that occurs with this process. In another study, refining dog housing, husbandry and training techniques reduced the variability of urine sodium excretion by reducing stress-related vomiting, so increasing the study's validity[39]. Finally, as we saw in Chapter 6, refined mouse handling techniques have been shown to result in mice that perform in less anxious ways in subsequent behavioural testing, so making them better experimental subjects (unless the aim in the experiment is to study a stressed animal)[40].

7.3.6.2 Enrichment and replicability

An argument sometimes raised against the use of enrichment is that variation in use of enrichment by different laboratories reduces standardisation so reducing replicability. The problem with this suggestion is that there is no such thing as truly standard housing anyway. Enclosure dimensions differ between jurisdictions, and even within jurisdictions some institutions voluntarily adopt standards that give more space than the minimum required. Different research institutions may also use different types of enclosure made of different materials and fabricated by different manufacturers, and this is particularly the case for the larger animals where housing is often designed to meet customers' individual specifications. Diet, ventilation and flooring type also all vary. Many institutions have now introduced barrier systems such as individually ventilated cages (IVCs) but others have not, and we know that IVCs can impact on mouse behaviour, including anxiety-like behaviour, in ways that vary by strain and gender[41]. Moreover, even if efforts are made to standardise conditions, standardisation seems to be very hard to achieve in practice. In a now famous study[42], researchers went to lengths that were far greater than normally employed to standardise strains of mouse used, apparatus, test protocols as well as housing and husbandry routines at three different sites. Despite these efforts, the animals still showed systematic site-related behavioural differences when tested in

---

[38] For example, Reinhardt (2003); Lambeth *et al.* (2006); Laule (2010); Lockhart *et al.* (2013).
[39] Hennessy *et al.* (1998); Ritskes-Hoitinga *et al.* (2006).
[40] Hurst and West (2010).
[41] Kallnik *et al.* (2007); Mineur and Crusio (2009).
[42] Crabbe *et al.* (1999).

open-field tests, elevated plus mazes and after cocaine injection. Anxiety-related behaviour in the elevated plus maze seems to be particularly prone to vary depending on environmental conditions and where the research is carried out[43].

Housing, husbandry, testing conditions and apparatus inevitably evolve over time, so standardisation over time, even within a research institute, is also limited. The issues involved in introducing enrichments to improve housing conditions are not therefore likely to be much greater than those involved in commissioning a new set of animal rooms or introducing new types of caging (e.g. in recent years IVCs have been widely introduced). A major problem with standardisation is that it can introduce a new cause of idiosyncratic results. This is because standardisation within organisations is generally easier to achieve than between them, so individual attempts to standardise tend to result in increasingly disparate and so less reproducible results. A better approach in drug development and testing would be to tackle environmental heterogeneity head on by using animals that are exposed to a variety of husbandry and environmental conditions in studies with appropriate designs and analysis. This would allow environmental heterogeneity to be controlled and studied so providing more information and reducing animal wastage[44].

7.3.6.3 Enrichment and variation

The last concern relates to the possibility that enrichment might increase variation within studies. On a-priori grounds, Garner argues that the opposite is likely to be the case as individual animals have different capacities to cope with stress. Under circumstances of inadequate housing this would result in higher interindividual variation as some animals develop greater physiological stress responses than others[45].

So is there any experimental evidence on the effects of enrichment on variation? A number of publications have addressed this issue, some suggesting that enriched environments reduce variation, while others assert that enrichment has the opposite effect[46]. The problem in interpreting these studies is that they were carried out at different laboratories and the enrichments varied in type. Moreover, it is quite likely that some of the enrichments studied in these papers were not truly beneficial, and if the enrichment increased stress, then interindividual variation would be expected to increase for the reasons that Garner has elaborated. However, one study[47] has properly looked at the effects of enrichment on variability and replicability between laboratories. The authors studied female mice of two different strains and provided a variety of different types of enrichment, including shelters, various materials that the animals could shred, pellets for which the animals could forage and an item that the animals could climb and swing on. They recorded the behaviour of the animals at three different laboratories using four commonly used behavioural

---

[43]   Wahlsten *et al.* (2006).
[44]   Würbel and Garner (2007); Richter *et al.* (2009).
[45]   Garner (2005).
[46]   Reviewed in Hubrecht (2010).
[47]   Wolfer *et al.* (2004). See also Richter *et al.* (2009).

tests, and concluded that for these animals the enrichment improved their well-being, but did not reduce the precision or the reproducibility of the data obtained from them, a case of good welfare being consistent with good science.

7.3.7 New possibilities for research

Enrichment can open up new possibilities for research. We have already discussed how enrichment interacts with various aspects of brain development and function, so perhaps it is not surprising that enrichment can ameliorate or slow the development of adverse effects in animal models of Alzheimer's and Huntington's disease. Like rodents, humans have evolved to interact with a complex environment so observation of these effects in animals has opened up new possibilities for approaches to treatment in the human population[48].

7.3.8 Nesting material for mice improves welfare, improves the quality of science, increases productivity and reduces costs

In Chapter 6, I described the importance of providing nesting material for mice to allow them to control their microenvironment, particularly when they are housed in temperatures below their thermoneutral zone. As one might expect, temperature control is harder for nude mice, as their lack of fur increases the rate of heat loss. Providing nesting material for nude mice has been shown to result in increased numbers of pups born and also improves feed conversion, so improving breeding performance. Furthermore, inter-subject variability was reduced when nesting material was provided. A further study of other (non-nude) mouse strains produced similar results. The authors calculated that providing nesting material could result in substantial savings of several hundred dollars per cage of breeding C57BL/6 mouse (the most widely used inbred strain). Hence, nesting material improves the animals' welfare by allowing them to avoid thermal stress, improves the quality of data collected from those animals and results in financial savings[49].

7.3.9 A dosing refinement that results in better science

Sometimes all the benefits of the refinement are not obvious at the planning stage. One group of researchers developed a refined method of dosing mice with a test substance. The traditional method was oral gavage, but this required handling that can be stressful to the mouse, as well as carrying risk of damage to the oral cavity and oesophagus, or accidental insertion of the gavage tube into the trachea with potentially fatal results. The researchers therefore developed an alternative method in which the substance to be administered was mixed with chocolate and sugar into a pellet. After a few adjustments they developed a pellet of a size that the mice would voluntarily eat completely, which was presumably a pleasant experience rather than the reverse. Even better, they found that significantly reduced doses

---

[48]   Effects of enrichment are reviewed in Hubrecht (2010) and Young (2003). For implications for human research and treatment, see Hockly *et al.* (2002); Spires *et al.* (2004); Jankowsky *et al.* (2005).
[49]   Gaskill *et al.* (2013b,c).

were necessary, suggesting that the refined method increased the substance's bioavailability[50].

7.3.10 Refined monitoring techniques that improve the quality of the science

We have already discussed some ways in which imaging can reduce animal use. Advances in imaging technologies now allow some procedures in oncology research to be refined by setting and adjusting individual humane endpoints based on the size and development of tumours. Imaging can also provide more and better information on tumour growth and its response to treatment. Similarly, in tuberculosis research, imaging techniques such as MRI, and more recently CT and PET, have allowed researchers studying this disease in macaques to use lower and more clinically relevant doses of the tuberculosis pathogen[51]. The imaging allows lesions to be detected and counted at stages where there are no observable clinical effects, such as weight loss and respiratory problems. The researchers can therefore distinguish at an earlier stage between animals for which the vaccine has provided protection and those which will go on to develop clinical signs. Once again, better and more informative scientific techniques have gone hand in hand with more humane experimental technique.

7.3.11 A refinement in coagulation research that also improves data collection

A final example of a refinement that benefits the science and which has considerable potential to reduce animal suffering comes from the field of research on blood coagulation. A commonly used technique to assess the effect of coagulation factors has been to transect a tail vein, or the tail, of mice deficient in these factors, after which the mice are returned to their cages. When clotting is impaired, as happens in mice that have received only small doses of the test factor in dose–response studies, the traditional study endpoint has been when the mouse is moribund or dead. The refined and validated technique uses anaesthetised mice so eradicating any suffering produced by the fatal loss of blood, and provides a number of benefits for the researchers and the quality of their science. Firstly, single tail vein transection allows lower, clinically relevant doses of coagulating factor to be used compared with standard tail clipping where the dorsal and two lateral veins and the ventral artery of the tail are all completely transected. Secondly, the researchers are able to obtain quantitative data on bleeding time and blood loss over time, instead of only dichotomous scale data (alive or not alive), and these more detailed data could also lead to a reduction in the number of animals required in these survival studies. Finally, the refined model extends the scope of research possible as it can also be used to study the effects of coagulation factors on the arterial system by transecting the ventral artery[52].

---

[50] Cadot, S., in Hawkins *et al.* (2010).
[51] Sharpe *et al.* (2009, 2010); M.J. Dennis (personal communication).
[52] P.B. Johansen, Novo Nordisk (personal communication).

7.4 Changes in Laboratory Animal Science and Challenges for the Future?

In this book I have attempted to provide an overview of animal use in research and the ways in which this can be made more humane. However, the world is rapidly changing, and this includes where, why and how animals are used in research. Changes in animal use will have consequences for implementation of the 3Rs, and if we could predict these changes it would help in planning to maximise animal welfare and avoid adverse consequences. Unfortunately, predicting the future is a hazardous business. Winston Churchill has been quoted as saying wisely, if rather unhelpfully, 'I always avoid prophesying beforehand, because it is a much better policy to prophesy after the event has already taken place'. Nonetheless, I believe that there are some trends in biomedical research that have implications for animal welfare and which seem likely to continue, at least into the near future, and outline some of these below.

7.4.1 Changing models and techniques

It seems reasonable to assume that advances in replacement and reduction will indeed reduce the need for animals in some areas of research. Nevertheless, over the next few decades at least, animals will continue to be required, to meet existing and new needs. Research methods and objectives are always changing and the development of new models will bring new challenges for the 3Rs. Political decisions can also affect animal use. The introduction of the European REACH regulation (Registration, Evaluation, Authorisation and Restriction of Chemical substances) has placed increased responsibility on companies to provide information on the safety of the chemicals that they produce and this could lead to greater animal use. On the other hand, efforts are being made to reduce animal use in this area through data sharing, the use of alternatives, and initiatives such as Tox21 in which robotic techniques are used to screen and prioritise the chemicals that need testing[53].

Some trends are likely to continue. As interest in the genome and gene function (genomics) is maintained and widened to include studies in a growing host of biomics including transcriptomics, proteomics and metabolomics[54], it seems unlikely that the trend for increasing use of genetically altered animals will be reversed in the near future. On the other hand, again, there have been technical developments that have limited adverse effects. The ability to switch genes on and off as desired and advances in detecting adverse phenotypes with conditions that adversely affect welfare have both helped to reduce suffering. Another trend that seems unlikely to change is the increasing use of barrier and isolator systems which, while improving health status and reducing the risk of allergen exposure for staff, can hinder welfare

---

[53] US Environmental Protection Agency, http://epa.gov/ncct/Tox21/, accessed 9 April 2013.
[54] Transcriptome: the complete set of RNA transcripts produced by the genome at any one time, http://genome.wellcome.ac.uk/doc_WTD020758.html. Proteomics: knowledge of the interactions of cellular proteins with each other and with diseases and drugs (Harding et al., 2010). Metabolomics: the characterisation of small molecule metabolites found in an organism, http://www.metabolomics.ca/.

assessment and animal husbandry and, as discussed earlier, can have some adverse impacts on the animals housed within them.

The species used in research are also changing. The UK 2011 statistics show that the use of rats has been decreasing whereas that of mice and fish has been increasing due to their use in GM work (the use or breeding of GM fish increased by 10% over the previous year and of mice by 3%). Conversely, procedures on non-GM mice have fallen. The increased usage of small teleost fish species, such as the zebrafish (*Brachydanio rerio*) and medaka (*Oryzias latipes*), in GM research is partly because they have advantages that include a high reproduction rate, small size and a relatively low maintenance cost, and because it is relatively easy to obtain and manipulate the eggs and observe development. It may also be partly because many people are less concerned about using fish than mammalian species[55]. In Chapter 4, I criticised the idea that fish are less likely to suffer and argued that rather than making such general assumptions, it is better to examine the impact of the research and husbandry techniques on the species in question. So, is using fish in this type of research really more humane than using mice? The task of obtaining fertilised eggs from fish and growing them on after genetic modification does not involve any invasive techniques[56], whereas when mice are used invasive techniques are required (see Chapter 5). When we turn to the harms involved in taking samples to identify genetically altered individuals there is less to choose between the two species, as in fish the caudal fin is clipped for genetic analysis (fish fins contain motor and very probably sensory nerve fibres), while mice are tail tipped or ear punched. These considerations might suggest that, from the welfare point of view, it is better to use fish than mice for this work. On the other hand, we know much less about the care and husbandry needs of fish species. Individual fish are also often difficult or impossible to monitor, particularly when there are a number of small fish in a tank, and our knowledge of signs of pain and distress in fish lags behind that for rodents. In my view, the choice on welfare grounds between using a fish or a mouse is a hard one to call, but with our current knowledge it may be more humane to use the rodent.

The pharmaceutical industry's increasing reliance on monoclonal antibodies as potential therapeutics is likely to create an increased demand for non-human primates[57]. The reason that primates are often the species of choice for efficacy, safety and batch testing of these products is because monoclonal antibodies are extremely specific in the way in which they bind to their target molecules, although, following the sequencing of the pig genome, pigs might also be used in this type of research. Given the likely increased need for primates, it is surprising that Home Office statistics at the time of writing show a fall in primate use in the UK. It may be that any increase in primates used for monoclonal antibody work has been hidden by a bigger overall decrease driven by concerns about using these species. Nonetheless,

---

[55]   Olsson *et al.* (2010).
[56]   Eggs may be stripped from ripe fish by gently rubbing the abdomen.
[57]   Chapman *et al.* (2011).

even if the total number of non-human primates has fallen, the need for monoclonal antibody testing will require more to be used than would otherwise be the case.

7.4.2 Changes in where and how research is carried out
7.4.2.1 Globalisation of biomedical research
For many years, laboratory animal science expertise and international pharmaceutical research organisations and suppliers tended to be concentrated in developed western countries. A decade or so ago, I remember a pharmaceutical company employee telling me that this situation would continue because Western countries possess greater experience and produce higher-quality research. His argument was probably wrong at the time and now looks very outdated. The reality is that the situation is changing fast. Many other countries are now rapidly developing their laboratory animal science skills and capacity. So it is not surprising that a number of global pharmaceutical companies, contract research organisations and suppliers either have, or are, establishing research centres in developing countries, and particularly in the Far East. Increased globalisation of research is a factor in academia also, as there are increased opportunities for researchers to collaborate internationally.

This trend for research to become more globalised seems certain to continue and to escalate, particularly for pharmaceutical companies where there are strong financial drivers to move to countries where research costs are lower. Another factor that may encourage research to move away from the UK and some other developed countries is the difficulty of importing animals. Long-tailed macaques are the non-human primate species most commonly used in safety testing, and these are supplied from overseas, often the Far East. Long transport times between countries have financial costs as well as the welfare risks identified in Chapter 5, but pressure from animal rights groups has also restricted the numbers of carriers willing to transport research animals, making it harder and more expensive to carry out research in countries that import primates.

Challenges and opportunities arising from globalisation
Globalisation and changes in where research is carried out provide both challenges and opportunities for those concerned about animal welfare. Countries have different histories regarding concern for the welfare of animals and legislative provisions for the welfare of animals used in research varies considerably (see Chapter 2). If the motivation for moving research were solely to conduct it where costs and bureaucratic burdens are less, then anyone interested in animal welfare would have serious concerns. If, on the other hand, when companies move they have the intention of maintaining and disseminating high standards, then there are great opportunities for them to contribute to the rapid global spread of the 3Rs, good science and good welfare. A pharmaceutical company may be only one of several clients of a research facility but, if large, it may have considerable leverage and by requiring specific standards it can help to spread good practice. On balance, I believe that higher global animal welfare standards are the most likely eventual outcome of the

relocation of research by large pharmaceutical companies. These companies are concerned about their public profile, and set global policies to ensure minimum standards for their own institutions and for collaborative or contracted-out research. They also have the resources and financial clout to ensure that reasonable expectations are met.

In contrast, academic researchers may not have as much experience of welfare standards in other countries, may not have access to the technical and financial resources necessary to audit a collaborative laboratory, and may not be as aware of the ethical and public relations risks of working with a collaborating research laboratory. In my view, to deal with these risks and ensure that welfare issues are identified and dealt with, universities should ensure that the researcher and senior animal care staff make at least one visit to collaborating laboratories to check the care and use of the animals, and a report of the visit should be made available to the researcher's home institution.

Any organisation or individual collaborating or contracting out research abroad may need to make several, or even regular, visits to follow up on any issues. Training or other forms of support may also be necessary. Visits by animal care specialists, such as veterinary surgeons, animal care staff or animal welfare researchers, provide real opportunities to pass on experience gained in the application of the 3Rs. AAALAC accreditation provides some reassurance that the collaborating institution meets local, or as a minimum US, standards but judgement will be needed as to whether these standards are sufficient when, if the work were to be carried out at the home institution, standards would be higher.

7.4.2.2 Increased contracting out of research

Another driver for change is that, for some years now, the big pharmaceutical industry's spend on research and development has not produced the level of returns that existed during the era of the hugely profitable blockbuster drugs. As a consequence, the industry has, in addition to reassessing location, been re-examining its model for drug discovery and development. The most obvious change has been a move towards investing less in in-house research and instead exploring more open drug innovation strategies, in which the big pharmaceutical companies search for promising new drugs and ideas being developed at universities or small biotech or pharmaceutical companies. The net result is that more animal research will be carried out in establishments not under the direct control of the commissioning company, and these are likely to be universities or, in the case of small biotech start-up companies that are unlikely to have the resources to carry out animal research on site, contract research organisations.

Challenges arising from contracted-out research and purchase of other services

The trend towards increased contracting out of research, mentioned above, could have animal welfare consequences. I have already touched on issues relating to contracting out research abroad, but even when it is within the same country where the legal requirements are the same, animal facilities in contracted-out research establishments may not be as lavish as in a traditional big pharmaceutical company.

Another likely consequence is that there will be less money spent by big pharma on new animal facilities. This may matter. The large pharmaceuticals have comparatively large resources and when they have built new facilities they have often tried to build something that is better than has gone before. As a result, some of the facilities they have developed have helped to drive improved animal care standards worldwide. Of course, pharmaceutical companies are not the only ones who innovate, and I have seen truly groundbreaking new animal facilities in CROs and university establishments. Nonetheless, if the trend for large pharmaceutical companies to close their in-house animal facilities continues, one of the drivers for housing and husbandry evolution will have been weakened.

7.5 Maintaining the Momentum

There is no doubt in my mind that animal welfare and the importance of the 3Rs have a higher profile now than ever before and that standards are generally improving. Nonetheless, ensuring high standards of welfare for animals used in research is a continuous process.

Further advances in animal welfare require an acceptance that there is a task still to be undertaken. Understandably, biomedical publicity, because of concerns about the public acceptability of research, tends to emphasise the positive aspects of research and the way that it is carried out. For example, some statements have suggested that UK welfare standards are amongst the highest in the world[58]. I suspect that there is an element of nationalism in these pronouncements, as good facilities and standards certainly exist outside the UK. However, even if there is some truth in the assertion, using it as a public relations tool is, I think, misguided as it risks producing a too complacent attitude. In my view, a more realistic and more useful approach would be to accept that in the past things were not done as well as they are today, that we need further research into animal welfare, and that occasionally, as in all spheres of life, errors or bad practice occur. The use of the word 'journey' has been devalued by too many reality show participants, but in a sense those who use animals in research are on a journey with animal welfare scientists in which the aim is to achieve continual improvement through better implementation of the 3Rs.

Advances in the 3Rs depend on research. Fortunately, there are now more animal welfare scientists working in the field of laboratory animal science than ever before. My UK experience has been that collaborative studies between animal

---

[58] For example, in October 2012 statements to this effect were available online from several UK universities and the Association of Medical Research Charities (AMRC). A parliamentary briefing from Understanding Animal Research (UAR) also stated that 'The UK's regulatory framework for the use of animals in research is founded on the Animals (Scientific Procedures) Act 1986 (ASPA) and is among the most stringent in the world'.

welfare scientists and the biomedical research industry are also much more common than when I started my career. This is to a great extent due to initiatives such as those run by UFAW and the NC3Rs, which fund or facilitate 3Rs research with the biomedical industry. Within the UK, funding in this area dramatically increased following the establishment of the NC3Rs, which spent an average of £3.75 million per annum in its first 8 years. Nonetheless, resources for research remain an issue in many countries.

There is a need to improve links between animal welfare scientists and those who use animals in other areas of research (e.g. biomedicine). These contacts would help to ensure that those who use animals are up to speed on the 3Rs and provide opportunities for collaborative animal welfare research with those scientists. Meetings organised by laboratory animal science (LAS) organisations are aimed at providing CPD, but despite the fact that the membership of these organisations includes researchers as well as veterinarians, animal care staff and animal welfare scientists, the number of researchers who attend and take part in LAS meetings or workshops is often quite small. Researchers may be reluctant to spend time and resources on attending meetings that are not directly related to their research, so ways need to be found to make animal welfare science more accessible to this audience. The organisers of researchers' professional meetings could arrange slots for speakers on animal welfare issues, or arrange joint meetings with LAS or welfare organisations. Some activities like this are already occurring, but more are needed.

Those working in animal welfare science also have their part to play in improving links. The science behind meeting animals' needs is a fascinating one and much of it is of high quality. Nonetheless, it is important that animal welfare professionals (including those who are not scientists) are open about uncertainties in their field. It may never be possible to prove that some animals have feelings that matter to them, however strongly some may feel this to be true. For the same reason, it is necessary to be clear that welfare assessment is a judgement, although ideally one based on science. It is also important that the 'language' in animal welfare science is as rigorous as possible. Terms such as enrichment, stress and even animal welfare need to be clearly defined and used consistently by all involved. Finally, it goes without saying that animal welfare research must be of high quality to be influential.

7.6 Conclusion

Despite the progress that has been made in improving the welfare of animals used in research, much more remains to be done. I think it unlikely that research on animals will cease in the near future. It is hard to see how animal use in many areas of fundamental research could be replaced using the technology that we have at present and we cannot be sure what types of animal research there may be in the future. Of course we must continue to work to find replacements and it is possible that at some point human ingenuity may prevail so that live animal use becomes completely redundant. In the meantime, we must continue our efforts to reduce animal use and to refine methods so as to avoid or minimise animal pain and distress and ensure good animal welfare. This will require the continued development of better techniques and good-quality training for all those involved. We still know remarkably little about the needs of some species commonly used in research, so it is essential that the science of laboratory animal welfare remains an active discipline, and that it should further develop internationally. This will require the support of governments around the world, of charitable organisations, and of the research industry and academia. We have made much progress in improving the welfare of animals used in research in the half-century since Russell and Burch first published *The Principles*, but we should not rest on our laurels. The next 50 years will be as great a challenge.

References

Abelson, K.S.P. & Roughan, J.V. (2011) Animal models in pain research. In *Handbook of Laboratory Animal Science, Volume II: Animal Models*, 3rd edn, Hau, J. & Schapiro, S.J. (eds.), pp. 123–147. CRC Press, Boca Raton, FL.

Agresti, J.J., Antipov, E., Abate, A.R., Ahn, K., Rowat, A.C., Baret, J.-C., Marquez, M., Klibanov, A.M., Griffiths, A.D. & Weitz, D.A. (2010) Ultrahigh-throughput screening in drop-based microfluidics for directed evolution. *Proceedings of the National Academy of Sciences of the United States of America* 107, 4004–4009. doi:10.1073/pnas.0910781107.

Akhtar, S. (2011) Animal pain and welfare: can pain be worse for them than for us? In *The Oxford Handbook of Animal Ethics*, Beauchamp, T.L. & Frey, R.G. (eds.), pp. 495–518. Oxford University Press, New York.

Allen, M. (1998) US and UK control of lab animal experimentation. *Lab Animal* 27, 34–39.

American Veterinary Medical Association (2007) AVMA Guidelines on Euthanasia (Formerly Report of the AVMA Panel on Euthanasia). American Veterinary Medical Association, Schaumburg, IL. Available at http://www.avma.org/issues/animal_welfare/euthanasia.pdf.

Anderson, R.C. (2006) Smart Octopus? *The Festivus* 38, 7–9.

Andrews, N., Harper, S., Issop, Y. & Rice, A.S.C. (2011) Novel, nonreflex tests detect analgesic action in rodents at clinically relevant concentrations. *Annals of the New York Academy of Sciences* 1245, 11–13.

Animal Procedures Committee (1997) *Report of the Animal Procedures Committee for 1997*. The Stationery Office, London.

Animal Procedures Committee (2001) Animal Procedures Committee Report on Biotechnology June 2001. Home Office. Available at http://www.homeoffice.gov.uk/agencies-public-bodies/apc/, accessed 12 October 2012.

Animal Procedures Committee (2003) Review of Cost–Benefit Assessment in the Use of Animals in Research. Home Office, Communication Directorate. Available at http://www.homeoffice.gov.uk/agencies-public-bodies/apc/, accessed 12 October 2012.

The Welfare of Animals Used in Research: Practice and Ethics, First Edition. Robert C. Hubrecht.
© 2014 Universities Federation for Animal Welfare. Published 2014 by John Wiley & Sons, Ltd.

Animal Procedures Committee (2006a) Acceptance of Overseas Centre Supplying Non-Human Primates to UK Laboratories: A Report by the Primates Sub-Committee of the Animals Procedures Committee. Available at http://www.homeoffice.gov.uk/agencies-public-bodies/apc/, accessed 12 October 2012.

Animal Procedures Committee (2006b) Review of Schedule 1 of The Animals (Scientific Procedures) Act 1986: Appropriate Methods of Humane Killing. Available at http://www.homeoffice.gov.uk/agencies-public-bodies/apc/, accessed 12 October 2012.

Animal Procedures Committee (2007) Consideration of policy concerning standards of animal housing and husbandry for animals from overseas non-designated sources. In *Report of the Animal Procedures Committee for 2006*, pp. 36–42. Animal Procedures Committee, London.

Animal Procedures Committee (2009a) Report of the Suffering and Severity Working Group on the Strengths and Weaknesses of the Current System of Severity Limits as a Way of Prospectively Assessing Suffering and Severity. Available at http://www.homeoffice.gov.uk/agencies-public-bodies/apc/, accessed 12 October 2012.

Animal Procedures Committee (2009b) Supplementary Review of Schedule 1 of the Animals (Scientific Procedures) Act 1986: Appropriate Methods of Humane Killing for Fish. Available at http://www.homeoffice.gov.uk/agencies-public-bodies/apc/, accessed 12 October 2012.

Animals Scientific Procedures Inspectorate (2010) A review on the issues and concerns raised in the report, 'The Ugly Truth: a BUAV investigation at Wickham Laboratories', November 2010. Available at http://www.homeoffice.gov.uk/publications/science-research-statistics/animals/wickham-laboratories.

Apfelbach, R., Blanchard, C.D., Blanchard, R.J., Hayes, R.A. & McGregor, I.S. (2005) The effects of predator odors in mammalian prey species: a review of field and laboratory studies. *Neuroscience and Biobehavioral Reviews* 29, 1123–1144.

Armus, H.L. (1970) Conditioning of the sensitive plant, *Mimosa pudica*. In *Comparative Psychology. Research in Animal Behavior*, Denny, M.R. & Ratner, S.C. (eds.), pp. 597–600. Dorsey Press, Homewood, IL.

Arts, J.W.M., Kramer, K. & Ohl, F. (2011) The effect of transportation on the physiology and behaviour of rats. In *ALTEX Alternatives to Animal Experimentation. Abstracts of the 8th World Congress, Montreal, 2011*, 28 (Special Issue), 243.

ASAB (2006) Guidelines for the treatment of animals in behavioural research and teaching. *Animal Behaviour* 71, 245–273. doi:10.1016/j.anbehav.2005.10.001.

Auersperg, A.M.I., Kacelnik, A. & von Bayern, A.M.P. (2013) Explorative learning and functional inferences on a five-step means-means-end problem in Goffin's Cockatoos (*Cacatua goffini*). *PLoS ONE* 8, e68979. doi:10.1371/journal.pone.0068979.

Baars, B.J. (1997) *In the Theater of Consciousness: The Workspace of the Mind*. Oxford University Press, New York.

Bailey, G.S., Reddy, A.P., Pereira, C.B., Harttig, U., Baird, W., Spitsbergen, J.M., Hendricks, J.D., Orner, G.A., Williams, D.E. & Swenberg, J.A. (2009) Nonlinear cancer response at ultralow dose: a 40800-animal ED001 tumor and biomarker study. *Chemical Research in Toxicology* 22, 1264–1276.

Baker, J.R. (1948) *The Scientific Basis of Kindness to Animals*. UFAW, London.

Baker, K.C., Weed, J.L., Crockett, C.M. & Bloomsmith, M.A. (2007) Survey of environmental enhancement programs for laboratory primates. *American Journal of Primatology* 69, 377–394.

Balls, M. (2007) Professor W.M.S. Russell (1925–2006): doyen of the three Rs. Proceedings of the 6th World Congress on Alternatives and Animal Use in the Life Sciences, 21–25 August 2007, Tokyo, Japan. *AATEX* 14 (Special Issue), 1–7.

Balls, M. (2009) The origins and early days of the Three Rs concept. *ATLA, Alternatives to Laboratory Animals* 37, 255.

Balls, M., Bennett, A. & Kendall, D. (2012) Translation of new technologies in biomedicines: shaping the road from basic research to drug development and clinical application – and back again. In *Pharmaceutical Biotechnology: Drug Discovery and Clinical Applications*, 2nd edn, Kayser, O. & Warzecha, H. (eds.), pp. 113–152. Wiley-VCH, Weinheim, Germany.

Barnard, C.J. & Hurst, J.L. (1996) Welfare by design: the natural selection of welfare criteria. *Animal Welfare* 5, 405–433.

Barr, S., Laming, P.R., Dick, J.T.A. & Elwood, R.W. (2008) Nociception or pain in a decapod crustacean? *Animal Behaviour* 75, 745–751. doi:10.1016/j.anbehav.2007.07.004.

Bateson, M. (2004) Mechanisms of decision-making and the interpretation of choice tests. *Animal Welfare* 13, 115–120.

Bateson, M. & Matheson, S. (2007) Performance on a categorisation task suggests that removal of environmental enrichment induces 'pessimism' in captive European starlings (*Sturnus vulgaris*). *Animal Welfare* 16, 33–36.

Bateson, M., Desire, S., Gartside, S.E. & Wright, G.A. (2011) Agitated honeybees exhibit pessimistic cognitive biases. *Current Biology* 21, 1070–1073. doi:10.1016/jcub.2011.05.017.

Bateson, P. (1986) When to experiment on animals. *New Scientist* 109, 30–32.

Bateson, P. (1991) Assessment of pain in animals. *Animal Behaviour* 42, 827–839.

Bateson, P. (2005) Ethics and behavioral biology. *Advances in the Study of Behavior* 35, 211–233.

Bateson, P. (2011a) Ethical debates about animal suffering and the use of animals in research. *Journal of Consciousness Studies* 18, 186–208.

Bateson, P. (2011b) Review of Research Using Non-Human Primates: Report of a panel chaired by Professor Sir Patrick Bateson FRS. Biotechnology and Biological Sciences Research Council (BBSRC), Medical Research Council (MRC), The National Centre for the Replacement, Refinement and Reduction of Animals in Research (NC3Rs), Wellcome Trust.

Bayne, K. (2005) Potential for unintended consequences of environmental enrichment for laboratory animals and research results. *ILAR Journal* 46, 129–139.

Bayne, K. (2011) The IACUC's impact on advancing best practice and the Three RS. *ILAR Journal* 52 (Suppl.), 436–439.

Bayne, K., Howard, B.R., Tsutomu, M.K. & Eugenia, M.A.N. (2010a) An overview of global legislation, regulation, and policies. In *Handbook of Laboratory Animal Science, Volume I: Essential Principles and Practices*, 3rd edn, Hau, J. & Schapiro, S.J. (eds.), pp. 39–64. CRC Press, Boca Raton, FL.

Bayne, K., Morris, T.H. & France, M.P. (2010b) Legislation and oversight of the conduct of research using animals: a global overview. In *The UFAW Handbook on the Care and Management of Laboratory and Other Research Animals*, 8th edn, Hubrecht, R. & Kirkwood, J. (eds.), pp. 107–123. Wiley-Blackwell, Oxford.

Bayvel, D., Carsons, L. & Littin, K. (2007) Severity assessment: the New Zealand experience and perspective. Proceedings of the 6th World Congress on Alternatives and Animal Use in the Life Sciences, 21–25 August 2007, Tokyo, Japan. *AATEX* 14 (Special Issue), 711–714.

Beauchamp, T.L. & Frey, R.G. (2011) *The Oxford Handbook of Animal Ethics*. Oxford University Press, New York.

Beerda, B., Schilder, M.B.H. & van Hooff, J. (1997) Manifestations of chronic and acute stress in dogs. *Applied Animal Behaviour Science* 52, 307–319.

Beerda, B., Schilder, M.B.H., van Hooff, J.A.R.A.M., de Vries, H.W. & Mol, J.A. (1998) Behavioural, saliva cortisol and heart rate responses to different types of stimuli in dogs. *Applied Animal Behaviour Science* 58, 365–381.

Beerda, B., Schilder, M.B.H., Bernadina, W., Van Hooff, J.A.N., De Vries, H.W. & Mol, J.A. (1999a) Chronic stress in dogs subjected to social and spatial restriction. II. Hormonal and immunological responses. *Physiology and Behavior* 66, 243–254.

Beerda, B., Schilder, M.B.H., Van Hooff, J.A.R.A.M., De Vries, H.W. & Mol, J.A. (1999b) Chronic stress in dogs subjected to social and spatial restriction. I. Behavioral responses. *Physiology and Behavior* 66, 233–242. doi:10.1016/S0031-9384(98)00289-3.

Bekoff, M. & Meaney, C.A. (1998) *Encyclopedia of Animal Rights and Animal Welfare*. Greenwood Press, Westport, CT.

Berdoy, M. (2002) *The Laboratory Rat: A Natural History*. Film, 27 minutes. Available at http://www.ratlife.org, accessed 13 April 2013.

Bermond, B. (2001) A neuropsychological and evolutionary approach to animal consciousness and animal suffering. *Animal Welfare* 10, 47–62.

Beynon, R. & Hurst, J. (2003) Multiple roles of major urinary proteins in the house mouse, *Mus domesticus. Biochemical Society Transactions* 31, 142–146.

Blecha, F. (2000) Immune system response to stress. In *The Biology of Animal Stress: Basic Principles and Implications for Animal Welfare*, Moberg, G.P. & Mench, J.A. (eds.), pp. 111–121. CABI Publishing, Wallingford, UK.

Block, N. (2005) Two neural correlates of consciousness. *Trends in Cognitive Sciences* 9, 46–52.

Boissy, A., Manteuffel, G., Jensen, M.B., Moe, R.O., Spruijt, B., Keeling, L.J., Winckler, C., Forkman, B., Dimitrov, I. & Langbein, J. (2007) Assessment of positive emotions in animals to improve their welfare. *Physiology and Behavior* 92, 375–397.

Botham, P. (2004) Acute systemic toxicity: prospects for tiered testing strategies. *Toxicology in Vitro* 18, 227–230.

Boyd Group (2002) The use of non-human primates in research and testing. June 2002. Available at http://www.boyd-group.demon.co.uk/.

Boyer, S. (2009) The use of computer models in pharmaceutical safety evaluation. *ATLA, Alternatives to Laboratory Animals* 37, 467–475.

Bradshaw, J. (2011) *In Defence of Dogs*. Allen Lane, London

Bradshaw, J.W.S. & Casey, R.A. (2007) Anthropomorphism and anthropocentrism as influences in the quality of life of companion animals. *Animal Welfare* 16 (Suppl. 1), 149–154.

Bradshaw, J.W.S., Blackwell, E.J. & Casey, R.A. (2009) Dominance in domestic dogs: useful construct or bad habit? *Journal of Veterinary Behavior: Clinical Applications and Research* 4, 135–144.

Bradshaw, R.H. (2002) The ethical review process in the UK and Australia: the Australian experience of improved dialogue and communication. *Animal Welfare* 11, 141–156.

Braithwaite, V. (2010) *Do Fish Feel Pain?* Oxford University Press, Oxford.

Braithwaite, V.A. & Huntingford, F.A. (2004) Fish and welfare: do fish have the capacity for pain perception and suffering? *Animal Welfare* 13, 87–92.

Brambell, F.W.R. (Chairman) (1965) Report of the Technical Committee to Enquire into the Welfare of Animals Kept Under Intensive Livestock Husbandry Systems. Cmnd 2836. HMSO, London.

Bräuer, J., Kaminski, J., Riedel, J., Call, J. & Tomasello, M. (2006) Making inferences about the location of hidden food: social dog, causal ape. *Journal of Comparative Psychology* 120, 38–47.

Breuer, K., Hemsworth, P.H. & Coleman, G.J. (2003) The effect of positive or negative handling on the behavioural and physiological responses of nonlactating heifers. *Applied Animal Behaviour Science* 84, 3–22.

Broom, D.M. (2003) Transport stress in cattle and sheep with details of physiological, ethological and other indicators. *DTW. Deutsche tierärztliche Wochenschrift* 110, 83–89.

Broom, D.M. (2007) Cognitive ability and sentience: which aquatic animals should be protected? *Diseases of Aquatic Organisms* 75, 99–108.

Broom, D.M. (2008) The welfare of livestock during transport. In *Long Distance Transport and the Welfare of Farm Animals*, Appleby, M., Cussen, V., Garcés, L., Lambert, L. & Turner, J. (eds.), pp. 157–181. CAB International, Wallingford, UK.

Broom, D.M. (2010) Cognitive ability and awareness in domestic animals and decisions about obligations to animals. *Applied Animal Behaviour Science* 126, 1–11.

Broom, D.M. (2011) A history of animal welfare science. *Acta Biotheoretica* 59, 121–137.

Broom, D.M., Sena, H. & Moynihan, K.L. (2009) Pigs learn what a mirror image represents and use it to obtain information. *Animal Behaviour* 78, 1037–1041.

Browne, J. (2002) *Charles Darwin: The Power of Place*. Alfred A. Knopf, London.

Brydges, N.M., Leach, M., Nicol, K., Wright, R. & Bateson, M. (2011) Environmental enrichment induces optimistic cognitive bias in rats. *Animal Behaviour* 81, 169–175.

Bshary, R., Wickler, W. & Fricke, H. (2002) Fish cognition: a primate's eye view. *Animal Cognition* 5, 1–13.

Buchanan-Smith, H.M., Rennie, A.E., Vitale, A., Pollo, S., Prescott, M.J. & Morton, D.B. (2005) Harmonising the definition of refinement. *Animal Welfare* 14, 379–384.

Budellmann, B. (2010) Cephalopoda. In *The UFAW Handbook on the Care and Management of Laboratory and Other Research Animals*, 8th edn, Hubrecht, R. & Kirkwood, J. (eds.), pp. 787–793. Wiley-Blackwell, Oxford.

Burch, R.L. (1995) The progress of humane experimental technique since 1959: a personal view. *ATLA, Alternatives to Laboratory Animals* 23, 776–783.

Burch, R.L. & Balls, M. (2009) The progress of humane experimental technique since 1959: a personal view. *ATLA, Alternatives to Laboratory Animals* 37, 269–275.

Burkhardt, R.W. Jr (1997) The founders of ethology and the problem of animal subjective experience. In *Animal Consciousness and Animal Ethics: Perspectives from the Netherlands*, Dol, M., Kasanmoentalib, S., Lijmbach, S., Rivas, E. & van den Bos, R. (eds.), pp. 1–13. Van Gorcum, Assen, The Netherlands.

Burman, O.H.P., Parker, R.M.A., Paul, E.S. & Mendl, M. (2008). Sensitivity to reward loss as an indicator of animal emotion and welfare. *Biology Letters* 4, 330–333. doi:10.1098/rsbl.2008.0113.

Burn, C.C., Deacon, R.M.J. & Mason, G.J. (2008) Marked for life? Effects of early cage-cleaning frequency, delivery batch, and identification tail-marking on rat anxiety profiles. *Developmental Psychobiology* 50, 266–277.

Butler, A.B. (2008) Evolution of brains, cognition, and consciousness. *Brain Research Bulletin* 75, 442–449.

Butterworth, A., Weeks, C.A., Crea, P.R. & Kestin, S.C. (2002) Dehydration and lameness in a broiler flock. *Animal Welfare* 11, 89–94.

Butterworth, A., Mench, J.A. & Wielebnowski, N. (2011) Practical strategies to assess (and improve) welfare. In *Animal Welfare*, 2nd edn, Appleby, M.C., Mench, J.A., Olsson, I.A.S. & Hughes, B.O. (eds.), pp. 200–214. CAB International, Wallingford, UK.

Byrne, R.W. (1995a) Primate cognition: comparing problems and skills. *American Journal of Primatology* 37, 127–141.

Byrne, R.W. (1995b) *Thinking Ape: Evolutionary Origins of Intelligence*. Oxford University Press, Oxford.

Byrne, R.W. (1996) Relating brain size to intelligence in primates. In *Modelling the Early Human Mind*, Mellars, P. & Gibson, K.R. (eds.), pp. 49–56. McDonald Institute for Archaeological Research, Cambridge, UK.

Byrne, R.W. (1999) Primate cognition: evidence for the ethical treatment of primates. In *Attitudes to Animals: Views in Animal Welfare*, Dolins, F.L. (ed.), pp. 114–125. Cambridge University Press, Cambridge, UK.

Carbone, L., Carbone, E.T., Yi, E.M., Bauer, D.B., Lindstrom, K.A., Parker, J.M., Austin, J.A., Seo, Y., Gandhi, A.D. & Wilkerson, J.D. (2012) Assessing cervical dislocation as a humane euthanasia method in mice. *Journal of the American Association for Laboratory Animal Science: JAALAS* 51, 352–356.

Cardon, A.D., Bailey, M.R. & Bennett, B.T. (2012a) The Animal Welfare Act: from enactment to enforcement. *Journal of the American Association for Laboratory Animal Science: JAALAS* 51, 301–305.

Cardon, A.D., Bailey, M.R. & Bennett, B.T. (2012b) The effect of public disclosure laws on biomedical research. *Journal of the American Association for Laboratory Animal Science: JAALAS* 51, 306–310.

Carey, W.D. & Cleveland Clinic Foundation (2010) *Current Clinical Medicine E-Book*, 2nd edn. Saunders/Elsevier, Philadelphia, PA. Available at http://www.clevelandclinicmeded.com/partners/elsevier/.

Carruthers, P. (1989) Brute experience. *Journal of Philosophy* 86, 258–269.

Carruthers, P. (2004) Suffering without subjectivity. *Philosophical Studies* 121, 99–125.

Carruthers, P. (2005) Why the question of animal consciousness might not matter very much. *Philosophical Psychology* 18, 83–102.

Casey, W., Schmitt, M., McFarland, R., Isbrucker, R., Levis, R., Arciniega, J., Descamps, J., Finn, T., Hendriksen, C., Horiuchi, Y. *et al.* (2011) Improving animal welfare and reducing animal use for human vaccine potency testing: state of the science and future directions. *Procedia in Vaccinology* 5, 33–46.

Castelhano-Carlos, M.J. & Baumans, V. (2009) The impact of light, noise, cage cleaning and in-house transport on welfare and stress of laboratory rats. *Laboratory Animals* 43, 311–327.

Chamove, A.S. (1996) Reducing animal numbers: sequential sampling. *Animal Welfare Information Center Newsletter* 7, 3–6.

Chandroo, K.P., Duncan, I.J.H. & Moccia, R.D. (2004a) Can fish suffer? Perspectives on sentience, pain, fear and stress. *Applied Animal Behaviour Science* 86, 225–250.

Chandroo, K.P., Yue, S. & Moccia, R.D. (2004b) An evaluation of current perspectives on consciousness and pain in fishes. *Fish and Fisheries* 5, 281–295. doi:10.1111/j.1467–2679.2004.00163.x.

Chang, F.T. & Hart, L.A. (2002) Human–animal bonds in the laboratory: how animal behavior affects the perspectives of caregivers. *ILAR Journal* 43, 10–18.

Chapman, K.L., Andrews, L., Bajramovic, J.J., Baldrick, P., Black, L.E., Bowman, C.J., Buckley, L.A., Coney, L.A., Couch, J. & Maggie Dempster, A. (2011) The design of chronic toxicology studies of monoclonal antibodies: implications for the reduction in use of non-human primates. *Regulatory Toxicology and Pharmacology* 62, 347–354.

Churchland, P. (1996) *The Engine of Reason, the Seat of the Soul*. MIT Press, Cambridge, MA.

Cinelli, P., Rettich, A., Seifert, B., Burki, K. & Arras, M. (2007) Comparative analysis and physiological impact of different tissue biopsy methodologies used for the genotyping of laboratory mice. *Laboratory Animals* 41, 174–184. doi:10.1258/002367707780378113.

Clayton, N.S. & Dickinson, A. (1998) Episodic-like memory during cache recovery by scrub jays. *Nature* 395, 272–273.

Clutton-Brock, J. (1995) Origins of the dog: domestication and early history. In *The Domestic Dog: Its Evolution, Behaviour, and Interactions with People*, Serpell, J. (ed.), pp. 7–20. Cambridge University Press, Cambridge, UK.

Coates, M.M. (2003) Visual ecology and functional morphology of Cubozoa (Cnidaria). *Integrative and Comparative Biology* 43, 542–548.

Coates, M.M., Garm, A., Theobald, J.C., Thompson, S.H. & Nilsson, D.E. (2006) The spectral sensitivity of the lens eyes of a box jellyfish, *Tripedalia cystophora* (Conant). *Journal of Experimental Biology* 209, 3758–3765.

Cooper, J. & Mason, G. (2001) The use of operant technology to measure behavioral priorities in captive animals. *Behavior Research Methods* 33, 427–434. doi:10.3758/bf03195397.

Cooper, J.J., Ashton, C., Bishop, S., West, R., Mills, D.S. & Young, R.J. (2003) Clever hounds: social cognition in the domestic dog (*Canis familiaris*). *Applied Animal Behaviour Science* 81, 229–244.

Coppinger, R. & Schneider, R. (1995) Evolution of working dogs. In *The Domestic Dog: Its Evolution, Behaviour, and Interactions with People*, Serpell, J. (ed.), pp. 22–47. Cambridge University Press, Cambridge, UK.

Corrado, M. (2001) No-one likes us – or do they? *Science and Public Affairs* 4, 14–15.

Cottingham, J. (1978) 'A brute to the brutes?' Descartes' treatment of animals. *Philosophy* 53, 551–559.

Crabbe, J.C., Wahlsten, D. & Dudek, B.C. (1999) Genetics of mouse behavior: interactions with laboratory environment. *Science* 284, 1670–1672.

Cressey, D. (2011a) Animal research: battle scars. *Nature* 470, 452–453.

Cressey, D. (2011b) Imaging animals for better research: modern techniques can cut the number of animals used and improve data quality. *Nature* doi:10.1038/news.2011.391.

Crick, F. (1994) *The Astonishing Hypothesis: The Scientific Search for the Soul*. Simon & Schuster, London.

Cronin, M.T.D. (2010) Quantitative structure–activity relationships (QSARs): applications and methodology. In *Recent Advances in QSAR Studies: Methods and Applications*, Puzyn, T., Leszczynski, J. & Cronin, M.T.D. (eds.), pp. 3–11. Springer, Dordrecht.

Cronin, M.T.D., Jaworska, J.S., Walker, J.D., Comber, M.H.I., Watts, C.D. & Worth, A.P. (2003) Use of QSARs in international decision-making frameworks to predict health effects of chemical substances. *Environmental Health Perspectives* 111, 1391–1401.

Crump, J.A. & Collignon, P.J. (2000) Intravascular catheter-associated infections. *European Journal of Clinical Microbiology and Infectious Diseases* 19, 1–8.

Culik, B.M., Wilson, R.P. & Bannasch, R. (1993) Flipper-bands on penguins: what is the cost of a life-long commitment? *Marine Ecology Progress Series* 98, 209–214.

Curren, R. (2009) Animal use in the chemical and product manufacturing sectors: can the downtrend continue? *ATLA, Alternatives to Laboratory Animals* 37, 623–629.

Cuthill, I.C. (2007) Ethical regulation and animal science: why animal behaviour is not so special. *Animal Behaviour* 74, 15–22.

Dahlborn, K., Bugnon, P., Nevalainen, T., Raspa, M., Verbost, P. & Spangenberg, E. (2013) Report of the Federation of European Laboratory Animal Science Associations Working

Group on animal identification. *Laboratory Animals* 47, 2–11. doi:10.1177/002367712473290.

Daly, M. (1989) Vivisection in eighteenth-century Britain. *Journal for Eighteenth-Century Studies* 12, 57–67.

Darwin, C. (1871) *The Descent of Man, and Selection in Relation to Sex*. Penguin Classics edition 2004. Penguin, London.

Dawkins, M.S. (1980) *Animal Suffering: The Science of Animal Welfare*. Chapman & Hall, London.

Dawkins, M.S. (1993) *Through Our Eyes Only? The Search for Animal Consciousness*. W.H. Freeman, New York.

Dawkins, M.S. (2001) Who needs consciousness? *Animal Welfare* 10, S19–S29.

Dawkins, M.S. (2008) The science of animal suffering. *Ethology* 114, 937–945.

Dawkins, M.S. (2012) *Why Animals Matter: Animal Consciousness, Animal Welfare and Human Well-being*. Oxford University Press, Oxford.

Dawkins, R. (2004) *The Ancestor's Tale: A Pilgrimage to the Dawn of Life*. Weidenfeld & Nicolson, London.

Deaner, R.O., Isler, K., Burkart, J. & van Schaik, C. (2007) Overall brain size, and not encephalization quotient, best predicts cognitive ability across non-human primates. *Brain, Behavior and Evolution* 70, 115–124.

de Leeuw, W. (2004) The Council of Europe: what is it? In *The Development of Science-based Guidelines for Laboratory Animal Care: Proceedings of the November 2003 International Workshop*, pp. 23–29. National Academies Press, Washington, DC.

Dennett, D.C. (1992) *Consciousness Explained*. Allen Lane The Penguin Press, London.

Dennis, M.B. (2010) Special housing arrangements. In *The UFAW Handbook on the Care and Management of Laboratory and Other Research Animals*, 8th edn, Hubrecht, R. & Kirkwood, J. (eds.), pp. 147–162. Wiley-Blackwell, Oxford.

de Waal, F.B.M. (2008) The thief in the mirror. *PLoS Biology* 6, 1621–1622.

Dewhurst, D. (2008) Is it possible to meet the learning objectives of undergraduate pharmacology classes with non-animal models? Proceedings of the 6th World Congress on Alternatives and Animal Use in the Life Sciences, 21–25 August 2007, Tokyo, Japan. *AATEX* 14 (Special Issue), 207–212.

Dielenberg, R.A. & McGregor, I.S. (2001) Defensive behavior in rats towards predatory odors a review. *Neuroscience and Biobehavioral Reviews* 25, 597–609.

DiVincenti Jr, L. & Wyatt, J.D. (2011) Pair housing of macaques in research facilities a science-based review of benefits and risks. *Journal of the American Association for Laboratory Animal Science: JAALAS* 50, 856–863.

Dolan, K. (1999) *Ethics, Animals and Science*. Blackwell Science, Oxford.

Dolan, K. (2007) *Laboratory Animal Law: Legal Control of the Use of Animals in Research*. Blackwell Publishing, Oxford.

Dolan, R.J. (2002) Emotion, cognition, and behavior. *Science* 298, 1191.

Dontas, I., Hau, J., Marinou, K. & Nevalainen, T. (2011) Basic procedures: dosing, sampling and immunisation. In *The COST Manual of Laboratory Animal Care and Use: Refinement, Reduction, and Research*, Howard, B., Nevalainen, T. & Perretta, G. (eds.), pp. 257–285. CRC Press, Boca Raton, FL.

Draayer, H. (2011) Overview of currently approved veterinary vaccine potency testing methods and methods in development that do not require animal use. *Procedia in Vaccinology* 5, 171–174.

Dresser, R. (1989) Measuring merit in animal research. *Theoretical Medicine and Bioethics* 10, 21–34. doi:10.1007/bf00625756.

Driscoll, J.W. (1995) Attitudes toward animals: species ratings. *Society and Animals* 3, 139–150.

Duffee, N., Nevalainen, T. & Hau, J. (2010) Education and training. In *Handbook of Laboratory Animal Science, Volume I: Essential Principles and Practices*, 3rd edn, Hau, J. & Schapiro, S.J. (eds.), pp. 81–96. CRC Press, Boca Raton, FL.

Dunbar, R.I.M. (1995) Neocortex size and group size in primates: a test of the hypothesis. *Journal of Human Evolution* 28, 287–296.

Dunbar, R.I.M. (2003) The social brain: mind, language, and society in evolutionary perspective. *Annual Review of Anthropology* 32, 163–181.

Duncan, I.J.H. (1993) Welfare is to do with what animals feel. *Journal of Agricultural and Environmental Ethics* 6 (Suppl.), 8–14.

Duncan, I.J.H. (2005) Science-based assessment of animal welfare: farm animals. *Revue scientifique et technique (International Office of Epizootics)* 24, 483–492.

Duncan, I.J.H. & Petherick, J.C. (1991) The implications of cognitive processes for animal welfare. *Journal of Animal Science* 69, 5017–5022.

Duncan, I.J.H., Widowski, T.M., Malleau, A.E., Lindberg, A.C. & Petherick, J.C. (1998) External factors and causation of dustbathing in domestic hens. *Behavioural Processes* 43, 219–228.

Edelman, D.B. & Seth, A.K. (2009) Animal consciousness: a synthetic approach. *Trends in Neurosciences* 32, 476–484.

Edgar, J.L. & Seaman, S.C. (2010) The effect of mirrors on the behaviour of singly housed male and female laboratory rabbits. *Animal Welfare* 19, 461–471.

Eisele, P.H. (2001) A practical dog bed for environmental enrichment for geriatric beagles, with applications for puppies and other small dogs. *Journal of the American Association for Laboratory Animal Science: JAALAS* 40, 36–38.

Ekins, S., Mestres, J. & Testa, B. (2007a) In silico pharmacology for drug discovery: methods for virtual ligand screening and profiling. *British Journal of Pharmacology* 152, 9–20.

Ekins, S., Mestres, J. & Testa, B. (2007b) In silico pharmacology for drug discovery: applications to targets and beyond. *British Journal of Pharmacology* 152, 21–37.

Elwood, R.W. & Appel, M. (2009) Pain experience in hermit crabs? *Animal Behaviour* 77, 1243–1246.

Elwood, R.W., Barr, S. & Patterson, L. (2009) Pain and stress in crustaceans? *Applied Animal Behaviour Science* 118, 128–136.

Emery, N. (2004) Are corvids feathered apes? Cognitive evolution in crows, jays, rooks and jackdaws. In *Comparative Analysis of Minds*, Watanabe, S. (ed.), pp. 181–213. Keio University Press, Tokyo.

Emery, N. (2006) Cognitive ornithology: the evolution of avian intelligence. *Philosophical Transactions of the Royal Society of London Series B: Biological Sciences* 361, 23–43.

Emery, N.J. & Clayton, N.S. (2005) Evolution of the avian brain and intelligence. *Current Biology* 15, R946–R950.

Esling, R.W.J. (1981) European animal experimentation law. In *Animals in Research: New Perspectives in Animal Experimentation*, Sperlinger, D. (ed.), pp. 39–62. John Wiley & Sons Ltd, Chichester.

European Commission (2012) National Competent Authorities for the Implementation of Directive 2010/63/EU on the Protection of Animals Used for Scientific Purposes: Working

Document on a Severity Assessment Framework, 11–12 July 2012. http://ec.europa.eu/environment/chemicals/lab_animals/home_en.htm.

European Commission (2013) Examples to illustrate the process of severity classification, day-to-day assessment and actual severity assessment. http://ec.europa.eu/environment/chemicals/lab_animals/pdf/examples.pdf.

Farm Animal Welfare Council (2009) Farm Animal Welfare in Great Britain: Past, Present and Future. Available at http://www.fawc.org.uk/pdf/ppf-report091012.pdf.

FELASA Working Group on Pain and Distress (1994) Pain and distress in laboratory rodents and lagomorphs. *Laboratory Animals* 28, 97–112.

FELASA Working Group on Standardization of Enrichment (2006) FELASA Working Group Standardization of Enrichment Report. Available at http://www.felasa.eu/recommendations/reports/Standardization-of-enrichment/, accessed 26 April 2013.

Fenwick, N. & Fraser, D. (2005) The Three Rs in the pharmaceutical industry: perspectives of scientists and regulators. *Animal Welfare* 14, 367–377.

Fenwick, N., Ormandy, E., Gauthier, C. & Griffin, G. (2011) Classifying the severity of scientific animal use: a review of international systems. *Animal Welfare* 20, 281–301.

Fernström, A.L., Sutian, W., Royo, F., Westlund, K., Nilsson, T., Carlsson, H.E., Paramastri, Y., Pamungkas, J., Sajuthi, D. & Schapiro, S.J. (2008) Stress in cynomolgus monkeys (*Macaca fascicularis*) subjected to long-distance transport and simulated transport housing conditions. *Stress: The International Journal on the Biology of Stress* 11, 467–476.

Festing, M.F.W. (2010) The design of animal experiments. In *The UFAW Handbook on the Care and Management of Laboratory and Other Research Animals*, 8th edn, Hubrecht, R. & Kirkwood, J. (eds.), pp. 23–36. Wiley-Blackwell, Oxford.

Festing, M.F.W. (2011) Reduction by careful design and statistical analsysis. In *The COST Manual of Laboratory Animal Care and Use: Refinement, Reduction, and Research*, Howard, B., Nevalainen, T. & Perretta, G. (eds.), pp. 131–149. CRC Press, Boca Raton, FL.

Festing, S. & Wilkinson, R. (2007) The ethics of animal research. Talking point on the use of animals in scientific research. *EMBO Reports* 8, 526–530. doi:10.1038/sj.embor.7400993.

Festing, M.F.W., Overend, P., Das, R.G., Borja, M.C. & Manuel, B. (2002) *The Design of Animal Experiments: Reducing the Use of Animals in Research Through Better Experimental Design*. Laboratory Animal Handbooks No. 14. Royal Society of Medicine Press Ltd, London.

Figgis, H. & Griffiths, G. (1998) Animal Experimentation: Background paper no 3/98. Parliamentary Library Research Service, Sydney, NSW. Available at http://preview.tinyurl.com/buy3y2v, accessed 26 April 2013.

Fleagle, J.G. (1988) *Primate Adaptation and Evolution*. Academic Press, San Diego, CA.

Flecknell, P. (2001) Recognition and assessment of pain in animals. In *Pain: Its Nature and Management in Man and Animals*, Lord Soulsby & Morton, D. (eds.), pp. 63–78. Royal Society of Medicine Press Ltd, London.

Flecknell, P.A. (2009) *Laboratory Animal Anaesthesia*. Academic Press, London.

Flecknell, P.A. & Liles, J.H. (1991) The effects of surgical procedures, halothane anaesthesia and nalbuphine on locomotor activity and food and water consumption in rats. *Laboratory Animals* 25, 50–60.

Flecknell, P.A. & Roughan, J.V. (2004) Assessing pain in animals: putting research into practice. *Animal Welfare* 13, S71–S75.

Flombaum, J.I. & Santos, L.R. (2005) Rhesus monkeys attribute perceptions to others. *Current Biology* 15, 447–452.

Francione, G.L. & Garner, R. (2010) *The Animal Rights Debate: Abolition or Regulation?* Columbia University Press, New York.

Francis, K.P. (2005) Biophotonic imaging and the reduction of animal usage in experimental research. Available at http://www.nc3rs.org.uk/news.asp?id=111.

Franco, N.H., Nunes, J.A. & Olsson, I.A.S. (2010) Shifting minds? The effect of training in laboratory animal science on researchers' attitudes to animal use in biomedical research. In *Global Food Security: Ethical and Legal Challenges. EurSafe 2010, Bilbao, Spain, 16–18 September 2010*, Casabona, C.M.R., San Epifanio, L.E. & Cirión, A.E. (eds.), pp. 456–458. Wageningen Academic Publishers, Wageningen, Netherlands.

Fraser, D. (1995) Science, values and animal welfare: exploring the 'inextricable connection'. *Animal Welfare* 4, 103–117.

Fraser, D. (1999) Animal ethics and animal welfare science: bridging the two cultures. *Applied Animal Behaviour Science* 65, 171–189.

Fraser, D. (2008) *Understanding Animal Welfare: The Science in its Cultural Context*. UFAW Animal Welfare Series. Wiley-Blackwell, Oxford.

Fraser, D., Weary, D.M., Pajor, E.A. & Milligan, B.N. (1997) A scientific conception of animal welfare that reflects ethical concerns. *Animal Welfare* 6, 187–205.

Friend, T.H. (2000) Dehydration, stress, and water consumption of horses during long-distance commercial transport. *Journal of Animal Science* 78, 2568–2580.

Gaeta, T.J., Fiorini, M., Ender, K., Bove, J. & Diaz, J. (2002) Potential drug–drug interactions in elderly patients presenting with syncope. *Journal of Emergency Medicine* 22, 159–162.

Gaines Das, R.E. (2004) 'Statistics' is not a sausage machine: a statistician's viewpoint and some comments on experimental design. *ATLA, Alternatives to Laboratory Animals* 32, 5.

Gaines Das, R.E., Fry, D., Preziosi, R. & Hudson, M. (2009) Planning for reduction. *ATLA, Alternatives to Laboratory Animals* 37, 27–32.

Garcia, J. & Koelling, R.A. (1996) Relation of cue to consequence in avoidance learning. *Psychonomic Science* 4, 123–124.

Garner, J.P. (2005) Stereotypies and other abnormal repetitive behaviors: potential impact on validity, reliability, and replicability of scientific outcomes. *ILAR Journal* 46, 106–117.

Gaskell, G., Stares, S., Allansdottir, A., Allum, N., Corchero, C., Fischler, C., Hampel, J., Jackson, J., Kronberger, N., Mejlgaard, N. *et al.* (2005) Europeans and Biotechnology in 2005: Patterns and Trends: Final report on Eurobarometer 64.3. A report to the European Commission's Directorate-General for Research. http://ec.europa.eu/research/biosociety/pdf/eb_64_3_final_report_second_edition_july_06.pdf.

Gaskill, B.N., Rohr, S.A., Pajor, E.A., Lucas, J.R. & Garner, J.P. (2009) Some like it hot: mouse temperature preferences in laboratory housing. *Applied Animal Behaviour Science* 116, 279–285.

Gaskill, B.N., Gordon, C.J., Pajor, E.A., Lucas, J.R., Davis, J.K. & Garner, J.P. (2012) Heat or insulation: behavioral titration of mouse preference for warmth or access to a nest. *PLoS ONE* 7, e32799. doi:10.1371/journal.pone.0032799.

Gaskill, B.N., Gordon, C.J., Pajor, E.A., Lucas, J.R., Davis, J.K. & Garner, J.P. (2013a) Impact of nesting material on mouse body temperature and physiology. *Physiology and Behavior* 110–111, 87–95. doi:10.1016/j.physbeh.2012.12.018.

Gaskill, B.N., Winnicker, C., Garner, J.P. & Pritchett-Corning, K.R. (2013b) The naked truth: breeding performance in nude mice with and without nesting material. *Applied Animal Behaviour Science* 143, 110–116.

Gaskill, B.N., Pritchett-Corning, K.R., Gordon, C.J., Pajor, E.A., Lucas, J.R., Davis, J.K. & Garner, J.P. (2013c) Energy reallocation to breeding performance through improved nest building in laboratory mice. *PLoS ONE* 8, e74153.

Geyer, M.A. & Markou, A. (1995) Animal models of psychiatric disorders. In *Psychopharmacology: The Fourth Generation of Progress*, Bloom, F.E. & Kupfer, D.J. (eds.), pp. 787–798. Raven Press, New York.

Geyer, M.A. & Markou, A. (2002) The role of preclinical models in the development of psychotropic drugs. In *Psychopharmacology: Fifth Generation of Progress*, Davis, K.L., Charney, D., Coyle, J.T. & Nemeroff, C. (eds.), pp. 445–455. American College of Neuropsychopharmacology, Nashville, TN. Available at http://www.acnp.org/publications/neuro5thgeneration.aspx.

Gherardi, F. (2009) Behavioural indicators of pain in crustacean decapods. *Annali dell'Istituto Superiore di Sanità* 45, 432–438.

Gonder, J.C., Smeby, R.R. & Wolfle, T.L. (eds.) (2001) *Performance Standards and Animal Welfare: Definition, Application and Assessment, Parts I and II*. Scientists Center for Animal Welfare, Bend, OR.

Gottlieb, D.H., Coleman, K. & McCowan, B. (2013) The effects of predictability in daily husbandry routines on captive rhesus macaques (*Macaca mulatta*). *Applied Animal Behaviour Science* 143, 117–127. doi:10.1016/j.applanim.2012.10.010.

Grandin, T. (ed.) (1998) *Genetics and the Behavior of Domestic Animals*. Academic Press, San Diego, CA.

Grandin, T. & Johnson, C. (2009) *Making Animals Happy: Creating the Best life for Animals*. Bloomsbury, London.

Grayson, L. (2000) *Animals in Research: For and Against*. British Library, London.

Gregory, N.G. (2004) *Physiology and Behaviour of Animal Suffering*. UFAW Animal Welfare Series. Blackwell Science, Ames, IA.

Griffin, D.R. (1976) *The Question of Animal Awareness: Evolutionary Continuity of Mental Experience*. Rockefeller University Press, New York.

Griffin, D.R. (1998) From cognition to consciousness. *Animal Cognition* 1, 3–16.

Griffin, D.R. (2001) *Animal Minds: Beyond Cognition to Consciousness*. University of Chicago Press, Chicago.

Griffin, D.R. & Speck, G.B. (2004) New evidence of animal consciousness. *Animal Cognition* 7, 5–18.

Güntürkün, O. & Durstewitz, D. (2001) Multimodal areas of the avian forebrain: blueprints for cognition. In *Brain Evolution and Cognition*, Roth, G. & Wullimann, M.F. (eds.), pp. 431–450. John Wiley & Sons, New York/Spektrum, Heidelberg.

Haemisch, A. & Gärtner, K. (1994) The cage design affects intermale aggression in small groups of male laboratory mice: strain specific consequences on social organization, and endocrine activations in two inbred strains (DBA/2J and CBA/J). *Journal of Experimental Animal Science* 36, 101.

Haemisch, A., Voss, T. & Gärtner, K. (1994) Effects of environmental enrichment on aggressive behavior, dominance hierarchies, and endocrine states in male DBA/2J mice. *Physiology and Behavior* 56, 1041–1048.

Hagelin, J., Carlsson, H.E. & Hau, J. (2003) An overview of surveys on how people view animal experimentation: some factors that may influence the outcome. *Public Understanding of Science* 12, 67–81. doi:10.1177/0963662503012001247.

Hall, M. (1847) On experiments in physiology as a question of medical ethics. *The Lancet* 49 (1220), 58–60.

Hanlon, R.T. & Messenger, J.B. (1996) *Cephalopod Behaviour*. Cambridge University Press, Cambridge, UK.

Hansen, A.K. (2010) Health status and health monitoring. In *Handbook of Laboratory Animal Science, Volume I: Essential Principles and Practices*, 3rd edn, Hau, J. & Schapiro, S.J. (eds.), pp. 251–305. CRC Press, Boca Raton, FL.

Harding, E.J., Paul, E.S. & Mendl, M. (2004) Animal behaviour: cognitive bias and affective state. *Nature* 427, 312.

Harding, J.D., Van Hoosier Jr, G.L. & Grieder, F.B. (2010) The contribution of laboratory animals to medical progress: past present and future. In *Handbook of Laboratory Animal Science, Volume I: Essential Principles and Practices*, 3rd edn, Hau, J. & Schapiro, S.J. (eds.), pp. 1–19. CRC Press, Boca Raton, FL.

Hardy, P. & Wolfensohn, S. (2011) Animal models: selecting and preparing animals for study. In *The COST Manual of Laboratory Animal Care and Use: Refinement, Reduction, and Research*, Howard, B., Nevalainen, T. & Perretta, G. (eds.), pp. 151–178. CRC Press, Boca Raton, FL.

Hare, B., Brown, M., Williamson, C. & Tomasello, M. (2002) The domestication of social cognition in dogs. *Science* 298, 1634–1636. doi:10.1126/science.1072702.

Hasiwa, N., Bailey, J., Clausing, P., Daneshian, M., Eileraas, M., Farkas, S., Gyertyan, I., Hubrecht, R., Kobel, W. & Krummenacher, G. (2011) Critical evaluation of the use of dogs in biomedical research and testing in Europe. *ALTEX* 28, 326–340.

Hau, J. & Schapiro, S.J. (eds.) (2011) *Handbook of Laboratory Animal Science, Volume II: Animal Models*. CRC Press, Boca Raton, FL.

Hawkins, P. (2002) Recognizing and assessing pain, suffering and distress in laboratory animals: a survey of current practice in the UK with recommendations. *Laboratory Animals* 36, 378–395.

Hawkins, P., Playle, L., Golledge, H., Leach, M., Banzett, R., Coenen, A., Cooper, J., Danneman, P., Flecknell, P., Kirkden, R. *et al.* (2006) Newcastle consensus meeting on carbon dioxide euthanasia of laboratory animals, 27–28 February 2006, Newcastle upon Tyne, UK. Available at http://www.nc3rs.org.uk/downloaddoc.asp?id=416&page= 292&skin=0, accessed 22 October 2012.

Hawkins, P., Fry, D., Wells, S., Honess, P., Yeates, J., Farmer, A.M., Main, D., Osborne, N., Jennings, M. & Hubrecht, R. (2010) Report of the 2009 RSPCA/UFAW Rodent Welfare Group meeting. Focusing on fifty years of the Three Rs and welfare assessment. *Animal Technology and Welfare* 9, 87–98.

Hawkins, P., Burn, C., Deacon, R., Dennison, N., Heath, M., Mackin, P., Tremoleda, J., Webb, A., Woodhall, G., Jennings, M. *et al.* (2012) Report of the 2011 RSPCA/UFAW Rodent Welfare Group meeting: harms and benefits of new technologies. *Animal Technology and Welfare* 11, 95–104.

Hawkins, P., Gimpel, J., Rice, A.S.C., Wells, D.J., Latcham, J., Heath, K., Gardiner, M., Wattam, T., Lilley, E., Jennings, M. *et al.* (2013) Report of the 2012 RSPCA/UFAW Rodent Welfare Group meeting. *Animal Technology and Welfare* 12, 49–58.

Healy, S.D. & Rowe, C. (2007) A critique of comparative studies of brain size. *Proceedings of the Royal Society of London Series B: Biological Sciences* 274, 453–464. doi:10.1098/rspb.2006.3748.

Healy, S.D. & Tovée, M.J. (1999) Environmental enrichment and impoverishment: neurophysiological effects. In *Attitudes to Animals: Views in Animal Welfare*, Dolins, F.L. (ed.), pp. 54–76. Cambridge University Press, Cambridge, UK.

Hedenqvist, P. & Flecknell, P. (2011) Anaesthesia and analgesia. In *The COST Manual of Laboratory Animal Care and Use: Refinement, Reduction, and Research*, Howard, B., Nevalainen, T. & Perretta, G. (eds.), pp. 313–322. CRC Press, Boca Raton, FL.

Hellebrekers, L.J. & Henenqvist, P. (2010) Laboratory animal analgesia, anesthesia, and euthanasia. In *Handbook of Laboratory Animal Science, Volume I: Essential Principles and Practices*, 3rd edn, Hau, J. & Schapiro, S.J. (eds.), pp. 485–534. CRC Press, Boca Raton, FL.

Hellyer, P.W., Frederick, C., Lacy, M., Salman, M.D. & Wagner, A.E. (1999) Attitudes of veterinary medical students, house officers, clinical faculty, and staff toward pain management in animals. *Journal of the American Veterinary Medical Association* 214, 238–244.

Hendriksen, C.F.M. & Morton, D.B. (eds.) (1999) *Humane Endpoints in Animal Experiments for Biomedical Research: Proceedings of the International Conference, 22–25 November 1998, Zeist, The Netherlands*. Royal Society of Medicine Press, London.

Hendriksen, C.F.M., Steen, B., Visser, J., Cussler, K., Morton, D. & Streijger, F. (1999) The evaluation of humane endpoints in pertussis vaccine potency testing. In *Humane Endpoints in Animal Experiments for Biomedical Research: Proceedings of the International Conference, 22–25 November 1998, Zeist, The Netherlands*, Hendriksen, C.F.M. & Morton, D.B. (eds.), pp. 5–12. Royal Society of Medicine Press, London.

Hendriksen, C.F.M., Morton, D.B. & Cussler, K. (2011) Use of humane endpoints to minimise suffering. In *The COST Manual of Laboratory Animal Care and Use: Refinement, Reduction, and Research*, Howard, B., Nevalainen, T. & Perretta, G. (eds.), pp. 333–351. CRC Press, Boca Raton, FL.

Hennessy, M.B., Williams, M.T., Miller, D.D., Douglas, C.W. & Voith, V.L. (1998) Influence of male and female petters on plasma cortisol and behaviour: can human interaction reduce the stress of dogs in a public animal shelter? *Applied Animal Behaviour Science* 61, 63–77.

Herzog, H. (2002) Ethical aspects of relationships between humans and research animals. *ILAR Journal* 43, 27–32.

Hessler, J.R. (2010) Laboratory animal facilities and equipment for conventional, barrier, and containment housing systems. In *Handbook of Laboratory Animal Science, Volume I: Essential Principles and Practices*, 3rd edn, Hau, J. & Schapiro, S.J. (eds.), pp. 145–207. CRC Press, Boca Raton, FL.

Heyes, C. (2008) Beast machines? Questions of animal consciousness. In *Frontiers of Consciousness*, Davies, M. & Weiskrantz, L. (eds.), pp. 259–274. Oxford University Press, Oxford.

Hobson-West, P. (2010) The role of 'public opinion' in the UK animal research debate. *Journal of Medical Ethics* 36, 46–49. doi:10.1136/jme.2009.030817.

Hockly, E., Cordery, P.M., Woodman, B., Mahal, A., Van Dellen, A., Blakemore, C., Lewis, C.M., Hannan, A.J. & Bates, G.P. (2002) Environmental enrichment slows disease progression in R6/2 Huntington's disease mice. *Annals of Neurology* 51, 235–242.

Hofman, M.A. (2001) Evolution and complexity of the human brain: some organizing principles. In *Brain Evolution and Cognition*, Roth, G. & Wullimann, M.F. (eds.), pp. 501–521. John Wiley & Sons, New York/Spektrum, Heidelberg.

Holmberg, A. & Pelletier, R. (2009) Automated blood sampling and the 3Rs. Available at http://www.nc3rs.org.uk/news.asp?id=1039.

Holmes, A.M., Rudd, J.A., Tattersall, F.D., Aziz, Q. & Andrews, P.L.R. (2009) Opportunities for the replacement of animals in the study of nausea and vomiting. *British Journal of Pharmacology* 157, 865–880. doi:10.1111/j.1476–5381.2009.00176.x.

Holmes, R. (2008) *The Age of Wonder: How the Romantic Generation Discovered the Beauty and Terror of Science*. HarperPress, London.

Home Office (2000) Guidance on the Operation of the Animals (Scientific Procedures) Act 1986. The Stationery Office, London.

Home Office (2012) *Statistics of Scientific Procedures on Living Animals Great Britain 2011*. The Stationery Office, London.

Honess, P. & Wolfensohn, S. (2010) The extended welfare assessment grid: a matrix for the assessment of welfare and cumulative suffering in experimental animals. *ATLA, Alternatives to Laboratory Animals* 38, 205–212.

Honess, P.E., Johnson, P.J. & Wolfensohn, S.E. (2004) A study of behavioural responses of non-human primates to air transport and re-housing. *Laboratory Animals* 38, 119–132.

Hooijmans, C., de Vries, R., Leenaars, M. & Ritskes-Hoitinga, M. (2010) The Gold Standard Publication Checklist (GSPC) for improved design, reporting and scientific quality of animal studies GSPC versus ARRIVE guidelines. *Laboratory Animals* 45, 61.

House of Lords (2002) Select Committee on Animals in Scientific Procedures, Volume I Report. HL Paper 150-I. Available at http://www.publications.parliament.uk/pa/ld200102/ldselect/ldanimal/150/15004.htm.

Howard, B. & Nevalainen, T. (2010) Attaining competence in the care of animals used in research. In *The UFAW Handbook on the Care and Management of Laboratory and Other Research Animals*, 8th edn, Hubrecht, R. & Kirkwood, J. (eds.), pp. 194–205. Wiley-Blackwell, Oxford.

Howard, B., Howard, K. & Sandøe, P. (2011) Education, training, and competence. In *The COST Manual of Laboratory Animal Care and Use: Refinement, Reduction, and Research*, Howard, B., Nevalainen, T. & Perretta, G. (eds.), pp. 369–390. CRC Press, Boca Raton, FL.

Hubrecht, R.C. (1993) A comparison of social and environmental enrichment methods for laboratory housed dogs. *Applied Animal Behaviour Science* 37, 345–361.

Hubrecht, R.C. (1995) Enrichment in puppyhood and its effects on later behavior of dogs. *Laboratory Animal Science* 45, 70.

Hubrecht, R.C. (1997) Current practice in maintaining marmosets: results of a UK survey In *Marmosets and Tamarins in Biological and Biomedical Research: Proceedings of a Workshop Organised by the European Marmoset Research Group*, Pryce, C., Scott, L. & Schnell, C. (eds.). DDSD Imagery, Salisbury, UK.

Hubrecht, R.C. (2002) Comfortable quarters for dogs in research institutions. In *Comfortable Quarters for Laboratory Animals*, Reinhardt, V. & Renhardt, A. (eds.), pp. 56–64. Animal Welfare Institute, Washington, DC.

Hubrecht, R.C. (2010) Enrichment: animal welfare and experimental outcomes. In *The UFAW Handbook on the Care and Management of Laboratory and Other Research Animals*, 8th edn, Hubrecht, R. & Kirkwood, J. (eds.), pp. 136–146. Wiley-Blackwell, Oxford.

Hubrecht, R.C. & Buckwell, A. (2004) The welfare of laboratory dogs. In *The Welfare of Laboratory Animals*, Kaliste, E. (ed.), pp. 245–273. Kluwer Academic Press, Dordrecht.

Hubrecht, R.C. & Kirkwood, J. (eds.) (2010) *The UFAW Handbook on the Care and Management of Laboratory and Other Research Animals*, 8th edn. Wiley-Blackwell, Oxford.

Hubrecht, R.C., Serpell, J.A. & Poole, T.B. (1992) Correlates of pen size and housing conditions on the behaviour of kennelled dogs. *Applied Animal Behaviour Science* 34, 365–383.

Hubrecht, R.C., Wickens, S. & Kirkwood, J. (in preparation) The welfare of dogs in human care. In *The Domestic Dog: Its Evolution, Behaviour and Interactions with People*, 2nd edn, Serpell, J. (ed.). Cambridge University Press, Cambridge, UK.

Hughes, J.R. (2007) A review of sleepwalking (somnambulism): the enigma of neurophysiology and polysomnography with differential diagnosis of complex partial seizures. *Epilepsy and Behavior* 11, 483–491.

Huh, D., Leslie, D.C., Matthews, B.D., Fraser, J.P., Jurek, S., Hamilton, G.A., Thorneloe, K.S., McAlexander, M.A. & Ingber, D.E. (2012) A human disease model of drug toxicity-induced pulmonary edema in a lung-on-a-chip microdevice. *Science Translational Medicine* 4, 159ra147. doi:10.1126/scitranslmed.3004249.

Hurst, J.L. & West, R.S. (2010) Taming anxiety in laboratory mice. *Nature Methods* 7, 825–826. doi:10.1038/nmeth.1500.

Inglis, I.R., Mathews, F. & Hudson, A. (2010) Wild mammals. In *The UFAW Handbook on the Care and Management of Laboratory and Other Research Animals*, 8th edn, Hubrecht, R. & Kirkwood, J. (eds.), pp. 231–245. Wiley-Blackwell, Oxford.

Izmirli, S., Aldavood, S.J., Yasar, A. & Phillips, C.J.C. (2010) Introducing ethical evaluation of the use of animals in experiments in the Near East. *ATLA, Alternatives to Laboratory Animals* 38, 331–336.

Jamieson, D. & Bekoff, M. (1992) Carruthers on nonconscious experience. *Analysis* 52, 23–28.

Jankowsky, J.L., Melnikova, T., Fadale, D.J., Xu, G.M., Slunt, H.H., Gonzales, V., Younkin, L.H., Younkin, S.G., Borchelt, D.R. & Savonenko, A.V. (2005) Environmental enrichment mitigates cognitive deficits in a mouse model of Alzheimer's disease. *Journal of Neuroscience* 25, 5217–5224.

Jarvis, E.D., Güntürkün, O., Bruce, L., Csillag, A., Karten, H., Kuenzel, W., Medina, L., Paxinos, G., Perkel, D. & Shimizu, T. (2005) Avian brains and a new understanding of vertebrate brain evolution. *Nature Reviews Neuroscience* 6, 151–159.

Jensen, P. (1993) Nest building in domestic sows: the role of external stimuli. *Animal Behaviour* 45, 351–358.

Jerison, H.J. (2001) The evolution of neural and behavioral complexity. In *Brain Evolution and Cognition*, Roth, G. & Wullimann, M.F. (eds.), pp. 523–553. John Wiley & Sons, New York/Spektrum, Heidelberg.

Joint Working Group on Refinement (1993) Removal of blood from laboratory mammals and birds. First Report of the BVA/FRAME/RSPCA/UFAW Joint Working Group on Refinement. *Laboratory Animals* 27, 1–22.

Joint Working Group on Refinement (2001) Refining procedures for the administration of substances. Report of the BVAAWF/FRAME/RSPCA/UFAW Joint Working Group on Refinement. *Laboratory Animals* 35, 1–41.

Joint Working Group on Refinement (2003a) Refinement and reduction in production of genetically modified mice. *Laboratory Animals* 37, S1–S49.

Joint Working Group on Refinement (2003b) Refinement and reduction in production of genetically modified mice. Sixth Report of the BVAAWF/FRAME/RSPCA/UFAW Joint Working Group on Refinement. *Laboratory Animals* 37 (Suppl. 1), S1–S51.

Joint Working Group on Refinement (2003c) Refinements in telemetry procedures. Seventh report of the BVAAWF/FRAME/RSPCA/UFAW Joint Working Group on Refinement, Part A. *Laboratory Animals* 37, 261–299. doi:10.1258/002367703322389861.

Joint Working Group on Refinement (2004) Refining dog husbandry and care. Eighth report of the BVAAWF/FRAME/RSPCA/UFAW Joint Working Group on Refinement. *Laboratory Animals* 38 (Suppl. 1), 1–94.

Joint Working Group on Refinement (2009) Refinements in husbandry, care and common procedures for non-human primates. Ninth report of the BVAAWF/FRAME/RSPCA/UFAW Joint Working Group on Refinement. *Laboratory Animals* 43, S1–S47.

Joint Working Group on Refinement (2011) A guide to defining and implementing protocols for the welfare assessment of laboratory animals. Eleventh report of the BVAAWF/FRAME/RSPCA/UFAW Joint Working Group on Refinement. *Laboratory Animals* 45, 1–13. doi:10.1258/la.2010.010031.

Jones, R.B., Duncan, I.J.H. & Hughes, B.O. (1981) The assessment of fear in domestic hens exposed to a looming human stimulus. *Behavioural Processes* 6, 121–133.

Jucker, M. (2010) The benefits and limitations of animal models for translational research in neurodegenerative diseases. *Nature Medicine* 16, 1210–1214.

Julian, R.J. (1998) Rapid growth problems: ascites and skeletal deformities in broilers. *Poultry Science* 77, 1773–1780.

Kallnik, M., Elvert, R., Ehrhardt, N., Kissling, D., Mahabir, E., Welzl, G., Faus-Kessler, T., de Angelis, M.H., Wurst, W. & Schmidt, J. (2007) Impact of IVC housing on emotionality and fear learning in male C3HeB/FeJ and C57BL/6J mice. *Mammalian Genome* 18, 173–186.

Kalman, R., Olsson, I., Bernadi, C., van den Broek, F., Brønstad, A., Marinou, K. & Zeller, W. (2011) Ethical evaluation of scientific procedures: recommendation for ethics committees. In *The COST Manual of Laboratory Animal Care and Use: Refinement, Reduction, and Research*, Howard, B., Nevalainen, T. & Perretta, G. (eds.), pp. 101–129. CRC Press, Boca Raton, FL.

Kaluzhny, Y., Kandárová, H., Hayden, P., Kubilus, J., D'Argembeau-Thornton, L. & Klausner, M. (2011) Development of the EpiOcular™ eye irritation test for hazard identification and labelling of eye irritating chemicals in response to the requirements of the EU Cosmetics Directive and REACH legislation. *ATLA, Alternatives to Laboratory Animals* 39, 339–364.

Kean, H. (1998) *Animal Rights: Political and Social Change in Britain since 1800*. Reaktion, London.

Kendrick, K.M. (1998) Intelligent perception. *Applied Animal Behaviour Science* 57, 213–231.

Kent, M.L., Harper, C. & Wolf, J.C. (2012) Documented and potential research impacts of subclinical diseases in zebrafish. *ILAR Journal* 53, 126–134.

Kilkenny, C., Parsons, N., Kadyszewski, E., Festing, M.F.W., Cuthill, I.C., Fry, D., Hutton, J. & Altman, D.G. (2009) Survey of the quality of experimental design, statistical analysis and reporting of research using animals. *PLoS ONE* 4, e7824.

Kilkenny, C., Browne, W.J., Cuthill, I.C., Emerson, M. & Altman, D.G. (2010) Improving bioscience research reporting: the ARRIVE Guidelines for Reporting Animal Research. *PLoS Biology* 8, e1000412.

Kirk, R.G.H. (2009) Between the clinic and the laboratory: ethology and pharmacology in the work of Michael Robin Alexander Chance, *c.*1946–1964. *Medical History* 53, 513–536.

Kirkden, R.D. & Pajor, E.A. (2006) Using preference, motivation and aversion tests to ask scientific questions about animals' feelings. *Applied Animal Behaviour Science* 100, 29–47.

Kirkwood, J.K. (2006) The distribution of the capacity for sentience in the animal kingdom. In *Animals, Ethics and Trade: The Challenge of Animal Sentience*, Turner, J. & D'Silva, J. (eds.), pp. 12–26. Earthscan, London.

Kirkwood, J.K. & Hubrecht, R.C. (2001) Animal consciousness, cognition and welfare. *Animal Welfare* 10, S5–S17.

Klein, H.J. & Bayne, K.A. (2007) Establishing a culture of care, conscience, and responsibility: addressing the improvement of scientific discovery and animal welfare through science-based performance standards. *ILAR Journal* 48, 3–11.

Knight, A. (2007) The effectiveness of humane teaching methods in veterinary education. *ALTEX* 24, 91–109.

Knight, A., Taylor, K., Gordon, N., Langley, G. & Higgins, W. (2008) 127 million non-human vertebrates used worldwide for scientific purposes in 2005. Authors' reply. *ATLA, Alternatives to Laboratory Animals* 36, 494–497.

Knowles, T.G. (1999) A review of the road transport of cattle. *Veterinary Record* 144, 197–201.

Knowles, T.G., Kestin, S.C., Haslam, S.M., Brown, S.N., Green, L.E., Butterworth, A., Pope, S.J., Pfeiffer, D. & Nicol, C.J. (2008) Leg disorders in broiler chickens: prevalence, risk factors and prevention. *PLoS ONE* 3, e1545. doi:10.1371/journal.pone.0001545.

Koch, C. (2004) *The Quest for Consciousness: A Neurobiological Approach*. Roberts and Co., Denver, CO.

Kola, I. & Landis, J. (2004) Can the pharmaceutical industry reduce attrition rates? *Nature Reviews Drug Discovery* 3, 711–716.

Kolle, S.N., Kandárová, H., Wareing, B., van Ravenzwaay, B. & Landsiedel, R. (2011) In-house validation of the EpiOcular™ eye irritation test and its combination with the Bovine Corneal Opacity and Permeability test for the assessment of ocular irritation. *ATLA, Alternatives to Laboratory Animals* 39, 365–387.

Koolhaas, J.M., Bartolomucci, A., Buwalda, B., de Boer, S.F., Flügge, G., Korte, S.M., Meerlo, P., Murison, R., Olivier, B., Palanza, P. *et al.* (2011) Stress revisited: a critical evaluation of the stress concept. *Neuroscience and Biobehavioral Reviews* 35, 1291–1301. doi:10.1016/j.neubiorev.2011.02.003.

Kottelat, M., Britz, R., Hui, T.H. & Witte, K.E. (2006) *Paedocypris*, a new genus of Southeast Asian cyprinid fish with a remarkable sexual dimorphism, comprises the world's smallest vertebrate. *Proceedings of the Royal Society of London Series B: Biological Sciences* 273, 895–899.

Kramer, K. & Hachtman, S. (2010) Applications of radiotelemetry in small laboratory animals. In *Handbook of Laboratory Animal Science, Volume I: Essential Principles and Practices*, 3rd edn, Hau, J. & Schapiro, S.J. (eds.), pp. 447–466. CRC Press, Boca Raton, FL.

Kramer, K. & Kinter, L.B. (2003) Evaluation and applications of radiotelemetry in small laboratory animals. *Physiological Genomics* 13, 197–205.

Lambeth, S.P., Hau, J., Perlman, J.E., Martino, M. & Schapiro, S.J. (2006) Positive reinforcement training affects hematologic and serum chemistry values in captive chimpanzees (*Pan troglodytes*). *American Journal of Primatology* 68, 245–256.

Landis, S.C., Amara, S.G., Asadullah, K., Austin, C.P., Blumenstein, R., Bradley, E.W., Crystal, R.G., Darnell, R.B., Ferrante, R.J., Fillit, H. *et al.* (2012) A call for transparent reporting to optimize the predictive value of preclinical research. *Nature* 490, 187–191.

Lane, N. & Jennings, M. (2004) *Supplementary Resources for Lay Members of Local Ethical Review Processes: Projects Involving Genetically Modified Animals*. RSPCA, Horsham, UK.

Langford, D.J., Bailey, A.L., Chanda, M.L., Clarke, S.E., Drummond, T.E., Echols, S., Glick, S., Ingrao, J., Klassen-Ross, T., LaCroix-Fralish, M.L. *et al.* (2010) Coding of facial expressions of pain in the laboratory mouse. *Nature Methods* 7, 447–449. doi:10.1038/nmeth.1455.

Lapage, G. (1960) *Achievement: Some Contributions of Animal Experiment to the Conquest of Disease*. Heffer & Sons Ltd, Cambridge.

LASA (1998) The Production and Disposition of Laboratory Rodents Surplus to the Requirements for Scientific procedures: The Report of a LASA Task Force Meeting held on 12 June 1998. Available at http://www.lasa.co.uk/publications.html

LASA (2010) Guiding Principles for Preparing and Undertaking Aseptic Surgery. A report by the LASA Education, Training and Ethics Section, Jennings, M. & Berdoy, M. (eds.). Available at http://www.lasa.co.uk/publications.html.

LASA/APC (2008) Final report of a LASA/APC Working Group to examine the feasibility of reporting data on the severity of scientific procedures on animals. Available at http://www.lasa.co.uk/publications.html, accessed 26 April 2013.

Latham, N. (2010) Brief introduction to welfare assessment: a 'toolbox' of techniques. In *The UFAW Handbook on the Care and Management of Laboratory and Other Research Animals*, 8th edn, Hubrecht, R. & Kirkwood, J. (eds.), pp. 76–91. Wiley-Blackwell, Oxford.

Latham, N. & Mason, G. (2004) From house mouse to mouse house: the behavioural biology of free-living *Mus musculus* and its implications in the laboratory. *Applied Animal Behaviour Science* 86, 261–289.

Laule, G. (2010) Positive reinforcement training for laboratory animals. In *The UFAW Handbook on the Care and Management of Laboratory and Other Research Animals*, 8th edn, Hubrecht, R. & Kirkwood, J. (eds.), pp. 206–218. Wiley-Blackwell, Oxford.

Leach, M.C., Coulter, C.A., Richardson, C.A. & Flecknell, P.A. (2011) Are we looking in the wrong place? Implications for behavioural-based pain assessment in rabbits (*Oryctolagus cuniculi*) and beyond? *PLoS ONE* 6, e13347.

Leach, M.C., Klaus, K., Miller, A.L., di Perrotolo, M.S., Sotocinal, S.G. & Flecknell, P.A. (2012) The assessment of post-vasectomy pain in mice using behaviour and the mouse grimace scale. *PLoS ONE* 7, e35656.

Leuven, J. & Višak, T. (2013) Ryder's painism and his criticism of utilitarianism. *Journal of Agricultural and Environmental Ethics* 26, 409–419. doi:10.1007/s10806-012-9381-3.

Liberg, O. & Sandell, M. (1988) Spatial organisation and reproductive tactics in the domestic cat and other felids. In *The Domestic Cat: The Biology of its Behaviour*, Turner, D.C. & Bateson, P. (eds.), pp. 83–98. Cambridge University Press, Cambridge, UK.

Liles, J.H. & Flecknell, P.A. (1993) The effects of surgical stimulus on the rat and the influence of analgesic treatment. *British Veterinary Journal* 149, 515–525.

Lockhart, J., Wilson, K. & Lanman, C. (2013) The effects of operant training on blood collection for domestic cats. *Applied Animal Behaviour Science* 143, 128–134. doi:10.1016/j.applanim.2012.10.011.

Lutz, C.K. & Novak, M.A. (2005) Environmental enrichment for nonhuman primates: theory and application. *ILAR Journal* 46, 178–191.

Lynne, U.S., Victoria, A.B. & Michael, J.G. (2003) Novel object test: examining nociception and fear in the rainbow trout. *Journal of Pain* 4, 431–440.

McFarland, R., Verthelyi, D., Casey, W., Arciniega, J., Isbrucker, R., Schmitt, M., Finn, T., Descamps, J., Horiuchi, Y., Sesardic, D. *et al.* (2011) Non-animal replacement methods for human vaccine potency testing: state of the science and future directions. *Procedia in Vaccinology* 5, 16–32.

McGlone, J.J. (1993) What is animal welfare? *Journal of Agricultural and Environmental Ethics* 6 (Suppl. 2), 26–36.

McKinley, J., Buchanan-Smith, H.M., Bassett, L. & Morris, K. (2003) Training common marmosets (*Callithrix jacchus*) to cooperate during routine laboratory procedures: ease of training and time investment. *Journal of Applied Animal Welfare Science* 6, 209–220.

Mackta, J. (2010) Monkey in the middle. *ALN World* 3, 23–25.

McMillan, F.D. (2005a) Do animals experience true happiness? In *Mental Health and Well-being in Animals*, McMillan, F.D. (ed.), pp. 221–233. Wiley-Blackwell, Ames, IA.

McMillan, F.D. (ed.) (2005b) *Mental Health and Well-being in Animals*. Wiley-Blackwell, Ames, IA.

Macnaghten, P. (2004) Animals in their nature: a case study on public attitudes to animals, genetic modification and 'Nature'. *Sociology* 38, 533–551.

Macphail, E.M. (2001) Conservation in the neurology and psychology of cognition in vertebrates. In *Brain Evolution and Cognition*, Roth, G. & Wullimann, M.F. (eds.), pp. 401–430. John Wiley & Sons, New York/Spektrum, Heidelberg.

Maehle, A.H. (1990) Literary responses to animal experimentation in seventeenth- and eighteenth-century Britain. *Medical History* 34, 27–51.

Maehle, A.H. (1994) Cruelty and kindness to the 'brute creation': stability and change in the ethics of the man–animal relationship, 1600–1850. In *Animals and Human Society: Changing Perspectives*, Manning, A. & Serpell, J. (eds.), pp. 81–105. Routledge, London.

Maehle, A.H. & Tröhler, U. (1987) Animal experimentation from antiquity to the end of the eighteenth century: attitudes and arguments. In *Vivisection in Historical Perspective*, Rupke, N.A. (ed.), pp. 14–47. Croom Helm, London.

Maestripieri, D., Hoffman, C.L., Fulks, R. & Gerald, M.S. (2008) Plasma cortisol responses to stress in lactating and nonlactating female rhesus macaques. *Hormones and Behavior* 53, 170–176.

Main, D.C.J., Kent, J.P., Wemelsfelder, F., Ofner, E. & Tuyttens, F.A.M. (2003) Applications for methods of on-farm welfare assessment. *Animal Welfare* 12, 523–528.

Manser, C.E., Elliott, H., Morris, T.H. & Broom, D.M. (1996) The use of a novel operant test to determine the strength of preference for flooring in laboratory rats. *Laboratory Animals* 30, 1–6.

Martignoni, M., Groothuis, G.M. & de Kanter, R. (2006) Species differences between mouse, rat, dog, monkey and human CYP-mediated drug metabolism, inhibition and induction. *Expert Opinion on Drug Metabolism and Toxicology* 2, 875–94. doi:10.1517/17425255.2.6.875.

Martin, B., Ji, S., Maudsley, S. & Mattson, M.P. (2010) 'Control' laboratory rodents are metabolically morbid: why it matters. *Proceedings of the National Academy of Sciences of the United States of America* 107, 6127–6133. doi:10.1073/pnas.0912955107.

Martin, P.R. & Bateson, P.P.G. (2007) *Measuring Behaviour: An Introductory Guide*, 3rd edn. Cambridge University Press, Cambridge, UK.

Marx, U., Walles, H., Hoffmann, S., Lindner, G., Horland, R., Sonntag, F., Klotzbach, U., Sakharov, D., Tonevitsky, A. & Lauster, R. (2012) 'Human-on-a-chip' developments: a translational cutting-edge alternative to systemic safety assessment and efficiency evaluation of substances in laboratory animals and man? *ATLA, Alternatives to Laboratory Animals* 40, 235–257.

Mason, G. & Latham, N.R. (2004) Can't stop, won't stop: is stereotypy a reliable animal welfare indicator? *Animal Welfare* 13, 57–69.

Mason, G. & Mendl, M. (1993) Why is there no simple way of measuring animal welfare? *Animal Welfare* 2, 301–319.

Mason, G. & Rushen, J. (eds.) (2006) *Stereotypic Animal Behaviour: Fundamentals and Applications to Welfare*, 2nd edn. CAB International, Wallingford, UK.

Mason, G., Wilson, D., Hampton, C. & Würbel, H. (2004) Non-invasively assessing disturbance and stress in laboratory rats by scoring chromodacryorrhoea. *ATLA, Alternatives to Laboratory Animals* 32, 153–159.

Matfield, M. (1996) Laboratory animal welfare around the globe: introduction. *Lab Animal* 25, 23.

Mather, J. (1992) Understimating the octopus. In *The Inevitable Bond: Examining Scientist–Animal Interactions*, Davis, H. & Balfour, D. (eds.), pp. 240–249. Cambridge University Press, Cambridge, UK.

Mather, J.A. (2008) Cephalopod consciousness: behavioural evidence. *Consciousness and Cognition* 17, 37–48.

Matsumiya, L.C., Sorge, R.E., Sotocinal, S.G., Tabaka, J.M., Wieskopf, J.S., Zaloum, A., King, O.D. & Mogil, J.S. (2012) Using the mouse grimace scale to reevaluate the efficacy of postoperative analgesics in laboratory mice. *Journal of the American Association for Laboratory Animal Science: JAALAS* 51, 42–49.

Matteri, R.L., Carroll, J.A. & Dyer, C.J. (2000) Neuroendocrine responses to stress. In *The Biology of Animal Stress: Basic Principles and Implications for Animal Welfare*, Moberg, G.P. & Mench, J.A. (eds.), pp. 43–76. CABI Publishing, Wallingford, UK.

Medawar, P. & Medawar, J. (1985) *Aristotle to Zoos*. Oxford, Oxford University Press.

Mellor, D.J. (2004) Comprehensive assessment of harms caused by experimental, teaching and testing procedures on live animals. *ATLA, Alternatives to Laboratory Animals* 32 (Suppl. 1), 453–457.

Melzack, R. & Wall, P.D. (1988) *The Challenge of Pain*. Penguin, London.

Mendl, M. & Harcourt, R. (1988) Individuality in the domestic cat. In *The Domestic Cat: The Biology of its Behaviour*, Turner, D.C. & Bateson, P. (eds.), pp. 41–54. Cambridge University Press, Cambridge, UK.

Mendl, M. & Paul, E.S. (2008) Do animals live in the present? Current evidence and implications for welfare. *Applied Animal Behaviour Science* 113, 357–382. doi:10.1016/j.applanim.2008.01.013.

Mendl, M., Burman, O.H.P., Parker, R.M.A. & Paul, E.S. (2009) Cognitive bias as an indicator of animal emotion and welfare: emerging evidence and underlying mechanisms. *Applied Animal Behaviour Science* 118, 161–181. doi:10.1016/j.applanim.2009.02.023.

Mendl, M., Brooks, J., Basse, C., Burman, O., Paul, E., Blackwell, E. & Casey, R. (2010a) Dogs showing separation-related behaviour exhibit a 'pessimistic' cognitive bias. *Current Biology* 20, R839–R840.

Mendl, M., Burman, O.H.P. & Paul, E.S. (2010b) An integrative and functional framework for the study of animal emotion and mood. *Proceedings of the Royal Society of London Series B: Biological Sciences* 277, 2895–2904.

Mendl, M., Paul, E.S. & Chittka, L. (2011) Animal behaviour: emotion in invertebrates? *Current Biology* 21, R463–R465.

Miklósi, A. (2007) *Dog Behaviour, Evolution, and Cognition.* Oxford University Press, Oxford.

Mineur, Y.S. & Crusio, W.E. (2009) Behavioral effects of ventilated micro-environment housing in three inbred mouse strains. *Physiology and Behavior* 97, 334–340. doi:10.1016/j.physbeh.2009.02.039.

Moberg, G.P. & Mench, J.A. (2000) *The Biology of Animal Stress: Basic Principles and Implications for Animal Welfare.* CABI Publishing, Wallingford, UK.

Monamy, V. (2009) *Animal Experimentation: A Guide to the Issues.* Cambridge University Press, Cambridge.

Morrison, M. (1981) Legislation and practice in the United States. In *Animals in Research: New Perspectives in Animal Experimentation,* Sperlinger, D. (ed.), pp. 63–78. John Wiley & Sons Ltd, Chichester.

Morton, D.B. (1999) Humane endpoints in animal experiments for biomedical research: ethical legal and practical aspects. In *Humane Endpoints in Animal Experiments for Biomedical Research: Proceedings of the International Conference, 22–25 November 1998, Zeist, The Netherlands,* Hendriksen, C.F.M. & Morton, D.B. (eds.), pp. 5–12. Royal Society of Medicine Press, London.

Morton, D.B. (2000) A systematic approach for establishing humane endpoints. *ILAR Journal* 41, 80–86.

Morton, D.B. & Griffiths, P.H. (1985) Guidelines on the recognition of pain, distress and discomfort in experimental animals and an hypothesis for assessment. *Veterinary Record* 116, 431–436. doi:10.1136/vr.116.16.431.

Morton, D.B. & Hau, J. (2010) Welfare assessment and humane endpoints. In *Handbook of Laboratory Animal Science, Volume I: Essential Principles and Practices,* 3rd edn, Hau, J. & Schapiro, S.J. (eds.), pp. 535–572. CRC Press, Boca Raton, FL.

Morton, D.M. (1998) Importance of species selection in drug toxicity testing. *Toxicology Letters* 102–103, 545–550.

Muzur, A., Pace-Schott, E.F. & Hobson, J.A. (2002) The prefrontal cortex in sleep. *Trends in Cognitive Sciences* 6, 475–481.

Nagel, T. (1974) What is it like to be a bat? *The Philosophical Review* 83, 435–450.

National Institutes of Health (2002) *ARENA/OLAW Institutional Animal Care and Use Committee Guidebook,* 2nd edn. Office of Laboratory Animal Welfare, National Institutes of Health, Bethesda, MD.

NC3Rs (2011) Research Review 2011. National Centre for the Replacement, Refinement and Reduction of Animals in Research, London. Available at http://www.nc3rs.org.uk/document.asp?id=1640.

Nelson, R.J. & Mandrell, T.D. (2005) Enrichment and nonhuman primates: first, do no harm. *ILAR Journal* 46, 171–177.

Nevison, C.M., Hurst, J.L. & Barnard, C.J. (1999) Strain-specific effects of cage enrichment in male laboratory mice (*Mus musculus*). *Animal Welfare* 8, 361–379.

Newcomer, C.E. (2012) The evolution and adoption of standards used by AAALAC. *Journal of the American Association for Laboratory Animal Science: JAALAS* 51, 293–297.

Newton, J.E.O. & Lucas, L.A. (1982) Differential heart-rate responses to person in nervous and normal pointer dogs. *Behavior Genetics* 12, 379–393.

NHMRC (2004) Australian Code of Practice for the Care and Use of Animals for Scientific Purposes, 7th edn. National Health and Medical Research Council, Canberra, Australia. Available at http://www.nhmrc.gov.au/publications/synopses/ea16syn.htm.

NHMRC (2008) Guidelines to Promote the Wellbeing of Animals used for Scientific Purposes. National Health and Medical Research Council, Canberra, Australia. Available at http://www.nhmrc.gov.au/_files_nhmrc/publications/attachments/ea18.pdf.

Nicol, C.J. (1996) Farm animal cognition. *Animal Science* 26, 375–391.

Nicol, C.J., Caplen, G., Edgar, J. & Browne, W.J. (2009) Associations between welfare indicators and environmental choice in laying hens. *Animal Behaviour* 78, 413–424.

Niklas, W., Baneux, P., Boot, R., Decelle, T., Deeny, A.A., Fumanelli, M. & Illgen-Wilcke, B. (2002) Recommendations for the health monitoring of rodent and rabbit colonies in breeding and experimental units: recommendations of the Federation of European Laboratory Animal Science Associations (FELASA) Working Group on Health Monitoring of Rodent and Rabbit Colonies accepted by the FELASA Board of Management, 9 June 2001. *Laboratory Animals* 36, 20–42.

Novak, M.A., Meyer, J.S., Lutz, C. & Tiefenbacher, S. (2006) Deprived environments: developmental insights from primatology. In *Stereotypic Animal Behaviour: Fundamentals and Applications to Welfare*, 2nd edn, Mason, G. & Rushen, J. (eds.), pp. 19–57. CAB International, Wallingford, UK.

Novak, M.A., Hamel, A.F., Kelly, B.J. & Dettmer, A.M. (2013) Stress, the HPA axis, and nonhuman primate well-being: a review. *Applied Animal Behaviour Science* 143, 135–149.

NRC (National Research Council) (1996) *Guide for the Care and Use of Laboratory Animals*. National Academy Press, Washington, DC.

NRC (National Research Council) (1998) *The Psychological Well-being of Nonhuman Primates*. National Academy Press, Washington, DC.

NRC (National Research Council) (2003) *Guidelines for the Care and Use of Mammals in Neuroscience and Behavioral Research*. National Academies Press, Washington, DC.

NRC (National Research Council) (2008) *Recognition and Alleviation of Distress in Laboratory Animals*. National Academies Press, Washington, DC.

NRC (National Research Council) (2009) *Recognition and Alleviation of Pain in Laboratory Animals*. National Academies Press, Washington, DC.

NRC (National Research Council) (2010) *Guide for the Care and Use of Laboratory Animals*. National Academies Press, Washington, DC.

Nuffield Council on Bioethics (2005) *The Ethics of Research Involving Animals*. Latimer Trend & Company Ltd, London.

OECD (2000) Guidance Document on the Recognition, Assessment, and Use of Clinical Signs as Humane Endpoints for Experimental Animals Used in Safety Evaluation. OECD, Paris. doi:10.1787/9789264078376-en.

Olson, H., Betton, G., Robinson, D., Thomas, K., Monro, A., Kolaja, G., Lilly, P., Sanders, J., Sipes, G. & Bracken, W. (2000) Concordance of the toxicity of pharmaceuticals in humans and in animals. *Regulatory Toxicology and Pharmacology* 32, 56–67.

Olsson, I.A.S. & Dahlborn, K. (2002) Improving housing conditions for laboratory mice: a review of environmental enrichment. *Laboratory Animals* 36, 243–270.

Olsson, I.A.S., Hansen, A.K. & Sandøe, P. (2008) Animal welfare and the refinement of neuroscience research methods: a case study of Huntington's disease models. *Laboratory Animals* 42, 277–283. doi:10.1258/la.2008.007147.

Olsson, I.A.S., Robinson, P. & Sandøe, P. (2010) Ethics of animal research. In *Handbook of Laboratory Animal Science, Volume I: Essential Principles and Practices*, 3rd edn, Hau, J. & Schapiro, S.J. (eds.), pp. 21–37. CRC Press, Boca Raton, FL.

Olsson, I.A.S., Franco, N.H., Weary, D.M. & Sandøe, P. (2012) The 3Rs principle: mind the ethical gap! In *ALTEX Proceedings 8th World Congress, Montreal, 2011*, Vol. 1, pp. 333–336. Available at http://www.animalethics.net/Nyheder/2012/Altex.aspx

Osborne, N.J., Payne, D. & Newman, M.L. (2009) Journal editorial policies, animal welfare, and the 3Rs. *American Journal of Bioethics* 9, 55–59. doi:10.1080/15265160 903318343.

Panksepp, J. (2011) The basic emotional circuits of mammalian brains: do animals have affective lives? *Neuroscience and Biobehavioral Reviews* 35, 1791–1804. doi:10.1016/j.neubiorev.2011.08.003.

Panksepp, J. & Northoff, G. (2009) The trans-species core SELF: the emergence of active cultural and neuro-ecological agents through self-related processing within subcortical–cortical midline networks. *Consciousness and Cognition* 18, 193–215.

Parker, K.J. & Maestripieri, D. (2011) Identifying key features of early stressful experiences that produce stress vulnerability and resilience in primates. *Neuroscience and Biobehavioral Reviews* 35, 1466–1483.

Paskal, C. (2009) Finding the courage: the unconventional lives of WMS and Claire Russell. *ATLA, Alternatives to Laboratory Animals* 37 (Suppl. 2), 101–106.

Patronek, G.J. & Rauch, A. (2007) Systematic review of comparative studies examining alternatives to the harmful use of animals in biomedical education. *Journal of the American Veterinary Medical Association* 230, 37–43.

Patterson, F. & Gordon, W. (1993) The case for the personhood of gorillas. In *The Great Ape Project: Equality beyond Humanity*, Cavalieri, P. & Singer, P. (eds.), pp. 58–77. Fourth Estate, London.

Patterson-Kane, E.G. (2004) Enrichment of laboratory caging for rats: a review. *Animal Welfare* 13 (Suppl. 1), 209–214.

Patterson-Kane, E.G., Hunt, M. & Harper, D. (2002) Rats demand social contact. *Animal Welfare* 11, 327–332.

Paul, E.S. & Podberscek, A.L. (2000) Veterinary education and students' attitudes towards animal welfare. *Veterinary Record* 146, 269–272. doi:10.1136/vr.146.10.269.

Paul, S.M., Mytelka, D.S., Dunwiddie, C.T., Persinger, C.C., Munos, B.H., Lindborg, S.R. & Schacht, A.L. (2010) How to improve R&D productivity: the pharmaceutical industry's grand challenge. *Nature Reviews Drug Discovery* 9, 203–214

Perlman, J.E., Bloomsmith, M.A., Whittaker, M.A., McMillan, J.L., Minier, D.E. & McCowan, B. (2012) Implementing positive reinforcement animal training programs at primate laboratories. *Applied Animal Behaviour Science* 137, 114–126.

Philippi, C.L., Feinstein, J.S., Khalsa, S.S., Damasio, A., Tranel, D., Landini, G., Williford, K. & Rudrauf, D. (2012) Preserved self-awareness following extensive bilateral brain damage to the insula, anterior cingulate, and medial prefrontal cortices. *PLoS ONE* 7, e38413.

Pietersen, C.Y., Bosker, F.J., Postema, F. & den Boer, J.A. (2006) Fear conditioning and shock intensity: the choice between minimizing the stress induced and reducing the number of animals used. *Laboratory Animals* 40, 180–185.

Pifer, L., Shimizu, K. & Pifer, R. (1994) Public attitudes toward animal research: some international comparisons. *Society and Animals* 2, 95–113.

Pintando, B. & Roon, M.V. (2011) Creation of genetically modified animals. In *The COST Manual of Laboratory Animal Care and Use: Refinement, Reduction, and*

Research, Howard, B., Nevalainen, T. & Perretta, G. (eds.), pp. 179–204. CRC Press, Boca Raton, FL.

Plotnik, J.M., De Waal, F. & Reiss, D. (2006) Self-recognition in an Asian elephant. *Proceedings of the National Academy of Sciences of the United States of America* 103, 17053–17057.

Plous, S. & Herzog, H. (2001) Reliability of protocol reviews for animal research. *Science* 293, 608–609. doi:10.1126/science.1061621.

Poole, T. (1997) Happy animals make good science. *Laboratory Animals* 31, 116–124.

Popp, M.B. & Brennan, M.F. (1981) Long-term vascular access in the rat: importance of asepsis. *American Journal of Physiology* 241, H606–H612.

Porter, D.G. (1992) Ethical scores for animal experiments. *Nature* 356, 101–102.

Povinelli, D.J. (2000) *Folk Physics for Apes: The Chimpanzee's Theory of How the World Works*. Oxford University Press Oxford.

Prescott, M.J. (2010) Ethics of primate use. *Advances in Science and Research* 5, 11–22.

Prescott, M.J. & Buchanan-Smith, H.M. (2005) Training laboratory-housed non-human primates. Part 2. Resources for developing and implementing training programmes *Animal Technology and Welfare* 16, 133–148.

Prescott, M.J., Nixon, M.E., Farningham, D.A.H., Naiken, S. & Griffiths, M.-A. (2012) Laboratory macaques: when to wean? *Applied Animal Behaviour Science* 137, 194–207.

Price, E.O. (1998) Behavioural genetics and the process of animal domestication In *Genetics and the Behavior of Domestic Animals*, Grandin, T. (ed.), pp. 31–65. Academic Press, San Diego, CA.

Prior, H., Schwarz, A. & Güntürkün, O. (2008) Mirror-induced behavior in the magpie (*Pica pica*): evidence of self-recognition. *PLoS Biology* 6, 1642–1650.

Pullen, A.J., Merrill, R.J.N. & Bradshaw, J.W.S. (2010) Preferences for toy types and presentations in kennel housed dogs. *Applied Animal Behaviour Science* 125, 151–156.

Radford, M. (2001) *Animal Welfare Law in Britain: Regulation and Responsibility*. Oxford University Press, Oxford.

Rajala, A.Z., Reininger, K.R., Lancaster, K.M. & Populin, L.C. (2010) Rhesus monkeys (*Macaca mulatta*) do recognize themselves in the mirror: implications for the evolution of self-recognition. *PLoS ONE* 5, e12865.

Rands, S.A. (2011) Inclusion of policies on ethical standards in animal experiments in biomedical science journals. *Journal of the American Association for Laboratory Animal Science: JAALAS* 50, 901–903.

Rasmussen, S., Miller, M.M., Filipski, S.B. & Tolwani, R.J. (2011) Cage change influences serum corticosterone and anxiety-like behaviors in the mouse. *Journal of the American Association for Laboratory Animal Science: JAALAS* 50, 479.

Regan, T. (1982) *All That Dwell Therein: Animal Rights and Environmental Ethics*. University of California Press, Berkeley, CA.

Regan, T. (1998) Animal rights. In *Encyclopedia of Animal Rights and Animal Welfare*, Bekoff, M. & Meaney, C.A. (eds.), pp. 42–43. Greenwood Press, Westport, CT.

Rehkämper, G., Frahm, H.D. & Mann, M.D. (2001) Evolutionary constraints of large telencephala. In *Brain Evolution and Cognition*, Roth, G. & Wullimann, M.F. (eds.), pp. 265–293. John Wiley & Sons, New York/Spektrum, Heidelberg.

Reinhardt, V. (2003) Working with rather than against macaques during blood collection. *Journal of Applied Animal Welfare Science* 6, 189–197.

Reiss, D. & Marino, L. (2001) Mirror self-recognition in the bottlenose dolphin: a case of cognitive convergence. *Proceedings of the National Academy of Sciences of the United States of America* 98, 5937–5942.

Rennie, A.E. & Buchanan-Smith, H.M. (2006) Refinement of the use of non-human primates in scientific research. Part III: refinement of procedures. *Animal Welfare* 15, 239–261.

Rice, A.S.C. (2010) Predicting analgesic efficacy from animal models of peripheral neuropathy and nerve injury: a critical view from the clinic. In *Pain 2010 – An Updated Review: Refresher Course Syllabus*, Mogil, J.S. (ed.), pp. 415–426. IASP Press, Seattle, WA.

Rice, A.S.C., Cimino-Brown, D., Eisenach, J.C., Kontinen, V.K., Lacroix-Fralish, M.L., Machin, I., Mogil, J.S. & Stohr, T. (2008) Animal models and the prediction of efficacy in clinical trials of analgesic drugs: a critical appraisal and call for uniform reporting standards. *Pain* 139, 243–247.

Richardson, C.A. & Flecknell, P.A. (2005) Anaesthesia and post-operative analgesia following experimental surgery in laboratory rodents: are we making progress. *ATLA, Alternatives to Laboratory Animals* 33, 119–127.

Richmond, J. (2004) A review and comparison of processes to change regulatory guidelines: a European perspective. In *The Development of Science-based Guidelines for Laboratory Animal Care: Proceedings of the November 2003 International Workshop*, pp. 46–49. National Academies Press, Washington, DC.

Richmond, J. (2010) The Three Rs. In *The UFAW Handbook on the Care and Management of Laboratory and Other Research Animals*, 8th edn, Hubrecht, R. & Kirkwood, J. (eds.), pp. 5–22. Wiley-Blackwell, Oxford.

Richter, S.H., Garner, J.P. & Würbel, H. (2009) Environmental standardization: cure or cause of poor reproducibility in animal experiments? *Nature Methods* 6, 257–261.

Ritskes-Hoitinga, M., Gravesen, L.B. & Jegstrup, I.M. (2006) Refinement benefits animal welfare and quality of science. Available at http://www.nc3rs.org.uk/news.asp?id=212.

Robb, J.W. (1993) The role and value of the unaffiliated member and the nonscientist member of the Institutional Animal Care and Use Committee. *ILAR Journal* 35, 50–53.

Robery, S., Mukanowa, J., Percie du Sert, N., Andrews, P.L.R. & Williams, R.S.B. (2011) Investigating the effect of emetic compounds on chemotaxis in *Dictyostelium* identifies a non-sentient model for bitter and hot tastant research. *PLoS ONE* 6, e24439.

Roi, A.J. & Grune, B. (2011) *THE ECVAM Search Guide: Data Retrieval Procedures. Basic Principles*. EUR 24391 EN-2011. European Commission, Luxembourg. doi:10.2788/98187.

Rollin, B.E. (1989) *The Unheeded Cry: Animal Consciousness, Animal Pain and Science*. Oxford University Press, Oxford.

Rollin, B.E. (2006) The regulation of animal research and the emergence of animal ethics: a conceptual history. *Theoretical Medicine and Bioethics* 27, 285–304. doi:10.1007/s11017-006-9007-8.

Rose, J.D. (2002) The neurobehavioral nature of fishes and the question of awareness and pain. *Reviews in Fisheries Science* 10, 1–38.

Rose, J.D. (2007) Anthropomorphism and 'mental welfare' of fishes. *Diseases of Aquatic Organisms* 75, 139–154.

Rose, M. (2011) Challenges to the development and implementation of public policies to achieve animal welfare outcomes. *Animals* 1, 69–82. doi:10.3390/ani1010069.

Roth, B. & Øines, S. (2010) Stunning and killing of edible crabs (*Cancer pagurus*). *Animal Welfare* 19, 287–294.

Roth, G. (2001) The evolution of consciousness. In *Brain Evolution and Cognition*, Roth, G. & Wullimann, M.F. (eds.), pp. 555–582. John Wiley & Sons, New York/Spektrum, Heidelberg.

Roth, G. & Dicke, U. (2005) Evolution of the brain and intelligence. *Trends in Cognitive Sciences* 9, 250–257. doi:10.1016/j.tics.2005.03.005.

Roughan, J.V. & Flecknell, P.A. (2006) Training in behaviour-based post-operative pain scoring in rats: an evaluation based on improved recognition of analgesic requirements. *Applied Animal Behaviour Science* 96, 327–342.

RSPCA & LASA (2012) Guiding principles on good practice for Ethical Review Processes. A report by the RSPCA Research Animals Department and LASA Education, Training and Ethics Section, 2nd edn, Jennings, M. (ed.). Available at http://www.rspca.org.uk/science-group/researchanimals/ethicalreview

Rushen, J. (2000) Some issues in the interpretation of behavioural responses to stress. In *The Biology of Animal Stress: Basic Principles and Implications for Animal Welfare*, Moberg, G.P. & Mench, J.A. (eds.), pp. 23–42. CABI Publishing, Wallingford, UK.

Russell, W.M.S. (2005) The Three Rs: past, present and future. *Animal Welfare* 14, 279–286.

Russell, W.M.S. & Burch, R.L. (1959) *The Principles of Humane Experimental Technique*. Special Edition 1992, Universities Federation for Animal Welfare. Methuen & Co. Ltd, Potters Bar.

Ryder, R.D. (1975) *Victims of Science: The Use of Animals in Research*. Davis-Poynter, London.

Ryder, R.D. (2009) Painism versus Utilitarianism. *Think. Philosophy for Everyone* 8, 85–89. doi:10.1017/S1477175608000420.

Sambrook, T.D. & Buchanan-Smith, H.M. (1997) Control and complexity in novel object enrichment. *Animal Welfare* 6, 207–216.

Samuels, G. (2000) Medicines: tried and tested in animals? In *Medicines: Tried And Tested – Or An Unknown Risk?* Medicines for Health Series. http://www.abpi.org.uk/Details.asp?ProductID=120, accessed 24 March 2006, page no longer available.

Sandøe, P. (1994) Involving the public in ethical decisions. In *Welfare and Science. Proceedings of the Fifth Symposium of the Federation of European Laboratory Animal Science Associations, 8–11 June 1993, Brighton, UK*, Bunyan, J. (ed.), pp. 331–334. Royal Society of Medicine Press, London.

Sarko, D., Marino, L. & Reiss, D. (2002) A bottlenose dolphin's (*Tursiops truncatus*) responses to its mirror image: further analysis. *International Journal of Comparative Psychology* 15, 69–76.

Schapiro, S.J. & Lambeth, S.P. (2010) Chimpanzees. In *The UFAW Handbook on the Care and Management of Laboratory and Other Research Animals*, 8th edn, Hubrecht, R. & Kirkwood, J. (eds.), pp. 618–633. Wiley-Blackwell, Oxford.

Schneiderman, N. & McCabe, P.M. (1985) Biobehavioral responses to stressors. In *Stress and Coping*, Field, T., McCabe, P.M. & Schneiderman, N. (eds.), pp. 13–62. Lawrence Erlbaum Associates, Hillsdale, NJ.

Scott, J.P. & Fuller, J.L. (1965) *Genetics and the Social Behavior of the Dog*. University of Chicago Press, Chicago.

Seaman, S.C., Waran, N.K., Mason, G. & D'Eath, R.B. (2008) Animal economics: assessing the motivation of female laboratory rabbits to reach a platform, social contact and food. *Animal Behaviour* 75, 31–42.

Sechzer, J.A. (1983) The ethical dilemma of some classical animal experiments. *Annals of the New York Academy of Sciences* 406, 5–12.

Sedcole, J.R. (2006) Experimental design: minimizing the number of subjects that suffer may not mean minimizing total suffering. *Animal Behaviour* 71, 735–738.

Sena, E.S., van der Worp, H.B., Bath, P.M.W., Howells, D.W. & Macleod, M.R. (2010) Publication bias in reports of animal stroke studies leads to major overstatement of efficacy. *PLoS Biology* 8, e1000344.

Serpell, J. & Paul, E. (1994) Pets and positive attitudes to animals. In *Animals and Human Society: Changing Perspectives*, Manning, A. & Serpell, J. (eds.), p. 127–144. Routledge, London.

Seth, A.K. & Baars, B.J. (2005) Neural Darwinism and consciousness. *Consciousness and Cognition* 14, 140–168.

Seth, A.K., Baars, B.J. & Edelman, D.B. (2005) Criteria for consciousness in humans and other mammals. *Consciousness and Cognition* 14, 119–139.

Sharpe, S.A., Eschelbach, E., Basaraba, R.J., Gleeson, F., Hall, G.A., McIntyre, A., Williams, A., Kraft, S., Clark, S. & Gooch, K. (2009) Determination of lesion volume by MRI and stereology in a macaque model of tuberculosis. *Tuberculosis* 89, 405–416.

Sharpe, S.A., McShane, H., Dennis, M.J., Basaraba, R.J., Gleeson, F., Hall, G., McIntyre, A., Gooch, K., Clark, S. & Beveridge, N.E.R. (2010) Establishment of an aerosol challenge model of tuberculosis in rhesus macaques and an evaluation of endpoints for vaccine testing. *Clinical and Vaccine Immunology* 17, 1170–1182.

Shaw, R. (2004) Reduction in laboratory animal use by factorial design. *ATLA, Alternatives to Laboratory Animals* 32, 49.

Sherwin, C.M. (1996) Laboratory mice persist in gaining access to resources: a method of assessing the importance of environmental features. *Applied Animal Behaviour Science* 48, 203–213.

Sherwin, C.M. (2001) Can invertebrates suffer? Or, how robust is argument-by-analogy? *Animal Welfare* 10, 103–118.

Sherwin, C.M. & Olsson, I.A.S. (2004) Housing conditions affect self-administration of anxiolytic by laboratory mice. *Animal Welfare* 13, 33–38.

Shettleworth, S.J. (2010) *Cognition, Evolution, and Behavior*. Oxford University Press, Oxford.

Shimizu, T. (2001) Evolution of the forebrain in tetrapods. In *Brain Evolution and Cognition*, Roth, G. & Wullimann, M.F. (eds.), pp. 135–184. John Wiley & Sons, New York/Spektrum, Heidelberg.

Short, C.E. (1998) Fundamentals of pain perception in animals. *Applied Animal Behaviour Science* 59, 125–133.

Singer, P. (1990) *Animal Liberation*, revised edition. Jonathan Cape, London.

Smith, A.J. & Allen, T. (2005) The use of databases, information centres and guidelines when planning research that may involve animals. *Animal Welfare* 14, 347–359.

Smith, J.A. & Boyd, K.M. (1991) *Lives in the Balance: The Ethics of Using Animals in Biomedical Research*. Oxford University Press, Oxford.

Smith, J.A. & Jennings, M. (2009) *A Resource Book for Lay Members of Local Ethical Review Processes*, 2nd edn. RSPCA, Horsham, UK.

Smith, J.A., van den Broek, F.A.R., Martorell, J.C., Hackbarth, H., Ruksenas, O. & Zeller, W. (2005) Principles and practice in ethical review of animal experiments across Europe: summary of the report of a FELASA working group on ethical evaluation of animal experiments. Available at http://www.felasa.eu/media/uploads/Principles-practice-ethical-review_full%20report%20.pdf.

Smith, J.A., van den Broek, F.A.R., Martorell, J.C., Hackbarth, H., Ruksenas, O. & Zeller, W. (2007) Principles and practice in ethical review of animal experiments across Europe: summary of the report of a FELASA working group on ethical evaluation of animal experiments. *Laboratory Animals* 41, 143–160. doi:10.1258/002367707780378212.

Sømme, L.S. (2005) Sentience and pain in invertebrates: Report to Norwegian Scientific Committee for Food Safety. Norwegian University of Life Sciences, Department of Animal and Aquacultural Sciences. Available at http://www.vkm.no/dav/413af9502e.pdf, accessed 26 April 2013.

Sørensen, D.B. (2010) Never wrestle with a pig. *Laboratory Animals* 44, 159–161.

Sotocinal, S.G., Sorge, R.E., Zaloum, A., Tuttle, A.H., Martin, L.J., Wieskopf, J.S., Mapplebeck, J.C.S., Wei, P., Zhan, S. & Zhang, S. (2011) The Rat Grimace Scale: a partially automated method for quantifying pain in the laboratory rat via facial expressions. *Molecular Pain* 7, 55.

Sparrow, S.S., Robinson, S., Bolam, S., Bruce, C., Danks, A., Everett, D., Fulcher, S., Hill, R.E., Palmer, H., Scott, E.W. *et al.* (2011) Opportunities to minimise animal use in pharmaceutical regulatory general toxicology: a cross-company review. *Regulatory Toxicology and Pharmacology* 61, 222–229. doi:10.1016/j.yrtph.2011.08.001.

Spires, T.L., Grote, H.E., Varshney, N.K., Cordery, P.M., van Dellen, A., Blakemore, C. & Hannan, A.J. (2004) Environmental enrichment rescues protein deficits in a mouse model of Huntington's disease, indicating a possible disease mechanism. *Journal of Neuroscience* 24, 2270–2276.

Spitsbergen, J.M., Buhler, D.R. & Peterson, T.S. (2012) Neoplasia and neoplasm-associated lesions in laboratory colonies of zebrafish emphasizing key influences of diet and aquaculture system design. *ILAR Journal* 53, 114–125.

Stafleu, F.R., Tramper, R., Vorstenbosch, J. & Joles, J.A. (1999) The ethical acceptability of animal experiments: a proposal for a system to support decision-making. *Laboratory Animals* 33, 295–303.

Stallard, N., Price, C., Creton, S., Indans, I., Guest, R., Griffiths, D. & Edwards, P. (2011) A new sighting study for the fixed concentration procedure to allow for gender differences. *Human and Experimental Toxicology* 30, 239–249.

Stephens, M.L. (2009) Personal reflections on Russell and Burch, FRAME, and the HSUS. *ATLA, Alternatives to Laboratory Animals* 37, 29–33.

Stewart, M., Foster, T.M. & Waas, J.R. (2003) The effects of air transport on the behaviour and heart rate of horses. *Applied Animal Behaviour Science* 80, 143–160.

Stickings, P., Rigsby, P., Coombes, L., Hockley, J., Tierney, R. & Sesardic, D. (2011) Animal refinement and reduction: alternative approaches for potency testing of diphtheria and tetanus vaccines. *Procedia in Vaccinology* 5, 200–212.

Stockman, C.A., Collins, T., Barnes, A.L., Miller, D., Wickham, S.L., Beatty, D.T., Blache, D., Wemelsfelder, F. & Fleming, P.A. (2011) Qualitative behavioural assessment and quantitative physiological measurement of cattle naive and habituated to road transport. *Animal Production Science* 51, 240–249. doi:10.1071/AN10122.

Stokes, E.L., Flecknell, P.A. & Richardson, C.A. (2009) Reported analgesic and anaesthetic administration to rodents undergoing experimental surgical procedures. *Laboratory Animals* 43, 149–154.

Strausfeld, N.J. (2001) Insect brain. In *Brain Evolution and Cognition*, Roth, G. & Wullimann, M.F. (eds.), pp. 367–400. John Wiley & Sons, New York/Spektrum, Heidelberg.

Suresh, K.P. & Chandrashekara, S. (2012) Sample size estimation and power analysis for clinical research studies. *Journal of Human Reproductive Sciences* 5, 7.

Swallow, J., Anderson, D., Buckwell, A.C., Harris, T., Hawkins, P., Kirkwood, J., Lomas, M., Meacham, S., Peters, A., Prescott, M. *et al.* (2005) Guidance on the transport of laboratory animals. *Laboratory Animals* 39, 1–39.

Syversen, E., Pineda, F.J. & Watson, J. (2008) Temperature variations recorded during inter-institutional air shipments of laboratory mice. *Journal of the American Association for Laboratory Animal Science: JAALAS* 47, 31–36.

Tannenbaum, J. (1989) *Veterinary Ethics.* Williams & Wilkins, Baltimore.

Tannenbaum, J. (2001) The paradigm shift towards animal happiness: what it is, why it is happening, and what it portends for medical research. In *Why Animal Experimentation Matters: The Use of Animals in Medical Research*, Paul, E.F. & Paul, J. (eds.), pp. 93–130. Transaction Publishers, Piscataway, NJ.

Taylor, J. (2009) *Not a Chimp: The Hunt to Find the Genes that Make us Human.* Oxford University Press, Oxford.

Taylor, K., Gordon, N., Langley, G. & Higgins, W. (2008) Estimates for worldwide labora-tory animal use in 2005. *ATLA, Alternatives to Laboratory Animals* 36, 327–342.

ter Riet, G., Korevaar, D.A., Leenaars, M., Sterk, P.J., Van Noorden, C.J.F., Bouter, L.M., Lutter, R., Elferink, R.P.O. & Hooft, L. (2012) Publication bias in laboratory animal research: a survey on magnitude, drivers, consequences and potential solutions. *PLoS ONE* 7, e43404. doi:10.1371/journal.pone.0043404.

Thon, R.W., Ritskes-Hoitinga, M., Gates, H. & Prins, J.-B. (2010) Phenotyping of geneti-cally modified mice. In *The UFAW Handbook on the Care and Management of Laboratory and Other Research Animals*, 8th edn, Hubrecht, R. & Kirkwood, J. (eds.), pp. 61–75. Wiley-Blackwell, Oxford.

Tinbergen, N. (1963) On aims and methods of ethology. *Zeitschrift für Tierpsychologie* 20, 410–433.

Toates, F.M. (1995) *Stress: Conceptual and Biological Aspects.* John Wiley & Sons Ltd, Chichester.

Toth, L.A., Kregel, K., Leon, L. & Musch, T.I. (2011) Environmental enrichment of laboratory rodents: the answer depends on the question. *Comparative Medicine* 61, 314–321.

Tricklebank, M.D. & Garner, J.P. (2012) The possibilities and limitations of animal models for psychiatric disorders. In *Drug Discovery for Psychiatric Disorders*, Rankovic, Z., Bingham, M., Nestler, E.J. & Hargreaves, R. (eds.), pp. 534–556. RSC Drug Discovery Series 28. Royal Society of Chemistry, London.

Tsaioun, K. & Jacewicz, M. (2009) De-risking drug discovery with ADDME: avoiding drug development mistakes early. *ATLA, Alternatives to Laboratory Animals* 37, 47–55.

Turner, P.V., Brabb, T., Pekow, C. & Vasbinder, M.A. (2011) Administration of substances to laboratory animals: routes of administration and factors to consider. *Journal of the American Association for Laboratory Animal Science: JAALAS* 50, 600–613.

Udell, M.A.R., Dorey, N.R. & Wynne, C.D.L. (2010) The performance of stray dogs (*Canis familiaris*) living in a shelter on human-guided object-choice tasks. *Animal Behaviour* 79, 717–725.

Van de Weerd, H., Van Loo, P., Van Zutphen, L., Koolhaas, J. & Baumans, V. (1998) Strength of preference for nesting material as environmental enrichment for laboratory mice. *Applied Animal Behaviour Science* 55, 369–382.

van der Harst, J.E. & Spruijt, B.M. (2007) Tools to measure and improve animal welfare: reward-related behaviour. *Animal Welfare* 16 (Suppl. 1), 67–73.

van der Harst, J.E., Baars, A.M. & Spruijt, B.M. (2003) Standard housed rats are more sensitive to rewards than enriched housed rats as reflected by their anticipatory behaviour. *Behavioural Brain Research* 142, 151–156.

van der Worp, H.B., Howells, D.W., Sena, E.S., Porritt, M.J., Rewell, S., O'Collins, V. & Macleod, M.R. (2010) Can animal models of disease reliably inform human studies? *PLoS Medicine* 7, e1000245.

van Rooijen, J. (2009) Do dogs and bees possess a theory of mind? *Animal Behaviour* 79, e7–e8. doi:10.1016/j.anbehav.2009.11.016.

Varner, G.E. (1994) The prospects for consensus and convergence in the animal rights debate. *The Hastings Center Report* 24, 24–28.

Vergara, P. & Demers, G. (2010) Laboratory animal science and service organisations. In *Handbook of Laboratory Animal Science, Volume I: Essential Principles and Practices*, 3rd edn, Hau, J. & Schapiro, S.J. (eds.), pp. 97–113. CRC Press, Boca Raton, FL.

Vila, C., Savolainen, P., Maldonado, J.E., Amorim, I.R., Rice, J.E., Honeycutt, R.L., Crandall, K.A., Lundeberg, J. & Wayne, R.K. (1997) Multiple and ancient origins of the domestic dog. *Science* 276, 1687–1689.

Wahlsten, D., Bachmanov, A., Finn, D.A. & Crabbe, J.C. (2006) Stability of inbred mouse strain differences in behavior and brain size between laboratories and across decades. *Proceedings of the National Academy of Sciences of the United States of America* 103, 16364–16369.

Wallace, J. (2008) Attitudes to rodent euthanasia techniques: views of trainee research workers. *ALN Europe* 1, 14–15.

Wang, Y.-X.J. & Yan, S.-X. (2008) Biomedical imaging in the safety evaluation of new drugs. *Laboratory Animals* 42, 433–441. doi:10.1258/la.2007.007022.

Warburton, H. & Mason, G. (2003) Is out of sight out of mind? The effects of resource cues on motivation in mink, *Mustela vison*. *Animal Behaviour* 65, 755–762.

Ward, L.M. (2011) The thalamic dynamic core theory of conscious experience. *Consciousness and Cognition* 20, 464–486.

Watanabe, S., Sakamoto, J. & Wakita, M. (1995) Pigeons' discrimination of paintings by Monet and Picasso. *Journal of the Experimental Analysis of Behavior* 63, 165–174.

Waterton, J.C. (2000) Reduced animal use in efficacy testing in disease models by the use of sequential experimental designs. In *Progress in the Reduction, Refinement and Replacement of Animal Experimentation*, Balls, M., van Zeller, A.-M. & Halder, M. (eds.), pp. 737–745. Elsevier Science, Amsterdam.

Weary, D.M. & Fraser, D. (1995) Signalling need: costly signals and animal welfare assessment. *Applied Animal Behaviour Science* 44, 159–169.

Weary, D.M., Ross, S. & Fraser, D. (1997) Vocalizations by isolated piglets: a reliable indicator of piglet need directed towards the sow. *Applied Animal Behaviour Science* 53, 249–257.

Weisbroth, S.H. (1996) Post-indigenous disease: changing concepts of disease in laboratory rodents. *Lab Animal* 25, 25–34.

Weiskrantz, L. (1997) *Consciousness Lost and Found: A Neuropsychological Exploration*. Oxford University Press, Oxford.

Wells, D.J., Playle, L.C., Enser, W.E.J., Flecknell, P.A., Gardiner, M.A., Holland, J., Howard, B.R., Hubrecht, R., Humphreys, K.R. & Jackson, I.J. (2006) Assessing the welfare of genetically altered mice. *Laboratory Animals* 40, 111–114.

Wells, M.J. (1962) *Brain and Behaviour in Cephalopods*. Heinemann, London.

Wemelsfelder, F. (2007) How animals communicate quality of life: the qualitative assessment of behaviour. *Animal Welfare* 16, 25–31.

Wemelsfelder, F. & Lawrence, A.B. (2001) Qualitative assessment of animal behaviour as an on-farm welfare-monitoring tool. *Acta Agriculturae Scandinavica Section A* 51, 21–25.

Wemelsfelder, F., Hunter, T.E.A., Mendl, M.T. & Lawrence, A.B. (2001) Assessing the 'whole animal': a free choice profiling approach. *Animal Behaviour* 62, 209–220.

White, W.J., Chou, S.T., Kole, C.B. & Sutcliffe, R. (2010) Transportation of laboratory animals. In *The UFAW Handbook on the Care and Management of Laboratory and Other Research Animals*, 8th edn, Hubrecht, R. & Kirkwood, J. (eds.), pp. 169–182. Wiley-Blackwell, Oxford.

Wiedenmayer, C. (1997) Causation of the ontogenetic development of stereotypic digging in gerbils. *Animal Behaviour* 53, 461–470.

Wildman, D.E., Uddin, M., Liu, G., Grossman, L.I. & Goodman, M. (2003) Implications of natural selection in shaping 99.4% nonsynonymous DNA identity between humans and chimpanzees: enlarging genus *Homo*. *Proceedings of the National Academy of Sciences of the United States of America* 100, 7181.

Willner, P., Towell, A., Sampson, D., Sophokleous, S. & Muscat, R. (1987) Reduction of sucrose preference by chronic unpredictable mild stress, and its restoration by a tricyclic antidepressant. *Psychopharmacology* 93, 358–364.

Wimpenny, J.H., Weir, A.A.S., Clayton, L., Rutz, C. & Kacelnik, A. (2009) Cognitive processes associated with sequential tool use in New Caledonian crows. *PLoS ONE* 4, e6471. doi:10.1371/journal.pone.0006471.

Wolfensohn, S. (2010) Euthanasia and other fates for laboratory animals. In *The UFAW Handbook on the Care and Management of Laboratory and Other Research Animals*, 8th edn, Hubrecht, R. & Kirkwood, J. (eds.), pp. 219–226. Wiley-Blackwell, Oxford.

Wolfensohn, S. & Anderson, D. (2012) Lifetime experience and cumulative severity: a report on assessment of lifetime experience and cumulative severity under EU Directive 2010/63. *The Forum* Summer 2012, 4–11.

Wolfensohn, S. & Honess, P. (2005) *Handbook of Primate Husbandry and Welfare*. Blackwell, Oxford.

Wolfensohn, S. & Lloyd, M. (2008) *Handbook of Laboratory Animal Management and Welfare*. Blackwell, Oxford.

Wolfer, D.P., Litvin, O., Morf, S., Nitsch, R.M., Lipp, H.P. & Würbel, H. (2004) Laboratory animal welfare: cage enrichment and mouse behaviour. *Nature* 432, 821–822.

Wong, D., Makowska, I.J. & Weary, D.M. (2013) Rat aversion to isoflurane versus carbon dioxide. *Biology Letters* 9(1), doi:10.1098/rsbl.2012.1000.

Workman, P., Aboagye, E.O., Balkwill, F., Balmain, A., Bruder, G., Chaplin, D.J., Double, J.A., Everitt, J., Farningham, D.A.H., Glennie, M.J. *et al.* (2010) Guidelines for the welfare and use of animals in cancer research. *British Journal of Cancer* 102, 1555–1577.

World Organisation for Animal Health (2010) Chapter 7.8. The use of animals in research and education. OIE Terrestrial Animal Health Code. World Organisation for Animal Health (OIE), Paris. Available at http://www.oie.int/animal-welfare/key-themes/.

Wright-Williams, S.L., Courade, J.-P., Richardson, C.A., Roughan, J.V. & Flecknell, P.A. (2007) Effects of vasectomy surgery and meloxicam treatment on faecal corticosterone levels and behaviour in two strains of laboratory mouse. *Pain* 130, 108–118.

Würbel, H. (2009) Ethology applied to animal ethics. *Applied Animal Behaviour Science* 118, 118–127.

Würbel, H. & Garner, J.P. (2007) Refinement of rodent research through environmental enrichment and systematic randomization. Available at http://www.nc3rs.org.uk/news.asp?id=395.

Yang, S.Y. (2010) Pharmacophore modeling and applications in drug discovery: challenges and recent advances. *Drug Discovery Today* 15, 444–450.

Young, J.Z. (1964) *A Model of the Brain*. Clarendon Press, Oxford.

Young, R.J. (2003) *Environmental Enrichment for Captive Animals*. UFAW Animal Welfare Series. Blackwell, Oxford.

Zahs, K.R. & Ashe, K.H. (2010) 'Too much good news': are Alzheimer mouse models trying to tell us how to prevent, not cure, Alzheimer's disease? *Trends in Neurosciences* 33, 381–389.

Zhuo, M. (1998) NMDA receptor-dependent long term hyperalgesia after tail amputation in mice. *European Journal of Pharmacology* 349, 211–220.

Ziemann, A. E., Allen, J. E., Dahdaleh, N. S., Drebot, I. I., Coryell, M. W., Wunsch, A. M., Lynch, C. M., Faraci, F. M., Howard, M. A., Welsh, M. J., *et al.* (2009) The amygdala Is a chemosensor that detects carbon dioxide and acidosis to elicit fear behavior. *Cell* 139, 1012–1021.

Zimmermann, M. (1983) Ethical guidelines for investigations of experimental pain in conscious animals. *Pain* 16, 109–110.

Glossary

AAALAC	International Association for Assessment and Accreditation of Laboratory Animal Care International.
AALAS	American Association for Laboratory Animal Science.
ACLAM	American College of Laboratory Animal Medicine.
ACTH	Adrenocorticotropic hormone.
ADMET	A group of assays used to predict absorption, distribution, metabolism, excretion and toxicity characteristics of a potential drug.
APC	Animal Procedures Committee (the national UK advisory committee now superseded by the Animals in Science Committee).
APHIS	Animal and Plant Health Information Service (USA).
ASRU	Animals in Science Regulation Unit, a department of the Home Office that is responsible for inspection and licensing of research using animals in the UK.
AWERB	Animal Welfare and Ethical Review Body, required at institutions licensed to carry out research under UK legislation.
AWIC	Animal Welfare Information Center (United States Department of Agriculture National Agricultural Library).
BUAV	British Union for the Abolition of Vivisection. Also the trading name for the Campaign to End All Animal Experiments.
CAAT	Center for Alternatives to Animal Testing.
CALAS	Chinese Association for Laboratory Animal Sciences
CCAC	Canadian Council on Animal Care.
CIOMS	Council for International Organizations of Medical Sciences.
CNS	Central nervous system.
CPD	Continuous personal development.

The Welfare of Animals Used in Research: Practice and Ethics, First Edition. Robert C. Hubrecht.
© 2014 Universities Federation for Animal Welfare. Published 2014 by John Wiley & Sons, Ltd.

CRO	Contract research organisation.
Dystocia	Difficult or abnormal birth.
ECLAM	European College of Laboratory Animal Medicine.
ECVAM	European Centre for the Validation of Alternative Methods.
EQ	Encephalisation quotient.
ERP	Ethical Review Process, required under UK legislation (replaced by AWERB) in January 2013.
EUPRIM-net	European Primate Breeders Network.
FDA	Food and Drug Administration (USA).
FELASA	Federation of European Laboratory Animal Science Associations.
FRAME	Fund for the Replacement of Animals in Medical Experiments.
GA (animal)	Genetically altered. This is used as a catch-all term to include animals that have had their genome modified using transgenic techniques, as well as natural mutants and cloned animals.
GM (animal)	Genetically modified. This refers to animals that have had their genome deliberately modified using techniques such as pronuclear microinjection, embryo stem cell targeting, or as the result of a mutagen such as radiation, etc. In my view, the distinction between GM and GA animals is somewhat artificial.
HPA axis	Hypothalamic–pituitary–adrenocortical axis. The hypothalamus, anterior pituitary, adrenal cortex and associated hormones that together form a major component of the neuroendocrine stress response system.
IACUC	Institutional Animal Care and Use Committee.
ICLAS	International Council for Laboratory Animal Science.
ILAR	Institute for Laboratory Animal Research, a unit in the Division on Earth and Life Studies of the National Research Council (USA).
IVC	Individually ventilated cage.
JALAM	Japanese Association for Laboratory Animal Medicine.
LAL	*Limulus* amoebocyte lysate (test).
LASA	Laboratory Animal Science Association.
NACWO	Named Animal Care and Welfare Officer under UK legislation.
NC3Rs	National Centre for the Replacement, Refinement and Reduction of Animals in Research (UK).
NIH	National Institutes of Health, Department of Health and Human Services (USA).
NORECOPA	Norwegian Reference Centre for Laboratory Animal Science and Alternatives.
NRC	National Research Council (National Academies) (USA).
OECD	Organisation for Economic Cooperation and Development.
OIE	World Organisation for Animal Health.
OLAW	Office of Laboratory Animal Welfare (part of the NIH in the USA).
PHS	Public Health Service (USA).

PIT tags	Passive inductive transponder tags, used for marking animals. Information can be read from the microchip using an electronic reader.
Procedure as defined by European Directive 2010/63/EU	Any use, invasive or non-invasive, of an animal for experimental or other scientific purposes, with known or unknown outcome, or educational purposes, which may cause the animal a level of pain, suffering, distress or lasting harm equivalent to, or higher than, that caused by the introduction of a needle in accordance with good veterinary practice. This includes any course of action intended, or liable, to result in the birth or hatching of an animal or the creation and maintenance of a genetically modified animal line in any such condition, but excludes the killing of animals solely for the use of their organs or tissues.
Project as defined by European Directive 2010/63/EU	A programme of work having a defined scientific objective and involving one or more procedures.
PRT	Positive reinforcement training.
QSAR	Quantitative structure–activity relationship *in vitro* model.
RSPCA	Royal Society for the Prevention of Cruelty to Animals (UK).
SAM	Sympatho-adrenomedullary. The adrenal medulla and its links with the sympathetic nervous system.
SAR	Structure–activity relationship *in vitro* model.
SNS	Sympathetic nervous system.
UFAW	Universities Federation for Animal Welfare, International Animal Welfare Science Society.

Index

Page numbers in *italics* denote figures, those in **bold** denote tables.

The Welfare of Animals Used in Research: Practice and Ethics, First Edition. Robert C. Hubrecht.
© 2014 Universities Federation for Animal Welfare. Published 2014 by John Wiley & Sons, Ltd.

Keep up with critical fields

Printed and bound by CPI Group (UK) Ltd, Croydon, CR0 4YY

27/10/2024

14580387-0001